Expert Knowledge and Its Application in Landscape Ecology

Ajith H. Perera · C. Ashton Drew
Chris J. Johnson
Editors

Expert Knowledge and Its Application in Landscape Ecology

Springer

Editors
Ajith H. Perera
Ontario Forest Research Institute
Ontario Ministry of Natural Resources
Sault Ste. Marie, ON P6A 2E5, Canada
ajith.perera@ontario.ca

Chris J. Johnson
Ecosystem Science and
Management Program
University of Northern British Columbia
Prince George, BC V2N 4Z9, Canada
johnsoch@unbc.ca

C. Ashton Drew
Department of Zoology
North Carolina State University
Biodiversity & Spatial Information Center
Raleigh, NC 27695, USA
cadrew@ncsu.edu

ISBN 978-1-4614-1033-1 e-ISBN 978-1-4614-1034-8
DOI 10.1007/978-1-4614-1034-8
Springer New York Dordrecht Heidelberg London

Library of Congress Control Number: 2011938788

Printed on acid-free paper

Springer is part of Springer Science+Business Media (www.springer.com)

Foreword

In the years since its emergence as a widely recognized scientific discipline a quarter-century or so ago, landscape ecology has become increasingly quantitative and analytically rigorous. Technological advances have made it possible to obtain empirical information about landscape configuration, movements of animals through a landscape, human land uses, landscape change, and a host of other interesting things about landscapes. Landscape ecology, like other sciences, has become data-driven.

Yet, landscapes are much more complex than the simple patch-matrix diagrams some of us have become fond of. Landscape structure, function, and dynamics interact in myriad ways over multiple scales. We do not have, nor will we ever have, data on everything that is important or interesting. Gaps in data, and uncertainties accompanying the data we do have, pose particularly difficult problems when landscape ecology is applied to practical issues in urban planning, resource management, sustainable agriculture, fire ecology, and the like.

Of course, people knew things about landscapes long before landscape ecology came into being, and even now not everything landscape ecologists know is embodied in digital bytes. These sources of knowledge – expert knowledge – can help to fill the data gaps and reduce the uncertainties. That is why the approaches developed in this book are so important.

But the phrase "expert knowledge" immediately conjures up a variety of possibilities. Expert knowledge might be anything from "It's true because I'm an expert and I say so" to highly formalized systems of knowledge elicitation or expert systems software. An "expert" might be someone who knows more about something than someone else who wants to know about it. Some have suggested that an expert is someone who knows more about a topic than the average person, but this does not mean much because most people know nothing about the topic, bringing down the average. By this definition, even a passing knowledge of, say, quantum physics or the epidemiology of AIDS might qualify one as an expert. This is why "experts" are usually defined by the regard with which they are held by their peers. But there is a sociological element at play here: people who know a lot about a topic but challenge the conventional wisdom of a discipline may be called "iconoclasts" rather than "experts," and the value of their knowledge is often correspondingly diminished.

In the legal arena, expert witnesses are highly qualified people who are called upon to provide objective testimony about the state of knowledge related to an area of their expertise. Because of their expert status, their testimony may carry inordinate weight. But good lawyers know that it is not difficult to find well-credentialed experts who present diametrically opposed views of the same issue. The open, questioning nature of scientific investigation virtually assures this. As an example, expert witnesses for the plaintiffs and the defendants often gave conflicting statements about the effects of the *Exxon Valdez* oil spill on marine ecosystems in Prince William Sound. The jury hearing the case was unable to evaluate the merits of the arguments presented by the "dueling scientists" and ended up ignoring the experts on both sides in making their decision. The scientific evidence was largely ignored.

The point of this is that the knowledge of experts is not necessarily pure and unbiased. It is a product of their experiences and their training – the "facts" are colored by one's perceptions of the world from which they came. It is easy to see this if the individual is, say, a tribal elder or a long-time fisherman with deep knowledge gained from decades of experience and insights extending back for generations. Such knowledge can provide invaluable perspectives on landscape dynamics and history, but it is clearly influenced by the cultural context in which it was gained. We tend to think of scientific knowledge as somehow being less swayed by context, and perhaps it is. But science has its cultures, too, and scientists are susceptible to the judgments of their peers, which can influence how they interpret data as well as the kinds of data they collect. "Knowledge" always has cultural overtones.

All of this is to say that, although any discipline, perhaps especially landscape ecology, must draw knowledge from multiple sources, there is a real need to ensure that this knowledge is as accurate and reliable as possible. How we accomplish that is the focus of this book. It is much needed.

John A. Wiens
Chief Conservation Science Officer PRBO
Conservation Science Petaluma, CA
USA

Acknowledgments

We would like to recognize the contributions of many individuals who assisted us in producing this book.

Several colleagues reviewed chapter manuscripts and made valuable suggestions to improve the quality and clarity of contents. They are: Mark Boyce, University of Alberta; Mark Burgman, University of Melbourne; Lisa Buse, Ontario Forest Research Institute; Frederik Doyon, Université du Québec; Michael Drescher, University of Waterloo; Ioan Fazey, University of St. Andrews; Brad Hawkes, Canadian Forest Service; Hong He, University of Missouri; Elise Irwin, Auburn University; Carrie Kappel, University of California-Santa Barbara; Roxolana Kashuba, US Geological Survey; Bob Keane, US Forest Service; Rebecca Kennedy, US Forest Service; Bruce Marcot, US Forest Service; Marissa McBride, University of Melbourne; Scott McNay, Wildlife Infometrics, British Columbia; Bill Meades, Canadian Forest Service; Alison Moody, Auburn University; Grant Murray, Vancouver Island University; Justine Murray, CSIRO Ecosystem Sciences; Rebecca O'Leary, Australian Institute of Marine Science; Janet Silbernagel, University of Wisconsin; Brian Sturtevant, US Forest Service; Mark White, The Nature Conservancy; and John Wiens, PRBO Conservation Science. One peer-reviewer wished to remain anonymous.

Several others assisted in expediting the publication process: Geoff Hart style-edited chapter manuscripts and made the text readable; Lisa Buse coordinated the peer review and style edit processes and organized the editors; and Janet Slobodien and Melissa Higgs provided the liaison with Springer US.

We most gratefully thank them all.

Ajith H. Perera
C. Ashton Drew
Chris J. Johnson

Contents

Contributors

Clair Alston Faculty of Science and Technology, Queensland University of Technology, 2 George St., Brisbane, QLD 4001, Australia
c.alston@qut.edu.au

Mark A. Burgman Australian Centre of Excellence for Risk Analysis, School of Botany, University of Melbourne, Parkville, VIC 3010, Australia
markab@unimelb.edu.au

Lisa J. Buse Ontario Forest Research Institute, 1235 Queen Street East, Sault Ste. Marie, ON, Canada P6A 2E5
lisa.buse@ontario.ca

Jaime A. Collazo US Geological Survey, North Carolina Cooperative Fish and Wildlife Research Unit, North Carolina State University, David Clark Labs, Box 7617, Raleigh, NC 27695, USA
jaime_collazo@ncsu.edu

Roger M. Cooke Resources for the Future, 1616 P St. NW, Washington, DC 20036-1400, USA
cooke@rff.org

Frédérik Doyon Département de Sciences Sociales, Secteur Foresterie, Centre d'Étude de la Forêt (CEF), Université du Québec en Outaouais, C.P. 1250, Succursale Hull, Gatineau, QC J8X 3X7, Canada
frederik.doyon@uqo.ca

Michael Drescher School of Planning, University of Waterloo, 200 University Avenue West, Waterloo, ON, Canada, N2L 3G1
mdresche@uwaterloo.ca

C. Ashton Drew Department of Biology, North Carolina Cooperative Fish and Wildlife Research Unit, North Carolina State University, David Clark Labs, Box 7617, Raleigh, NC 27695, USA
cadrew@ncsu.edu

Andrew Fall Gowlland Technologies Ltd, Tucker Bay Road, Lasqueti Island, BC V0R-2J0, Canada and Department of Resource and Environmental Management, Simon Fraser University, 8888 University Drive, Burnaby, BC V5A 1S6, Canada
andrew@gowlland.ca

Marie-Josée Fortin Department of Ecology and Evolutionary Biology, University of Toronto, 25 Harbord St., Toronto, ON M5S 3G5, Canada
mjfortin@zoo.utoronto.ca

James B. Grand USGS, Alabama Cooperative Fish and Wildlife Research Unit, 3236 School of Forestry and Wildlife Sciences, Auburn University, 602 Duncan Drive, Auburn, AL 36849, USA
grandjb@auburn.edu

Benjamin S. Halpern National Center for Ecological Analysis and Synthesis, 735 State Street, Santa Barbara, CA 93101, USA
halpern@nceas.ucsb.edu

Michael Hurley The Natural Resources Institute, University of Manitoba, 303-70 Dysart Road, Winnipeg, MB R3T 2N2, Canada
umhurle5@cc.umanitoba.ca

Allan James High Performance Computing Unit, Queensland University of Technology, GPO Box 2434, Brisbane, QLD 4001, Australia
ar.james@qut.edu.au

Patrick M.A. James Department of Biological Sciences, University of Alberta, 11455 Saskatchewan Drive, Edmonton, AB T6G 2E9, Canada
pjames@ualberta.ca

Chris J. Johnson Ecosystem Science and Management Program, University of Northern British Columbia, 3333 University Way, Prince George, BC, Canada V2N 4Z9
johnsoch@unbc.ca

Carrie V. Kappel National Center for Ecological Analysis and Synthesis, 735 State Street, Santa Barbara, CA 93101, USA
kappel@nceas.ucsb.edu

Robert E. Keane Missoula Fire Sciences Laboratory, USDA Forest Service, Rocky Mountain Research Station, 5775 Highway 10 West, Missoula, MT 59808, USA
rkeane@fs.fed.us

Daniel D. Kneeshaw Département des Sciences Biologiques, Centre d'Étude de la Forêt (CEF), Institut des Sciences Environnementales, Université du Québec à Montréal, C.P. 8888, Succursale Centre-Ville, Montréal, QC H3C 3P8, Canada
daniel.kneeshaw@uqam.ca

Samantha Low-Choy Cooperative Research Centre for National Plant Biosecurity and Faculty of Science and Technology, Queensland University of Technology, 2 George Street, Brisbane, QLD 4001, Australia
s.lowchoy@qut.edu.au

Marissa F. McBride School of Botany, University of Melbourne, Parkville, VIC 3010, Australia
mcbridem@pgrad.unimelb.edu.au

R. Scott McNay Wildlife Infometrics, Inc., Box 308, Mackenzie, BC, Canada V0J 2C0
scott.mcnay@wildlifeinfometrics.com

Kerrie Mengersen Queensland University of Technology, GPO Box 2434, Brisbane, QLD 4001, Australia
k.mengersen@qut.edu.au

Christian Messier Département des Sciences Biologiques, Centre d'Étude de la Forêt (CEF), Institut des Sciences Environnementales, Université du Québec à Montréal, C.P. 8888, Succursale Centre-Ville, Montréal, QC H3C 3P8, Canada
Christian.messier@uqam.ca

Brian Miranda Institute for Applied Ecosystem Studies, Northern Research Station, USDA Forest Service, 5985 Hwy K, Rhinelander, WI 54501, USA
brmiranda@fs.fed.us

Allison T. Moody School of Forestry and Wildlife Sciences, Auburn University, 3301 Forestry and Wildlife Science Building, 602 Duncan Drive, Auburn, AL 36849, USA
atm0005@auburn.edu

Justine Murray Spatial Ecology Laboratory, The University of Queensland, Brisbane, QLD, Australia and Sustainable Ecosystems, CSIRO Ecosystem Sciences, PO Box 2583, Brisbane, QLD 4001, Australia
justine.murray@csiro.au

Marc R. Ouellette Ontario Forest Research Institute, 1235 Queen Street East, Sault Ste. Marie, ON, Canada P6A 2E5
marc.ouellette@ontario.ca

Michael J. Papaik Department of Biology, Sonoma State University, 1801 E. Cotati Ave., Rohnert Park, CA 94928, USA
papaikm@comcast.net

Ajith H. Perera Ontario Forest Research Institute, 1235 Queen Street East, Sault Ste. Marie, ON, Canada P6A 2E5
ajith.perera@ontario.ca

Michael Pullinger School of Geography and Environmental Studies, University of Tasmania, Private Bag 78, Hobart, TAS 7001, Australia
mpullinger@gmail.com

Eric Rapaport Faculty of Architecture and Planning, School of Planning, Dalhousie University, PO Box 1000, Halifax, NS, Canada B3J 2X4
Eric.Rapaport@dal.ca

Matt Reeves RMRS Forestry Sciences Laboratory, USDA Forest Service, Rocky Mountain Research Station, 800 E. Beckwith Ave. Missoula, MT 59801, USA
mreeves@fs.fed.us

Wayne Rochester CSIRO Marine and Atmospheric Research, G.P.O Box 2583, Brisbane, QLD 001, Australia
wayne.rochester@csiro.au

Kimberly A. Selkoe National Center for Ecological Analysis and Synthesis, 735 State Street, Santa Barbara, CA 93101, USA
selkoe@nceas.ucsb.edu

Brian R. Sturtevant Institute for Applied Ecosystem Studies, Northern Research Station, USDA Forest Service, 5985 Hwy K, Rhinelander, WI 54501, USA
bsturtevant@fs.fed.us

Kristen J. Williams CSIRO Ecosystem Sciences, G.P.O Box 1700, Canberra, ACT 2601, Australia
kristen.williams@csiro.au

Chapter 1
Experts, Expert Knowledge, and Their Roles in Landscape Ecological Applications

Ajith H. Perera, C. Ashton Drew, and Chris J. Johnson

Contents

1.1 Introduction

In an attempt to develop a forest succession model that simulates scenarios of future landscape patterns, researchers encounter many gaps in the published knowledge of forest succession trajectories. They resort to consulting local foresters and using their knowledge of forest succession to parameterize the model. In another situation, management of an elusive bird species requires estimates of the likelihood of its occurrence under specific sets of site conditions. Because the habitat characteristics

A.H. Perera (✉)
Ontario Forest Research Institute, 1235 Queen Street East,
Sault Ste. Marie, ON, Canada P6A 2E5
e-mail: ajith.perera@ontario.ca

C.A. Drew
Department of Biology, North Carolina Cooperative
Fish and Wildlife Research Unit, North Carolina State University,
Campus Box 7617, Raleigh, NC 27695, USA

C.J. Johnson
Ecosystem Science and Management Program, University of Northern
British Columbia, 3333 University Way, Prince George, BC, Canada V2N 4Z9

A.H. Perera et al. (eds.), *Expert Knowledge and Its Application in Landscape Ecology*,
DOI 10.1007/978-1-4614-1034-8_1, © Springer Science+Business Media, LLC 2012

of this species are not well studied or have not been published, the investigators seek the advice of specialist wildlife biologists to learn where these birds could potentially occur. In yet another case, natural resource and conservation professionals turn to expert knowledge to help them conserve or manage wildlife habitats in high-risk environments, and researchers investigate the relative merits of that expert knowledge in comparison with empirical data, as well the uncertainty and variability in expert-based predictions.

The scenarios described above are taken from our personal experience as landscape ecology researchers: forest succession (Perera), elusive birds (Drew), and wildlife habitat (Johnson). These situations undoubtedly appear familiar to most applied ecologists; in research and its application, such gaps and shortfalls are pervasive in the available data, information, and knowledge. Often, we must rely on expert knowledge to complement and supplement empirical data. But in so doing, we face a problem: the use of expert knowledge in ecological research and the application of that knowledge in practical situations are viewed with skepticism by many, since such knowledge is considered to be very different from, and even inferior to, traditional data gathered through rigorously designed sampling of ecosystem components (Drew and Perera 2011).

Reliance on expert knowledge in research and natural resource management has been formally acknowledged for many decades (Burgman 2005). Although its early uses in ecology were limited to using expert judgment to support decision-making, as in the examples of an expert-systems approach to decision-making and the use of artificial intelligence systems (e.g., Coulson et al. 1987; Rykiel 1989; Meyer and Booker 1991), the frequency and diversity of uses of expert knowledge are both increasing (Drew and Perera 2011). This growth is evident in the published literature: our brief search through the Web of Science database (http://www.images.isiknowledge.com/WOK46/help/WOS/h_database.html) showed that the terms *expert knowledge*, *expert opinion*, and *expert judgment* appeared 220 times in the ecological literature before 2000, versus more than 400 times since 2000.

In this chapter, we set the stage for the rest of this book by discussing some of the key issues involved in making better use of expert knowledge. These include clarifying the meaning of the word *expert*, defining the sources of expertise, characterizing the nature of expert knowledge, explaining its relevance, and describing some of the concerns that arise from using expert knowledge.

1.1.1 Who Do We Mean by Experts?

A wide variety of experts is mentioned in the literature with reference to using their knowledge in an array of professions, ranging from science to the arts and even sports (Ayyub 2001; Ericsson et al. 2006). In our view, there are several key categories of experts that pertain to ecological applications: scientists, who conduct research and publish their knowledge formally; practitioners, who apply scientific knowledge in management but typically do not conduct research and publish their knowledge formally; stakeholders, who have an interest in the outcome of applying ecological knowledge to inform conservation or resource extraction issues; and elders of local societies (aboriginal or other) who are rich sources of traditional knowledge. Of course, a given individual may belong to more than one of these categories, and no category is inherently more important than the others.

Our focus in this book is on experts who are practitioners (termed *expert practitioners* hereafter). We have done so because the greatest wealth of untapped latent knowledge that waits to be utilized rests with expert practitioners, including such professionals as ecologists, biologists, foresters, and geologists, who are highly knowledgeable in their technical domains. This does not exclude nor does it diminish the value of expert knowledge contributed by scientists, whose knowledge is typically published in scientific media and is therefore not latent, nor do we intend to lessen the value of the expertise and insights provided by traditional or local ecological knowledge, which is increasingly recognized in the literature (e.g., Johannes 1989; Huntingdon 2000; Anadon et al. 2009). As well, we acknowledge that stakeholders can provide expert knowledge on occasion while recognizing their inherent biases (e.g., O'Hagan et al. 2006) towards certain desired outcomes of resource management decisions.

We encounter expert practitioners in all facets of landscape ecological applications. Beyond generalists, this group occasionally includes topic specialists (for example, an authority on the feeding habits of cougars), skilled individuals (for example, a naturalist capable of identifying animal signs and tracking cougars in the wild), and sage practitioners (for example, an extraordinarily experienced cougar biologist). These expert practitioners do not always formally record their knowledge as researchers do (which is typically via a peer-review process), and their expertise is more likely to be local in scope than global.

Consequently, it is difficult to recognize the existence of that expertise and gauge the quality of the knowledge by consulting only the scientific media. It is equally tricky to formulate an objective and a globally measurable definition of what constitutes an expert practitioner. Still, expert practitioners have one common trait: they are recognized and held in high regard by their peers (other practitioners in the same technical domain) as being skilled and knowledgeable (Chi 2006). Although not all of those recognized by their peers may be true experts (Ericsson 2006), we feel that all experts are recognized by their peers, and that peer-recognition is thus a reliable initial filter for identifying experts.

1.1.2 How Do Experts Gain Expertise?

Recently, a considerable body of scientific support has developed for the belief that individuals become experts through "deliberate practice," a concept first suggested by Ericsson et al. (1993): after some level of formal education in a technical domain, individuals must engage in practice of their profession while actively attempting to improve their knowledge and skills. This period of practice is typically no less than 10,000 hours of intensively focused engagement with a subject (Ericsson 2006), which amounts to around 10 years for most experts (Horn and Masunaga 2006). We believe that the ideal expert practitioners in the context of this book are those who possess superior reasoning skills, have received a strong foundation of formal education, and (most importantly) have augmented that knowledge with years of deliberate practice. This combination of formal training, intelligence (the ability to reason and synthesize), and experience (a lengthy period of observation and practice) present numerous opportunities to encounter and solve professional problems, thereby gaining knowledge that may not exist in the published literature and recognition as an expert by their peers. From a strict scientific perspective, critics may view this learning process as implicit, subjective, and nonrepeatable. The implications and limitations of this process of gaining expertise are recognized (e.g., Chi 2006), and we expand on their consequences for an expert's knowledge and its use later in this chapter as well as Chapter 2.

1.1.3 What Is the Nature of Expert Knowledge?

Much of the knowledge amassed by experts remains informal, primarily because it is typically not documented and remains tacit until its expression is demanded in specific applications. Moreover, experts express and apply their knowledge in different forms (e.g., implicit, qualitative, equivocal) than what is typical for empirical science. As a result, expert knowledge may appear latent, fragmented, and nonunified, and properties such as the variability, uncertainty, and even veracity of the knowledge remain unassessed (e.g., Ayyub 2001; Burgman et al. 2011). By definition, expert knowledge is highly subjective and methods of its formulation are not explicit and repeatable unless someone makes an effort to rigorously elicit the knowledge. Although expert knowledge may include the processes of synthesis, testing, validation, and reformulation, these steps are not always explicit, unlike in the development of a body of scientific knowledge. This makes expert knowledge highly unattractive to scientists who have been trained to use the scientific method to obtain objective knowledge. We emphasize, however, that methods for eliciting and validating expert knowledge can be rigorous and consistent with the approaches used by scientists to collect and apply empirical data.

While others have categorized expertise, for example the Dreyfus scale (Dreyfus and Dreyfus 1986), we view degree of expertise as a continuum. Many dimensions can be used to quantify the degree of expertise, as illustrated in Fig. 1.1. Ideal experts are those who have a global view, with broad spatiotemporal and cross-scale understanding; base their knowledge on observations, critical thought, and deliberate practice; can understand complex systems and recognize detailed relationships; approach problem solving with parsimony; and acknowledge the variability and stochasticity in ecosystems and the resulting uncertainty of knowledge.

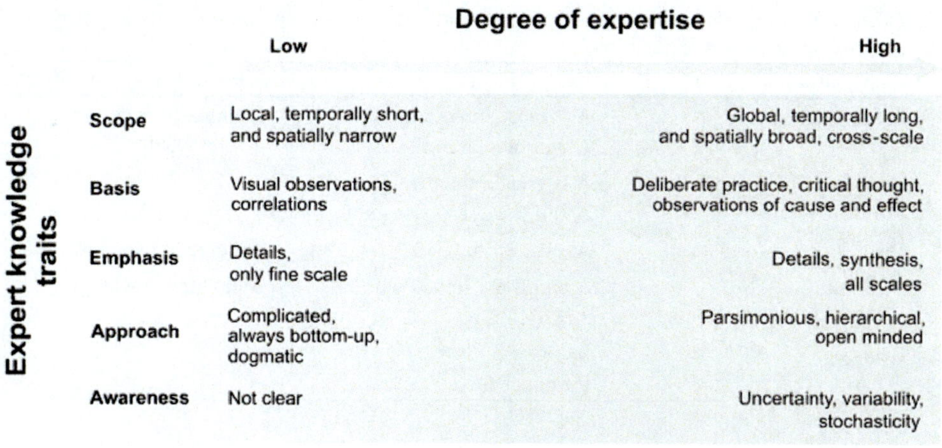

Fig. 1.1 Expert knowledge can be positioned along a continuum of expertise that can be quantified using multiple traits

1.1.4 Why Should We Care About Expert Knowledge?

The value of expert knowledge extends beyond traditional expert systems which focus mostly on an expert's decision-making ability in specific and narrow instances, such as policy development, strategic planning, or tactical management (e.g., Rykiel 1989; Cleaves 1994; Burgman 2005; Aspinall 2010). Given that expertise is not a single level, but rather a continuum, the utility of expert knowledge to landscape ecologists can differ according to an expert's position within the spectrum. At lower levels of expertise, experts are informative in providing data on local observations and specific local conditions; for example, they can provide knowledge of exceptions and rarities and can fill gaps in more formal data sets. Such knowledge is more of empirical and local value. These experts just "know" from their observations, reading, discussions, and other means, and may offer their opinions or specific knowledge of ecological patterns and processes. At higher levels, experts can potentially inform researchers about ecological processes and patterns; for example, they can offer insights, syntheses, and hypotheses. We consider these experts to be "sages"; in addition to just "knowing," they have thought deeply about an issue and have formulated their own syntheses based on synoptic knowledge. They may be able to bypass otherwise complex systems and provide parsimonious solutions that focus on the key aspects of a situation. Examples of the contributions of expert knowledge to landscape ecology research and application development range from judgments to decisions, and these expressions may be founded on different forms of expert knowledge (Table 1.1).

Landscape ecologists may find useful roles for expert knowledge in all aspects of research and development: conceiving and clarifying research ideas; hypothesis development and testing; model development, parameterization, and validation; and subsequent knowledge transfer (Fig. 1.2).

Table 1.1 Examples of the types of contributions experts have made to landscape ecology research and applications

Contribution	Example	Foundation
Judgment	A is more important than B A will increase in time	Opinion or synoptic knowledge
Qualitative information	A is greater than B A and B exist; C does not	
Quantitative information	A = 10, B = 5, C = 0	Specific knowledge
Synthesis	A and B are linked with C A = 2(B + C)	Synoptic knowledge
Decision	Increase B Introduce C	

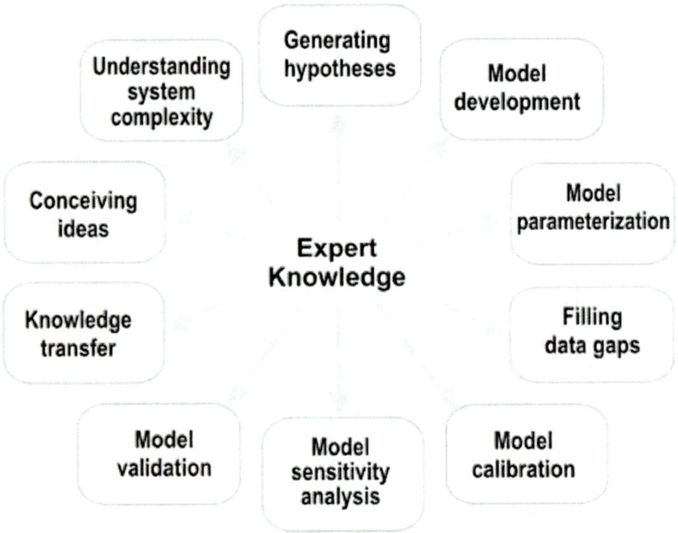

Fig. 1.2 Expert knowledge is useful at many levels in landscape ecology research and development

1.1.5 What Are the Concerns of Using Expert Knowledge?

Although expert knowledge is useful in most phases of landscape ecology research and development, users of that knowledge must be aware of the potential pitfalls. For example, given the many biases and complications involved in cognition, communication, behavior, and other human traits associated with the acquisition and expression of expert knowledge, it is possible to misrepresent, misinterpret, and improperly apply expert knowledge (Burgman 2005).

Credible and proper use of expert knowledge requires a rigorous scientific approach to selecting the experts; to eliciting, analyzing, and verifying their knowledge; and to applying it appropriately. The long-term credibility and utility of expert knowledge depends on the rigor of the method, not merely on the acceptability of the results of applying the knowledge. The general principles and detailed methods for engaging experts and using their knowledge are evolving steadily (e.g., Cooke 1991; Meyer and Booker 1991; Ayyub 2001; O'Hagan et al. 2006). Much of this work was pioneered in the social sciences, with relatively recent application and innovation in the natural and life sciences (e.g., Burgman 2005; Doswald et al. 2007 Kuhnert et al. 2010). We suggest, however, that most practitioners and researchers involved in landscape ecology have little training in the rigorous methods and approaches that must be used to engage experts and elicit their knowledge. Very few of us are exposed to the science of dealing with human subjects during our undergraduate or postgraduate training. This lack of awareness not only compromises the principles of scientific rigor, but potentially has large implications for landscape ecological applications if knowledge is not elicited and used in a rigorous manner (Johnson and Gillingham 2004; Burgman et al. 2011).

1.2 Road Map of the Book

Based on the abovementioned context, our goal in this book is to introduce landscape ecologists to the applicability, advantages, and pitfalls of expert knowledge. We have included methodological chapters that provide guidance in developing appropriate and defensible methods for the elicitation and use of expert knowledge. We have also included case studies that reveal additional methods and applications of this knowledge, as well as the costs, pitfalls, and benefits of applying expert knowledge to a wide range of problems and questions in landscape ecology. However, we do not intend for the text to serve as a comprehensive handbook of "best practices" for landscape ecologists on how to use expert knowledge. We anticipate that such a how-to volume, similar to the work of Cooke (1991), Meyer and Booker (1991), Ericsson et al. (2006), and O'Hagan et al. (2006) will be produced in time as the applications of expert knowledge in landscape ecology become more prevalent, as lessons are learned about its disadvantages, as elicitation and analytical methods are refined, and as further insights are gained into its applications.

The experts described in this book, and whose knowledge was elicited and used in ecological applications, ranged from expert practitioners to scientists. Here, we define expert knowledge as "Acquaintance with facts; state of being aware or informed; intellectual perception of fact or truth; clear and certain understanding or awareness," following the definition of *knowledge* provided by Oxford English Dictionary (2002, 5th edition). In doing so, we recognize its difference from *expert opinion*, which is defined by the Oxford English Dictionary as "a view held about a particular subject or point; a judgment formed; a belief; a formal statement by a member of an advisory body, an expert, etc., of what he or she judges or advises on a matter" (2002, 5th edition).

In Chapter 2, McBride and Burgman expand on the concept of experts and their knowledge. They especially focus on how that expertise is gained, limitations of

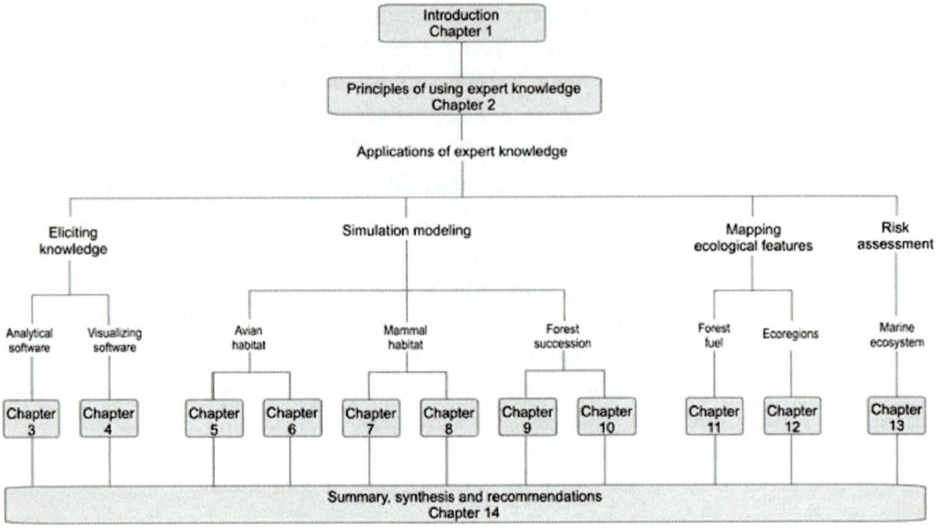

Fig. 1.3 Overview of the organization of this book

expert knowledge, and how others could rigorously elicit and use that knowledge in ecological applications; in so doing, they provide a conceptual foundation for knowledge elicitation. Subsequent chapters describe the application of expert knowledge to landscape ecology through case studies from Australia (Chaps. 3 and 12), Canada (Chaps. 4, 7–10), and the USA (Chaps. 5, 6, 11, and 13). They constitute a broad array of ecological contexts (Fig. 1.3), ranging from modeling of avian habitat (Chaps. 5 and 6), conservation of mammal habitat (Chaps. 7 and 8), forest succession modeling (Chaps. 9 and 10), and mapping of ecological features (Chaps. 11 and 12), to assessing risks in marine ecosystems (Chap. 13).

In Chapter 3, Low-Choy et al. describe analytical software they have designed to elicit and assess expert knowledge to support its application within Bayesian models. Drew and Collazo (Chap. 5), Moody and Grand (Chap. 6), McNay (Chap. 7), and Williams et al. (Chap. 12) relied on Bayesian methodologies in their case studies of expert knowledge. Johnson et al. (Chap. 8) describe the application of methods to evaluate the uncertainty and the relative accuracy of knowledge. Drescher et al. (Chap. 4) detail the development of a customized software tool that facilitated the elicitation of expert knowledge by visually simplifying the otherwise intractably complex details of forest ecosystems. Drescher and Perera (Chap. 9) discuss the assessment and verification of that expert knowledge. Doyon et al. (Chap. 10) present a similar case of using expert knowledge as a source of supplementary information for simulation models of forest succession.

Drew and Collazo (Chap. 5) provide an excellent example of using expert knowledge to model the habitat of bird species, where formal scientific data did not exist. Moody and Grand (Chap. 6) report their use of expert knowledge of bird habitat associations to develop a decision-support tool for avian conservation plans. McNay

(Chap. 7) presents a case study of using expert knowledge to develop long- and short-term recovery objectives for woodland caribou. Johnson et al. (Chap. 8) provide a summary of the lessons they learned from case studies of mammal habitat modeling using expert knowledge, with particular emphasis on the potential pitfalls of using expert knowledge in ecological applications. Keane and Reeves (Chap. 11) and Williams et al. (Chap. 12) classify and map the ecological features of landscapes in applications that rely on interpretation by experts using aspatial logic and implicit models. Kappel et al. (Chap. 13) describe how expert knowledge can be used to assess vulnerability risk of marine ecosystems. Finally, Johnson et al. (Chap. 14) provide a summary and a synthesis of the knowledge and insights presented in this volume, as well as recommendations for the rigorous use of expert knowledge in landscape ecological applications.

References

Anadon JD, Gimenez A, Ballestar R, Perez I (2009) Evaluation of local ecological knowledge as a method for collecting extensive data on animal abundance. Conserv Biol 23:617–625

Aspinall W (2010) A route to more tractable expert advice. Nature 463:294–295

Ayyub BM (2001) Elicitation of expert opinions for uncertainty and risks. CRC Press, Boca Raton

Burgman MA (2005) Risks and decisions for conservation and environmental management. Cambridge University Press, Cambridge

Burgman MA, Carr A, Godden L et al (2011) Redefining expertise and improving ecological judgement. Conserv Lett 4:81–87

Chi MTH (2006) Two approaches to the study of experts' characteristics. In: Ericsson KA, Charness N, Hoffman RR, Feltovich PJ (eds) The Cambridge handbook of expertise and expert performance. Cambridge University Press, New York, pp 20–30

Cleaves DA (1994) Assessing uncertainty in expert judgments about natural resources. USDA Forest Service, Southern Forest Experiment Station, New Orleans, Gen. Tech. Rep. SO-110

Cooke RM (1991) Experts in uncertainty: Opinion and subjective probability in science. Oxford University Press. New York

Coulson RN, Folse LJ, Loh DK (1987) Artificial intelligence and natural resource management. Science 237:262–267

Doswald N, Zimmerman F, Breitenmoser U (2007) Testing expert groups for a habitat suitability model for the lynx *Lynx lynx* in the Swiss Alps. Wildlife Biol 13:430–446

Drew CA, Perera AH (2011) Expert knowledge as a basis for landscape ecological predictive models. In: Drew CA, Wiersma Y, Huetmann F (eds) Predictive species and habitat modeling in landscape ecology: concepts and applications. Springer, New York, pp 229–248

Dreyfus HL, Dreyfus SE (1986) Mind over machine The power of human intuition and expertise in the era of the computer. The Free Press, New York

Ericsson KA (2006) An introduction to the *Cambridge Handbook of Expertise and Expert Performance:* its development, organization and content. In: Ericsson KA, Charness N, Hoffman RR, Feltovich PJ (eds) The Cambridge handbook of expertise and expert performance. Cambridge University Press, New York, pp 3–19

Ericsson, KA, Krampe RTh, Tesch-Roemer C (1993). The role of deliberate practice in the acquisition of expert performance. Psych Rev 100:363–406

Ericsson KA, Charness N, Hoffman RR, Feltovich PJ (eds) (2006) The Cambridge handbook of expertise and expert performance. Cambridge University Press, New York

Horn J, Masunaga H (2006) A merging theory of expertise and intelligence. In: Ericsson KA, Charness N, Hoffman RR, Feltovich PJ (eds) (2006) The Cambridge handbook of expertise and expert performance. Cambridge University Press, New York, pp 587–611
Huntingdon HP (2000) Using traditional ecological knowledge in science: methods and applications. Ecol Appl 10:1270–1274
Johannes RE (1989) Traditional ecological knowledge: a collection of essays. World Conservation Union (IUCN), Gland
Johnson CJ, Gillingham MP (2004) Mapping uncertainty: sensitivity of wildlife habitat ratings to variation in expert opinion. J Appl Ecol 41:1032–1041
Kuhnert P, Martin TG, Griffiths SP (2010) A guide to eliciting and using expert knowledge in Bayesian ecological models. Ecol Lett 13:900–914
Meyer MA, Booker JM (1991) Eliciting and analyzing expert judgment: A practical guide. Academic Press, San Diego, Knowledge-based systems, Volume 5
O'Hagan AO, Buck CE, Daneshkhah A, Eiser JR, et al (2006) Uncertain judgments: Eliciting experts' probabilities. John Wiley & Sons, Chichester
Rykiel EJ Jr (1989) Artificial intelligence and expert systems in ecology and natural resource management. Ecol Model 46:3–8

Chapter 2
What Is Expert Knowledge, How Is Such Knowledge Gathered, and How Do We Use It to Address Questions in Landscape Ecology?

Marissa F. McBride and Mark A. Burgman

Contents

2.1 Introduction: Why Use Expert Knowledge?

Expert knowledge plays an integral role in applied ecology and conservation (Burgman 2005). Environmental systems are characterized by complex dynamics, multiple drivers, and a paucity of data (Carpenter 2002). Action is often required before uncertainties can be resolved. Where empirical data are scarce or unavailable, expert knowledge is often regarded as the best or only source of information (Sutherland 2006; Kuhnert et al. 2010). Experts may be called upon to provide input for all stages of the modeling and management process, and specifically to inform

M.F. McBride (✉)
School of Botany, University of Melbourne, Parkville, VIC 3010, Australia
e-mail: mcbridem@pgrad.unimelb.edu.au

M.A. Burgman
Australian Centre of Excellence for Risk Analysis, School of Botany,
University of Melbourne, Parkville, VIC 3010, Australia

A.H. Perera et al. (eds.), *Expert Knowledge and Its Application in Landscape Ecology*, DOI 10.1007/978-1-4614-1034-8_2, © Springer Science+Business Media, LLC 2012

the definition and structuring of the problem (Cowling and Pressey 2003; Sutherland et al. 2008), to inform the selection of data or variables, model structures, and assumptions about functional relationships between variables (Pearce et al. 2001; Czembor and Vesk 2009), and to inform the analysis of data, estimation of parameters, interpretation of results, and the characterization of uncertainty (Alho and Kangas 1997; Martin et al. 2005).

Expert judgment is susceptible to a range of cognitive and motivational biases, to an expert's particular context, and to their personal beliefs and experiences (Shrader-Frechette 1996; Camerer and Johnson 1997; Slovic 1999; Ludwig et al. 2001; Campbell 2002). Formal elicitation methods anticipate and account for the most serious and predictable frailties of expert opinions (Morgan and Henrion 1990; Cooke 1991). These methods improve the quality of elicited knowledge by treating elicitation as formal data acquisition, using systematic, well-defined protocols that reduce the impact of extraneous factors on the results and that make assumptions and reasoning explicit (van Steen 1992; Burgman et al. 2011).

Expert knowledge incorporates uncertainty derived from multiple sources. Uncertainty may arise from incertitude (sometimes termed "epistemic uncertainty"), natural variation (sometimes termed "aleatory uncertainty"), and linguistic uncertainty (Anderson and Hattis 1999; Regan et al. 2002). Incertitude arises from incomplete knowledge and can be reduced by additional research and data collection. Natural variation results from inherent natural randomness, such as fluctuations in rainfall and temperature. It can be better understood but not reduced by additional study or measurement improvements (Burgman 2005). Linguistic uncertainty arises from imprecision in language, and results from ambiguous, vague, underspecified, and context-dependent terms. This form of uncertainty can be reduced by resolving meanings and clarifying context, terms, and expressions (Regan et al. 2002). For example, Whitfield et al. (2008) used expert judgment to quantify the flight initiation distance (FID) of breeding birds in response to an approaching human. Epistemic uncertainty arose in, for example, the average FID, as a result of the expert's lack of knowledge, and could be reduced by additional study. Natural variation arose because different individual birds exhibit different FID responses, and the same individuals exhibit different responses in different circumstances.

Different types of uncertainty have different implications for decision-makers, and ideally, experts will be given the opportunity to address different sources of uncertainty separately (Ferson and Ginzburg 1996; Regan et al. 2002). Incertitude may prompt further research, whereas natural variation may lead to the development of management strategies, such as a maximum approach distance in the FID example. However, in practice, clear distinctions between the different types of uncertainty do not always exist (Hofer 1996; O'Hagan 1998).

In this chapter we explore the capacity of experts to contribute to better management and decision-making in environmental systems. We look at what expertise is and how it is acquired. We outline the process involved in the formal elicitation of expert knowledge, including the selection of appropriate experts, deciding the form of knowledge to elicit, and verification of expert responses. Finally, we discuss more broadly the role for experts and expert knowledge when addressing questions in landscape ecology, including examples of problems for which expert knowledge can usefully contribute, problems and pitfalls, and areas for possible improvement.

2.2 What Is Expert Knowledge?

"Expert knowledge" is what qualified individuals know as a result of their technical practices, training, and experience (Booker and McNamara 2004). It may include recalled facts or evidence, inferences made by the expert on the basis of "hard facts" in response to new or undocumented situations, and integration of disparate sources in conceptual models to address system-level issues (Kaplan 1992). For a more detailed discussion of expert knowledge, see Perera et al. (Chap. 1). Experts are usually identified on the basis of qualifications, training, experience, professional memberships, and peer recognition (Ayyub 2001), although broader definitions of expertise may include untrained people who possess direct, practical experience (Burgman et al. 2011; see Table 2.1). For example, a typical expert in landscape ecology might be a practitioner who has formal training, years of deliberate practice, and whose ability to solve professional problems has led to their recognition as an "expert" by their peers.

Expert knowledge is a product of unique reasoning systems (Ericsson and Lehmann 1996; Fazey et al. 2005; Chi 2006). Skilled experts have acquired extensive knowledge and experience that affects how they perceive systems and how they are able to organize and interpret information. The cognitive basis for expert performance is recognition: experts develop organizational structures that allow them to recognize a situation and efficiently recall the most appropriate knowledge to solve a specific problem (Ericsson and Charness 1994). As a result, experts are skilled in determining the most relevant information for a given context, structuring the problem definition, and finding an appropriate solution method (Chi 2006). Their reasoning typically is characterized as being automatic, abstract, intuitive, tacit, and reflexive. An expert operating in their area of direct expertise is often able to perform tasks without being aware of exactly how or what they do (Kidd and Welbank 1984).

Table 2.1 A proficiency scale for expertise under a traditional approach to expertise (modified from Collins and Evans 2007; see also R.R. Hoffman 1998).

Type	Characteristics
Contributory expertise	Fully developed and internalized skills and knowledge, including an ability to contribute new knowledge or to teach.
Interactional expertise	Knowledge gained from learning the language of specialist groups, without necessarily obtaining practical competence.
Primary source knowledge	Knowledge gained from the primary literature, including basic technical competence.
Popular understanding	Knowledge from the media, with little detail and less complexity.
Specific instruction	Formulaic, rule-based knowledge, typically simple, context-specific, and local.

A domain (or substantive) expert is an individual familiar with the subject at hand and responsible for the analysis of the issue and providing judgments. The expert literature distinguishes between substantive expertise, which represents an expert's domain knowledge, and normative expertise, the expert's ability to accurately and clearly communicate beliefs in a particular format, such as probabilities (Ferrell 1994; Stern and Fineberg 1996). However, knowledge about a subject area does not translate into an ability to convey that knowledge. Similarly, experts are often required to convert incomplete knowledge into judgments for use in decision-making, or to extrapolate knowledge to new and unfamiliar circumstances. The degree to which they are able to extrapolate or adapt to new circumstances, referred to as "adaptive expertise" (Fazey et al. 2005), varies depending on the individual and not necessarily according to their substantive knowledge or training. As with substantive expertise, normative and adaptive expertise must be acquired through training and experience (Murphy and Winkler 1984; Ferrell 1994; Wilson 1994; Fazey et al. 2005).

2.2.1 Development of Expertise

Expert skill requires substantial domain knowledge and repeated experience with relevant tasks so that experts recognize the appropriate cues for future information demands (Ericsson and Kintsch 1995; Ericsson 2004). The traditional theory of expertise (Chase and Simon 1973; Richman et al. 1995) assumes that experts are trained appropriately, and then slowly accumulate knowledge over long periods through experience, and that this leads to a gradual improvement in their ability to estimate parameter values and make predictions (Ericsson and Towne 2010). However, experience and qualifications are often poor indicators of this kind of performance (Ericsson and Lehmann 1996; Camerer and Johnson 1997). Experience and training contribute to expertise, but their value depends on the characteristics of the task environment in which they are obtained (Shanteau 1992).

Where expertise is acquired in appropriate environments with adequate experience and feedback, it can be highly effective. In particular, when feedback quality is high (frequent, prompt, and diagnostic) and judgments are made in exacting environments (where mistakes are costly), expert knowledge is likely to be accurate. For example, chess players (Chase and Simon 1973), weather forecasters (Murphy and Winkler 1984), athletes (Ericsson et al. 2006), and physicists in textbook problem solving (Larkin et al. 1980) all display highly skilled expertise, developed through experience over an extended period in conjunction with consistent and diagnostic feedback.

When feedback quality is low, or when mistakes are not costly to those making the estimates, inaccurate beliefs are easily acquired. In such environments, experts are likely to have difficulty separating the influences of skill from those of chance and are likely to form superstitious beliefs (Kardes 2006). Delayed feedback, for example, makes it difficult for physicians to learn about the accuracy of their diagnoses (Christensen-Szalanski and Bushyhead 1981).

Sutherland et al. (2004) give several instances in which the failure to evaluate the outcomes of management actions resulted in the persistence of misperceptions about their effectiveness and suitability. For example, winter flooding of grasslands was considered by many experts to be beneficial for wading birds. However, an in-depth study by Ausden et al. (2001) revealed that although flooding of previously unflooded grasslands improved conditions for bird foraging, it also killed the invertebrates upon which the birds fed. Incorrect beliefs were propagated because appropriate diagnostic feedback about the effectiveness of grassland flooding was initially absent.

Adaptive expertise may be inhibited by knowledge within a narrow domain. Greater expert knowledge and more structured, automated reasoning processes can lead to more entrenched thinking that may be difficult to alter when circumstances change. For example, Chi (2006) noted that experts may perform worse than novices when adapting to new situations. This is particularly likely to arise when experts become complacent or do not recognize when a task lies outside their direct area of expertise.

2.2.2 Limitations of Expertise

The way in which expertise is acquired means that expert skill is limited to the tasks and domains in which it was acquired. Where experts deal with a known situation for which they have had repeated performance feedback, they give more accurate, better-calibrated information than nonexperts (Shanteau 1992; Hogarth 2001). Outside their specific sphere of expertise, experts fall back on the same reasoning processes as everyone else, and their judgments are subject to the same psychological and contextual frailties. The degree to which a person's unique set of experiences and training are relevant to a particular context is often difficult to determine (Bransford et al. 2000).

The seminal work by Tversky and Kahneman (Tversky and Kahneman 1974; Kahneman and Tversky 1982), and others (e.g., Fischhoff et al. 1982; Dawes and Kagan 1988; Gilovich et al. 2002; Slovic et al. 2004) has shown that experts rely on "heuristics" (shortcuts). Experts who make appropriate use of these shortcuts can make powerful inferences with limited time and data (Gigerenzer 1999, 2008). However, incorrect use of judgmental heuristics often leads to biases (Kahneman 1991; Shanteau and Stewart 1992; Wilson 1994).

Cognitive biases result from limitations on human processing ability and occur because of a failure to adequately process, aggregate, or integrate relevant information (Wilson 1994). For example, judgments from experts (and lay people) are undermined by overconfidence, with experts specifying narrower bounds than is warranted based on their knowledge or experience (Fischhoff et al. 1982; Speirs-Bridge et al. 2010). Overconfident experts fail to correctly process the full extent of uncertainty in their knowledge about a variable. For example, Baran (2000), as discussed by Burgman (2005), asked professional ecologists to estimate how many 0.1-ha quadrats would be necessary to sample 95% of the plant species within a 40-ha Australian dry temperate sclerophyll forest landscape. Field ecologists routinely perform this type of estimation task, and the respondents were familiar with the methodology and habitat. However, Baran (2000) found that only 2 of the 28 experts specified 90% credible bounds that included the true value.

Motivational biases arising from context, personal beliefs, and from what the expert stands to gain or lose personally from a decision may also color their judgments (Kunda 1990; Garthwaite et al. 2005). Motivational biases are "a conscious or subconscious adjustment in the subject's responses motivated by his [sic] perceived system of personal rewards for various responses" (Spetzler and Stael Von Holstein 1975). Other biases common among scientists include a tendency to treat model or experimental results as more reliable than they really are (Hora 1992), predicting the future based on past events ("hindsight" bias), overestimating their degree of control over an outcome, and underestimating the amount of variability in a system (Anderson 1998; Burgman 2000). Formal elicitation processes are motivated by the need to make experts aware of these potential biases, and to mitigate their effects (Morgan and Henrion 1990; Hokstad et al. 1998; Arnott 2006).

2.3 Gathering Expert Knowledge

Experts provide knowledge informally when they specify information "off the top of their heads". Informal, subjective judgments are often incorporated into scientific decisions through the selection of which problem needs to be analyzed, how the problem is to be structured, what data sources to draw upon, how results are interpreted, and what actions are recommended. Formal procedures have been developed to counter the cognitive and motivational biases prevalent in informal expert judgments (Morgan and Henrion 1990; Hokstad et al. 1998). They are employed with the aim of increasing the credibility, repeatability, and transparency of expert knowledge. Generally, they involve a protocol for elicitation; that is, a set of defined, repeatable steps that control the way in which information is elicited to reduce the effects of extraneous factors.

A successful elicitation is one that provides an accurate representation of an expert's true beliefs (Garthwaite et al. 2005). There is a particular emphasis on establishing a complete understanding of the reasoning and assumptions behind an expert's judgments, and ensuring that experts make judgments on the basis of all relevant information. Questions are formulated to help experts draw on appropriate data and relevant background information (Spetzler and Stael Von Holstein 1975). Feedback and verification stages are included to ensure that experts give fully reasoned responses and that the responses are internally (for the expert) and externally (with existing knowledge) consistent (Keeney and von Winterfeldt 1991). Although the specifics vary between protocols, there is general agreement on the key stages (Spetzler and Stael Von Holstein 1975; von Winterfeldt and Edwards 1986; Morgan and Henrion 1990; Cooke 1991; Keeney and von Winterfeldt 1991):

1. Preparation:
   - Problem definition and development of questions.
   - Definition and selection of experts.
2. Elicitation:
   - Training of experts before conducting the actual elicitation.
   - The actual elicitation.
3. Analysis:
   - Verification of responses.
   - Aggregation of expert responses.

Within this broad framework, there is scope for considerable variation at each of the stages. Key variables include the format for the elicitation, number of experts selected, kind and degree of interaction among the experts and between the elicitors and experts, format of the elicitation, and the way in which the elicited knowledge is combined. Often, details depend on the preferences of the researcher and the characteristics of the problem at hand. Key factors include the number and type of experts available, and the time and other resources available to the researcher (Kuhnert et al. 2010). The development of a tailored elicitation protocol for the requirements of a particular problem is referred to as elicitation design (Low-Choy et al. 2009).

Readers interested in eliciting expert knowledge must understand the distinct roles that are involved in a formal elicitation process (Rosqvist and Tuominen 2004; O'Hagan et al. 2006):

1. The *client* is the decision-maker or body that will use the results of the elicitation.
2. *Substantive experts* have the relevant domain knowledge about the parameters that will be elicited; most of these experts contribute judgments, but ideally one or two should inform the initial structuring of the problem and design of the questions.
3. *Analytical experts* have relevant quantitative knowledge and are responsible for analyzing the expert responses.
4. The *facilitator* manages the dialogue with or among the experts.

We refer to the individual who undertakes the elicitation as the researcher; there may be more than one. The researcher may also function as the analytical expert, facilitator, and even as the client. However, generally the steps in the elicitation are best performed by separate individuals with experience performing the necessary tasks (Hoffman and Kaplan 1999; Garthwaite et al. 2005; O'Hagan et al. 2006).

2.3.1 Preparation

The preparation stage is where the researcher decides the structure of the elicitation. Key tasks include definition of the problem, development of questions, and selection of experts. Adequate preparation is a key part of successful elicitation, since it will ensure a smoother process and maximize opportunities for identifying and countering possible biases. Experts respect and appreciate the effort a researcher has put into developing the elicitation documentation and the questions, and are generally inclined to reciprocate by devoting similar time and effort when making their judgments (van der Gaag et al. 1999).

2.3.1.1 Problem Definition and Question Development

The first step is to determine the purpose of the elicitation and define the objectives precisely. The elicitation designer must determine what information is required, the level of precision, and the appropriate selection of experts. For example, is the purpose to inform policy, support decision-making, determine research priorities, or characterize uncertainty about a particular model, analysis, or parameter? The researcher may need to work with decision-makers and stakeholders to develop goals if the objectives of the process are not already specified.

The scientific literature should be reviewed to determine the extent of relevant scientific knowledge and to identify information gaps. It is usually helpful to provide experts with documentation outlining the relevant evidence that has been compiled into an appropriate, accessible form (Cooke and Goossens 2000). Background materials usually provide information about the objectives of the elicitation, explain the motivations for the formal methodology, outline what the elicitation will involve, explain relevant statistical concepts, and document the questions (e.g., Hogarth 1987; Morgan and Henrion 1990; Rothlisberger et al. 2010). Experts should have time to review the materials, raise any potential concerns, and volunteer relevant information prior to the elicitation proper.

Having identified the requirements for the elicitation, the researcher then defines and structures the problem and identifies the variables for which knowledge is to be elicited. Problem structuring refers to the process of breaking down the problem into a set of variables or relationships for which knowledge will be elicited. Planning, often in conjunction with substantive experts, aims to ensure that the structure is straightforward and intuitive for experts (Keeney and von Winterfeldt 1991). The level of problem disaggregation is an important consideration. In general, researchers disaggregate complex questions into more manageable sub-problems, aiming to create knowledge environments that are more comfortable and familiar to experts. This strategy aims to create a set of variables that best allow experts to incorporate their knowledge, for example, about quantities that are observable or that the experts have experienced directly (Cooke and Goossens 2000). The variables should be sufficiently well defined that experts can answer questions without further specification (Morgan and Henrion 1990).

Habitat suitability indices are good example of disaggregation techniques in ecology. These indices provide a quantitative representation of the relative suitability of some part of a landscape for the survival and reproduction of a species (Reading et al. 1996; Cohen et al. 2004). Rather than asking experts to estimate the suitability outright for every point in the landscape, elicitation of these indices instead requires experts to nominate which variables are most important in determining suitable habitat for a species, and how measures of these variables should be combined into an overall measure of suitability. Thus, they represent a disaggregated model that links environmental data to the persistence of a species.

The draft protocol and background information should be carefully piloted (tested and revised before it is used to collect actual data) to ensure that the questions have been framed appropriately, to identify possible problems with biases or question phrasing, and to receive feedback about any potential ways to improve the quality of the process and of the knowledge that is being elicited. To some degree, all questions are biased, but careful development combined with testing and refinement of the protocol by substantive experts can minimize adverse effects considerably (Payne 1951). It should also be noted that experts used in testing the protocol should not be used to answer the questions; this is a formal technical requirement in Bayesian analysis.

2.3.1.2 Selection of Experts

The selection process involves identification of the expertise that will be relevant to the elicitation process, and selection of the subset of experts who best fulfill the requirements for expertise within the existing time and resource constraints. In some cases, the selection of appropriate experts is straightforward, but in other cases, an appropriate expert group will need to be defined by the researcher according to the experts' availability and the requirements of the elicitation. Experts should be selected using explicit criteria to ensure transparency, and to establish that the results represent the full range of views in the expert community. Common metrics for identifying experts include qualifications, employment, memberships in professional bodies, publication records, years of experience, peer nomination, and perceived standing in the expert community (e.g., Chuenpagdee et al. 2003; Drescher et al. 2008; Whitfield et al. 2008; Czembor and Vesk 2009). Additional considerations include the availability and willingness of the experts to participate, and the possibility of conflicts of interest.

The appropriate number of experts depends on the scope of the problem, the available time and other resources, and the level of independence between experts. Experts often share beliefs because of shared information sources and training. In such cases, the marginal benefits of including more than about five to eight experts decrease quickly (Winkler and Makridakis 1983; Clemen and Winkler 1985). As a result, researchers are encouraged to include as diverse a range of experts as possible. The literature on expert elicitation strongly recommends the use of multiple experts to buffer against individual mistakes and biases, and to allow for assessments that are representative of the whole expert community (Hokstad et al. 1998; Clemen and Winkler 1999; Armstrong 2006). Even in cases where one expert is considered substantially more knowledgeable than the others, a diversity of opinions from a group of "lesser" experts may outperform the opinion of a single "best" expert (Bates and Granger 1969; Dickinson 1973; 1975; Otway and von Winterfeldt 1992;

Clemen and Winkler 1999; Armstrong 2001; Fisher 2009). The combined judgment also tends to be more reliable, since *a priori* identification of a single best expert is not always straightforward.

In most ecological settings, the breadth of concerns means that no one individual will be expert for all aspects of the problem (e.g., Ludwig et al. 2001; Martin et al. 2005). For example, in the elicitation described by Martin et al. (2005), no single expert had the required expertise for all 20 bird species that were considered. Using multiple experts was an important strategy to obtain the required expert coverage. The use of larger expert groups may also be beneficial if it will increase the acceptance or perceived validity of the elicitation outcomes. This is particularly true in contexts such as a public consultation process, in which the stakeholders may include many groups of individuals who are not traditionally considered to be experts, but who nonetheless possess expertise in certain relevant domains.

2.3.2 *Elicitation*

2.3.2.1 Expert Pretraining

Substantive experts may be unfamiliar with expressing their beliefs numerically or in the format required by the elicitation protocol. Pretraining provides participants with appropriate experience, and where relevant, improves their understanding of the concepts involved in the elicitation. Given sufficient practice combined with adequate feedback, experts can substantially improve their performance, thereby becoming more reliable and accurate (Ferrell 1994; Renooij 2001). Inclusion of pretraining may be particularly important where elicitations involve the assessment of complex, unintuitive statistical formats such as quantiles or the moments of a probability distribution (see Hogarth 1987; Morgan and Henrion 1990; Cooke and Goossens 2000; Renooij 2001).

2.3.2.2 Elicitation

During this step, the experts respond to questions to assess the required variables, usually under the guidance of a facilitator. The expert performs four tasks during the elicitation (Meyer and Booker 1991):

1. Understands the question.
2. Searches for and recalls the relevant information.
3. Makes judgments.
4. Constructs and reports an answer.

Errors may enter the elicitation process at any of these stages. The process should, therefore, be viewed as one that helps an expert construct a set of carefully reasoned and considered judgments.

Five steps can help to counteract the psychological biases associated with elicitation: motivating, structuring, conditioning (i.e., defining any conditions that affect the problem definition), encoding, and verifying (Spetzler and Stael Von Holstein 1975; von Winterfeldt and Edwards 1986; Morgan and Henrion 1990; Shephard and Kirkwood 1994). We outline these steps in the remainder of this section. They involve ensuring that the expert has a complete understanding of each variable for which knowledge will be elicited and of any assumptions or conditioning factors, that they have had a chance to discuss and develop their reasoning and reflect on the relevant evidence, and having responded, that they have a chance to review and verify their responses.

2.3.2.3 Motivating

The facilitator works to develop an initial rapport or understanding with the experts and to establish their approval of the objectives of the elicitation. Facilitators explain the context and reasons for the elicitation and how the results will be used, the motivation for the experts' involvement, and how the expert's judgments will contribute (Walls and Quigley 2001). An introduction to the psychology of human judgment and bias in the elicitation will help the expert to understand the need for the formal elicitation process.

Experts are often wary of giving estimates that are not based on direct evidence. It is usually important to stress that there is no single correct response and that the aim of the process is only to elicit an accurate representation of the expert's true beliefs (Cooke 1991). The facilitator also identifies issues that may bias an expert's assessments, such as personal beliefs or conflicts of interest.

2.3.2.4 Structuring

At this stage, the facilitator goes through the details of each of the independent variables for which knowledge is to be elicited, including the model structure and conditions that constrain the expert's responses, and resolves any ambiguities. The aim is to ensure that each expert has a complete, unambiguous understanding of what information they are being asked to provide and what assumptions they are based on.

2.3.2.5 Conditioning

The facilitator and experts review the information and any assumptions on which the experts will base their assessments. The facilitator then questions the experts about their reasoning to ensure they have fully considered all possibilities, for example, by considering scenarios that may lead to unexpected outcomes.

2.3.2.6 Encoding

At this stage, the expert is asked to state their beliefs for each variable, for example, as probabilities or relative weights. Different techniques can be employed to encode the expert's beliefs, and we outline a number of the approaches commonly applied within landscape ecology in Sect. 3.2.8. In-depth coverage of different encoding techniques can be found in Spetzler and Stael Von Holstein (1975), von Winterfeldt and Edwards (1986), Morgan and Henrion (1990), Cooke (1991), Renooij (2001), Garthwaite et al. (2005), and the references therein.

2.3.2.7 Verifying

Following the assessment, the facilitator reviews the responses for signs of bias (e.g., experts who gave consistently high or low probabilities), and confirms that the responses are logical and consistent. Experts are asked to review their judgments, consider alternatives, and verify or change their judgments if they wish. Experts are rarely aware of the full implications of a set of judgments, and viewing their assessments in multiple formats (e.g., computer visualizations, statistical graphs, data tables) prompts a more rigorous reassessment of their beliefs. Experts should also be given an opportunity to review the outputs of any model or final representation, such as a graphical representation of the probability distribution constructed from their responses, to ensure that this result represents a reasonable reflection of their beliefs. The facilitator should actively question the expert, and should provide examples of their responses in multiple formats to prompt the expert to reconsider their statements in a new light.

2.3.2.8 Encoding Techniques

At the encoding stage, the expert is asked to state their knowledge using a particular response format. Experts can be asked to state their knowledge directly, for example, using questions such as "What is the probability of the event?" or "What is the value of variable *x*?". However, methods such as these do not help the expert to construct their beliefs. Experts are sensitive to the effects of the question framing and the response format in constructing their beliefs, and may benefit from assistance that reduces the cognitive strain in translating their beliefs into the required response format. Encoding techniques in the form of particular question formats have been developed to assist in estimating quantities and probability distributions that align with the expert's beliefs. We discuss these techniques in the remainder of this section.

Different encoding approaches can be used, depending on the type of information being elicited (e.g., probabilities, means, probability distributions; see Table 2.2). A complete enumeration of the full set of approaches is beyond the scope of this chapter. Below, we outline a few techniques that have been applied in landscape ecology.

Table 2.2 Summary of some approaches that have been commonly used for elicitation in landscape ecology (after Kuhnert et al. 2010, reproduced with permission from Wiley)

	Quantitative measures						Qualitative measures	
Criteria/methods	(1) Probability	(2) Frequency	(3) Quantity (e.g., means)	(4) Weighting	(5) Quantitative interval	(6) Probability distribution	(7) Categorical measure	(8) Relative measure
Type of elicitation	Direct	Indirect	Direct	Indirect	Direct/indirect	Direct	Indirect	Indirect
Benefits from multiple experts	✓	✓	✓	✓			✓	✓
Language easily interpreted		✓	✓	✓			✓	✓
High level of expertise required	✓	✓	✓		✓	✓		
Knowledge of probability theory required						✓		
Can be performed remotely	✓	✓	✓	✓	✓		✓	✓
Fast and easy	✓	✓	✓	✓			✓	✓
Uncertainty explicitly specified					✓	✓		
Example	$p=0.3$	x out of n trials	n	rank	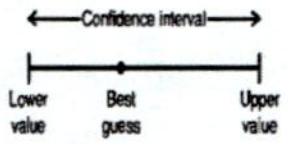	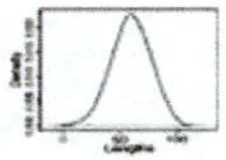	low/medium/high	↑ ↓
Key references	Griffiths et al. (2007)	Gigerenzer (2002)	Christen and Nakamura (2000)	MacNally (2007)	O'Neill et al. (2008)	Kynn (2004)	Yamada et al. (2003)	Martin et al. (2005)

Ranking methods can be used to elicit information indirectly. In the analytical hierarchy process (Saati 1980; Tavana et al. 1997), the expert is presented with pairs of events or criteria and asked to rank the relative importance of each pair. Rankings use a scale ranging from 1, which represents equal importance, to 9, which represents a situation in which one alternative is "absolutely" more important. The analytical hierarchy process can also be adapted to elicit information about relative likelihoods. Weights or probabilities are then fitted using matrix algebra. Experts find this process easy and intuitive. However, it is best suited to small numbers of discrete events because the number of assessments becomes impractically large for large numbers of events. Assessed probabilities may be anchored by including events for which the "true" probability is known.

Verbal qualifiers of uncertainty, which include words or phrases such as "highly likely" or "uncertain", can be used to qualify a probability or a degree of confidence, or to specify the incertitude associated with a concept. They are intuitive and are used as an alternative to numerical probabilities in eliciting information from experts (Wallsten et al. 1997). People often prefer to express their uncertainty with verbal phrases rather than numbers, though as experts gain experience with numerical techniques, this preference often lessens (Spetzler and Stael Von Holstein 1975; Cooke 1991; Walls and Quigley 2001).

Verbal qualifiers have the potential to introduce substantial linguistic uncertainty. Phrases do not correspond to a single numerical value, and individuals interpret them differently depending on the context (Beyth-Marom 1982; Budescu and Wallsten 1985; Wallsten et al. 1986; Wallsten and Budescu 1995; Windschitl and Wells 1996). For example, the phrase "very unlikely" may mean different things when referring to the possibility of a disease outbreak and the chance of rain tomorrow. Variance in the interpretation of such phrases between individuals can span almost the entire probability scale. People are usually unaware of the extent of these differences (Brun and Teigen 1988). Phrases such as "insignificant", "negligible", or "moderate" may also carry implied value judgments.

Probabilities are often difficult to elicit directly. Tools such as probability scales and probability wheels provide a straightforward visual representation for experts, though responses may be subject to scaling biases such as centering and spacing. Renooij (2001) recommended the use of such tools when experts are inexperienced with assessing probabilities. Presenting and eliciting information using natural frequencies (e.g., 13 out of 100), rather than percentages or probabilities (e.g., 13% or 0.13), can improve the accuracy of elicitation, particularly when experts are unfamiliar with probabilistic terms (Gigerenzer and Hoffrage 1995; Cosmides and Tooby 1996). For example, rather than assessing the probability that Hawaiian birds will become extinct in the next 10 years, we can ask experts to predict the number of bird species that will become extinct out of the number of original bird species. Frequency formats are easier to understand and may be less susceptible to mistakes such as overconfidence and base-rate neglect, in which an expert tends to ignore background frequencies when estimating probabilities (Tversky and Kahneman 1983; Tversky and Koehler 1994; Gigerenzer and Hoffrage 1995; Price 1998; Hertwig and Gigerenzer 1999). However, they may be less useful when experts find it difficult to imagine occurrences of a very rare event (e.g., Slovic et al. 2000; van der Gaag et al. 2002).

There are two main ways to elicit intervals: using a fixed probability (a quantile) or using a fixed value (Tallman et al. 1993). In the fixed-probability method, experts are asked to specify the value of a quantity within a specified quantile. It is common to elicit the 5, 50, 80, and 95% quantiles and to elicit quartiles (25, 50, and 75%). In the fixed-value method, the expert is asked to assign a probability that a quantity lies within a specific range of values, normally centered at the median. With both methods, experts typically display overconfidence, generating too-narrow intervals or assigning too-high levels of confidence.

O'Neill et al. (2008) were interested in estimating polar bear populations in the Arctic in the future. To elicit opinions about the relative changes in these populations, they asked experts to estimate the population in 2050 under current management regimes (based on the change in sea-ice distribution, which was shown using maps), expressed as percentage of today's population. The experts were asked to give their opinion and associated uncertainty using questions such as the following (adapted from O'Neill et al. 2008):

1. Please estimate the lower confidence bound for the total polar bear population in 2050.
2. Please estimate the upper confidence bound for the total polar bear population in 2050.
3. Please give your best estimate for total polar bear population in 2050.

Speirs-Bridge et al. (2010) reduced the level of overconfidence with a four-step question format. They recommended asking:

1. Realistically, what is the smallest the value could be?
2. Realistically, what is the largest the value could be?
3. What is your best guess for the true value?
4. How confident are you that the interval from lowest to highest contains the true value?

The most comprehensive form of elicitation is to elicit full probability distributions for each quantity. Parametric methods for eliciting distributions involve fitting expert assessments to a particular distribution or family of distributions (Garthwaite et al. 2005). Non-parametric distributions are usually constructed from a series of points or intervals elicited using graphical and numerical techniques, such as those described above. Points or intervals are elicited because the ability of experts to specify parameters such as the sample variance is poor (Peterson and Beach 1967). Eliciting four to five (well chosen) points allows a curve to be fitted that provides a reasonable approximation of the expert's beliefs (e.g., O'Hagan 1998; O'Hagan et al. 2006).

Methods have been developed for eliciting many of the commonly used parametric distributions, such as the normal and multivariate normal. We do not review these parametric methods here, but excellent overviews are given in, among others, Kadane et al. (1980), Al-Awadhi and Garthwaite (1998), Kadane and Wolfson (1998), Garthwaite et al. (2005), and O'Hagan et al. (2006).

2.3.3 *Analysis*

2.3.3.1 Verification

Following the elicitation, the researcher should perform a second, more rigorous verification process. In addition to checking for obvious errors or inconsistencies, the researcher compares the expert's responses to those of others in the group and against available information to establish the external validity of the expert responses. External validation is important, but is often limited by a lack of appropriate alternative sources of information with which to corroborate expert responses. In comparing an individual expert's responses with those of the rest of the group, the researcher looks for biases, anomalies, or strongly discordant opinions, as well as for varying interpretations of the information. The researcher should follow up on any interesting or problematic responses through further discussion with the expert. In some procedures, the verification stage includes a step in which experts see and may even question the responses of other experts before making their final judgment (Cooke 1991). If any calculations are performed using the expert's responses, the results should be provided for the expert to review and confirm. The aim of this stage is to arrive at a final set of judgments that the experts have approved. The responsibility rests with the researcher to ensure that the documented responses are consistent and that they faithfully reflect each expert's true beliefs.

2.3.3.2 Aggregation

Where judgments are elicited from two or more experts, it will usually be necessary to aggregate their opinions. Expert opinions often vary considerably and can often be contradictory or inconsistent. For example, it is not uncommon for experts to specify estimates that don't overlap.

Deciding how to aggregate the responses depends on why the expert judgments differ. Differences may arise as a result of (1) differing levels of knowledge or expertise, (2) different interpretations or weights assigned to pieces of evidence, (3) different theoretical models, and (4) differences in personal values or motivational biases (Morgan and Henrion 1990). In some cases, combining expert judgments may not be theoretically defensible or practical, or might lead to misrepresentations of the data (Keith 1996; Hora 2004; O'Hagan et al. 2006).

In some cases, differences in responses may lead the analyst to revisit earlier stages of the elicitation, or to consult experts further to understand the source of their beliefs. For example, it is possible that some of the experts failed to use information that others found to be influential, or weighed evidence differently. Alternatively, it may be clear from the responses that one of the experts misunderstood the question (or understood it differently). In these cases, it may be possible to ask the expert to revisit their response.

If there are wide differences in opinion, especially relative to intraexpert variability (i.e., the epistemic uncertainty in an expert's judgments), this is an important insight and should be communicated to decision-makers. Similarly, it may be important to know whether disagreements will have a significant impact on a decision. If differences of opinion persist and they could affect a decision, the analyst may elect to present a range of scenarios, each based on a different set of expert judgments (e.g., Crome et al. 1996).

If aggregation is appropriate, judgments may be combined using either behavioral or mathematical approaches (Clemen and Winkler 1999). Behavioral approaches involve interactions among experts, typically in a group setting, with opinions aggregated by the experts. Behavioral methods for resolving opinions may be structured, such as following a protocol for reaching agreement, or unstructured, by means of informal seeking of consensus (see Hogarth 1977; Crance 1987; Lock 1987; Burgman 2005; Macmillan and Marshall 2006).

Mathematical approaches involve combining the expert opinions using rules and do not involve any interactions between experts. Mathematical aggregation can be accomplished with Bayesian methods or opinion pools. Bayesian methods treat the resolution of differences among experts as a Bayesian inference problem (Morris 1974, 1977). A practical impediment is that the Bayesian approach requires the estimation of complex dependencies between experts (Jacobs 1995). Instead, in practice, opinion pools (typically the average or median for the group) are commonly implemented (Clemen 1989; Genest and McConway 1990; Armstrong 2001). Averaging is easy to implement, and more complicated methods may not provide better results (Clemen 1989). Methods may also combine elements from both behavioral and mathematical approaches (Cooke 1991). The theory and application of expert aggregation methods is reviewed in detail in Seaver (1978), Genest and Zidek (1986), and more recently by Clemen and Winkler (1999).

2.3.4 Trade-offs Between Cost and Accuracy

The use of a full formal elicitation protocol is neither necessary nor desirable for every analysis or decision (Pate-Cornell 1996). A tradeoff exists between time and precision, since methods that provide precise estimates by mitigating cognitive biases are also the most time-consuming. Interviews, for instance, are likely to result in better-quality responses than questionnaires, but make onerous time and resource demands. A full-scale elicitation process can involve dozens of people and last from 1 to 2 years, with estimated costs ranging from $100,000 to in excess of $1 million (e.g., Moss and Schneider 2000; Slottje et al. 2008). It is reasonable to assume that in many cases, decision analysts will not have access to, or wish to commit, this level of time and resources to elicitation.

Different formats and techniques will be appropriate, depending on the available time and resources and on the requirements of the problem. Particular considerations will include the number and types of experts who are available, the precision required, and the time and resources available to conduct the elicitation (Kuhnert et al. 2010). For example, Shephard and Kirkwood (1994) noted that the analyst must balance the desire for a probability distribution that more accurately represent the expert's knowledge against the need to retain their interest and attention throughout the elicitation process and to complete the elicitation efficiently. This tradeoff can require compromises, leading the analyst to forgo opportunities to iterate the estimation–validation–discussion process, or to use simpler question formats.

Less-intensive elicitations should still be guided by the principles outlined above. Researchers should always construct questions carefully, for example, and provide experts with the opportunity to revise their responses. In some cases, an expert may be reluctant to make estimates if they feel it is not scientifically appropriate. Morgan and Henrion (1990) suggest that there is a big difference between taking a position on what the answer might be and identifying what range of values might be correct. Indeed, scientists frequently advance their research using this type of reasoning.

2.4 Expert Knowledge in Landscape Ecology

In the previous sections, we examined expertise and techniques for the formal elicitation of expert knowledge. A core theme has been that both expert characteristics and appropriate elicitation practices vary with the task setting and requirements. In this section, we use this framework to critically examine current practices for employing expert knowledge in ecology, and make recommendations for future use of this knowledge.

The use of expert knowledge in landscape ecology is widespread. It is used regularly in problem characterization, model conceptualization, parameterization, and processing of data (Burgman 2005). Expert knowledge is frequently used as an alternative source of information when empirical data are not available (Burgman 2005; Sutherland 2006). The recourse to expert knowledge is particularly common for decision-makers operating in new, changing, or understudied systems. It is also valuable as a tool to supplement empirical information when the empirical information available is biased or incomplete, to corroborate model findings, to synthesize existing knowledge, and to correctly extrapolate, interpret, and apply knowledge to new situations (Pellikka et al. 2005; Teck et al. 2010).

Structured techniques and expert judgments have been used in scenario planning, species distribution modeling (Pearce et al. 2001; Johnson and Gillingham 2004), forest planning (Crome et al. 1996; Alho and Kangas 1997; Kangas and Kangas 2004), and the evaluation of conservation priorities (Sanderson et al. 2002; Marsh et al. 2007; Teck et al. 2010). The increasing use of Bayesian techniques, which provide a framework for the explicit inclusion of expert knowledge through the creation of a "prior" distribution for the problem parameters and subsequent improvement of the distribution using empirical knowledge, has contributed to a wider awareness of structured elicitation protocols (Kuhnert et al. 2010).

Despite the advances in and the advantages of structured elicitation methods, informal expert knowledge is more commonly deployed. For example, distances between bird nests and human habitations and analyses of breeding success in relation to distance to human habitations have been used to designate buffer zones for some species (e.g., Helander and Stjernberg 2003; Helander et al. 2003). It has become apparent that in many cases expert opinion had been used to recommend and designate buffer zones. Although such approaches are valid, this reliance on expert rather than empirical knowledge was rarely acknowledged explicitly (e.g., Grier et al. 1993; Currie and Elliott 1997). The problem this creates for decision-makers and subsequent researchers is that without knowing the sources of the knowledge or how it was elicited, it becomes difficult to know how much to rely on the knowledge. In addition, it becomes difficult to update the knowledge, since the assumptions and reasoning on which the previous knowledge was based are unknown.

Formal applications of expert knowledge in ecology and conservation typically omit many of the principles for structured elicitation outlined in Sect. 3. Only a handful of examples of elicitations have employed the principles of elicitation design (Low-Choy et al. 2009). Selection or development of an elicitation approach appears to have been primarily *ad hoc*, and documentation of the methodology was usually incomplete or absent. Experts are rarely trained before the elicitation. It is rare that clear explanations of the elicitation process and goals, or opportunities to verify or evaluate the elicited knowledge are provided (Roloff and Kernohan 1999).

2.5 Conclusions and Future Directions

Expert knowledge should be incorporated formally within a framework that is explicit and transparent, and both the experts and the researchers must be accountable to those who will use the elicited knowledge. Formal methods help to make knowledge available that otherwise might not have been accessible. As a result of a structured elicitation process, experts consider more facets of the problem, are interrogated more fully about their beliefs, and have opportunities to correct ambiguities and errors of understanding (Burgman et al. 2011).

The move in ecology toward more formal, structured processes for incorporating expert knowledge is promising (Martin et al. 2005; Low-Choy et al. 2009; Kuhnert et al. 2010; Burgman et al. 2011). The development of elicitation procedures should be informed by the characteristics of the task at hand and of the environment in which the experts have acquired their knowledge. Lessons from the formal paradigm include the importance of adequate preplanning and preparation (including pretesting of the protocol), of an opportunity to train experts, of appropriate tailoring of questions and elicitation formats to the expert's knowledge and experience, and of including a verification stage.

Table 2.3 summarizes what we view as the key decisions that characterize the development of an elicitation procedure. Design of an elicitation procedure may be viewed as a resource-allocation problem in which the analyst allocates limited

Table 2.3 Eight key decisions in the design of a formal elicitation procedure

Decision	Characterization	Guidelines
1. The format for the elicitation	Setting in which the elicitation will take place. For example, via e-mail survey, phone interview, or in person.	Interviews are preferable unless the expense or number of experts makes it infeasible. In person, it is easier to correct any misunderstandings, maintain expert motivation, provide training and feedback, and incorporate interactions between experts and between the expert and the facilitator.
2. The information that will be elicited	Involves decisions about how many variables will be elicited, in what form, and under what conditioning assumptions. Usually determined as a part of structuring the problem description and the conceptual models for the decision or processes of interest.	Ideally, experts should be able to state their knowledge directly, in a format that is as close as possible to the conditions under which the knowledge was acquired. This helps to remove any additional, unnecessary cognitive burdens. Research suggests that for complex problems, expert knowledge is best incorporated within a model or a broader conceptual framework (Armstrong 2001).
3. The experts who will be involved	How the experts will be identified and the number that will be included.	Multiple experts should be involved to provide corroboration and avoid simple errors. Diversity of experts may be more important than their number or years of experience because this helps to ensure that all aspects of the problem are considered, from multiple perspectives.
4. The level of pretraining to be provided	The number and type of practice questions that will be provided, and the level of feedback. Additional options include an introduction to cognitive and motivational biases, and to probability concepts if probabilities are to be elicited.	Practice accompanied by feedback on the expert's performance has been shown to improve performance for questions that are sufficiently similar to those used in the actual elicitation. This is particularly beneficial where experts are inexperienced with the question format. There is no evidence yet that providing information about cognitive and motivational biases help experts to avoid reasoning fallacies.
5. How uncertainty will be elicited	How uncertainty is to be incorporated and propagated through the analysis. For example, elicitation of a complete probability distribution versus definition of the upper and lower bounds around an estimate.	The choice of method with which to elicit uncertainty will depend on the level of precision required, the time available for elicitation, and the expert's knowledge. If uncertainty is not elicited, decision-makers will need to infer the precision of the responses.

(continued)

Table 2.3 (continued)

Decision	Characterization	Guidelines
6. The question format	Whether qualitative or quantitative information will be elicited and in what format, for instance as probabilities, probability distributions, ranks, or categorical measures.	Knowledge is available to inform the selection of appropriate response formats (see O'Hagan et al. 2006 and the references therein). Ranks and category formats are often preferred by experts over numerical responses, but are susceptible to linguistic uncertainty and confounding of knowledge with value judgments.
7. The degree to which experts will verify their responses	Whether and how experts will verify their responses, for example, in conjunction with graphical feedback, analysis of the output, or assisted by responses and reasoning from other experts.	Some minimum level of verification is important to catch errors and misunderstandings, particularly for less intensive protocols. Provision of feedback in multiple formats helps experts to check the coherence and accuracy of their responses more thoroughly.
8. How judgments from experts will be combined	Via mathematical or behavioral means, and the degree to which the experts will be given the opportunity to interact.	Empirical results suggest that mathematical methods outperform behavioral techniques. Use of measures such as the group average is a standard approach. Group discussions should be facilitated by a skilled facilitator, and may be most fruitful when combined with a final mathematical step to summarize the data that results from the discussions (Clemen and Winkler 1999).

available resources to achieve the greatest expected gains in response quality. Elicitation procedures should be developed with a view to how each feature will contribute to the elicitation as a whole. Improvement of existing practices within landscape ecology will require a greater awareness of the tools available to improve elicitation quality, and an understanding of how to select and tailor these techniques to best suit the decision problem at hand.

Ecological systems are complex and non-linear, with processes that unfold over long timescales and large spatial scales. In making predictions about future dynamics, experts are likely to be operating outside their direct area of expertise. Our guidelines (Table 2.3) suggest that expert knowledge may be most appropriately incorporated within a conceptual framework that avoids the need for experts to make predictions for complex, compound events. Use of multiple experts introduces more knowledge about a system and its dynamics, thereby creating a more detailed and comprehensive picture of the problem, and if the knowledge is deployed appropriately, it may lead ultimately to better decisions.

The primary focus of the methods presented in this chapter is on eliciting numerical information, which is a useful way of making tacit (implicit) knowledge more transparent, explicit, and useful to the decision-maker. The translation of expert knowledge into numbers is often difficult and requires care, but it is worthwhile making the effort to rigorously obtain these numbers, as they have considerable benefit for the decision-maker. In this chapter, we focus less on eliciting conceptual models or qualitative information, though many of the principles remain the same. The details of such elicitations are beyond the scope of the chapter, but they are nonetheless important in some contexts. For example, qualitative information may provide useful insight into the understanding of a system (e.g., McCoy et al. 1999), Yamada et al. 2003).

Expert knowledge is a necessary component in the analysis of any complex decision problem (Keeney and von Winterfeldt 1991). This knowledge represents a valuable resource for decision-makers, but as with any tool or resource, its value may be lessened by inappropriate or ill-informed application. Expert status alone is not enough to guarantee accurate responses, and traditional metrics of expertise such as the expert's age, rank, or experience, do not necessarily predict an expert's performance (Burgman et al. 2011). Structured elicitation techniques can be used to increase the reliability of expert opinions and counter some of the limitations associated with expert knowledge.

The use of formal practices within landscape ecology is increasing, but these uses would benefit from a greater emphasis on structured design. Steps such as the use of multiple, diverse experts and the inclusion of pretesting, training, and validation stages will contribute significantly to the elicitation of better-quality results. A move toward greater evaluation of both expert knowledge and the elicitation practices used to elicit that knowledge will improve the quality of knowledge available to inform future decisions, and improve expert and decision-maker accountability.

References

Al-Awadhi SA, Garthwaite PH (1998) An elicitation method for multivariate normal distributions. Commun Stat A-Theor 27:1123–1142

Alho JM, Kangas J (1997) Analyzing uncertainties in experts' opinions of forest plan performance. For Sci 43:521–528

Anderson EL, Hattis D (1999) A. Uncertainty and variability. Risk Anal 19:47–49

Anderson JL (1998) Embracing uncertainty: the interface of Bayesian statistics and cognitive psychology. Ecol Soc 2(1), article 2. Available from http://www.consecol.org/vol2/iss1/art2/ (accessed May 2011)

Armstrong JS (ed) (2001) Principles of forecasting: a handbook for researchers and practitioners. Kluwer Academic Publishers, Norwell

Armstrong JS (2006) Findings from evidence-based forecasting: methods for reducing forecast error. Int J Forecasting 22:583–598

Arnott D (2006) Cognitive biases and decision support systems development: a design science approach. Inform Syst J 16:55–78

Ausden M, Sutherland WJ, James R (2001) The effects of flooding lowland wet grassland on soil macroinvertebrate prey of breeding wading birds. J Appl Ecol 38:320–338

Ayyub BM (2001) Elicitation of expert opinions for uncertainty and risks. CRC Press, Boca Raton
Baran N (2000) Effective survey methods for detecting plants. MSc Thesis. University of Melbourne, Melbourne
Bates JM, Granger CWJ (1969) The combination of forecasts. Oper Res Q 20:451–468
Beyth-Marom R (1982) How probable is probable? A numerical translation of verbal probability expressions. J Forecasting 1:257–269
Booker JM, McNamara LA (2004) Solving black box computation problems using expert knowledge theory and methods. Reliab Eng Syst Safe 85:331–340
Bransford JD, Brown AL, Cocking RR (2000) How people learn: brain, mind, experience and school. National Academy Press, Washington
Brun W, Teigen KH (1988) Verbal probabilities: ambiguous, context-dependent, or both. Organ Behav Hum Dec 41:390–404
Budescu, DV, Wallsten TS (1985) Consistency in interpretation of probabilistic phrases. Organ Behav Hum Dec 36:391–405
Burgman MA (2000) Population viability analysis for bird conservation: prediction, heuristics, monitoring and psychology. Emu 100:347–353
Burgman MA (2005) Risks and decisions for conservation and environmental management. Cambridge University Press, Cambridge
Burgman MA, Carr A, Godden L et al (2011) Redefining expertise and improving ecological judgement. Conserv Lett 4:81–87
Camerer CF, Johnson EJ (1997) The process-performance paradox in expert judgment: how can experts know so much and predict so badly? In: Goldstein WM, Hogarth RM (eds) Research on judgment and decision making: currents, connections and controversies. Cambridge University Press, Cambridge, pp 342–364
Campbell LM (2002) Science and sustainable use: views of marine turtle conservation experts. Ecol Appl 12:1229–1246
Carpenter SR (2002) Ecological futures: building an ecology of the long now. Ecology 83:2069–2083
Chase WG, Simon HA (1973) The mind's eye in chess. In: Chase WG (ed) Visual information processing. Academic Press, New York, pp 215–281
Chi MTH (2006) Two approaches to the study of experts' characteristics. In: Ericsson KA, Charness N, Feltovich PJ, Hoffman, RR (eds) The Cambridge handbook of expertise and expert performance. Cambridge University Press, New York, pp 21–30
Christen JA, Nakamura M (2000) On the analysis of accumulation curves. Biometrics 56:748–754
Christensen-Szalanski JJJ, Bushyhead JB (1981) Physicians' use of probabilistic information in a real clinical setting. J Exp Psychol Human Percept Perform 7:125–126
Chuenpagdee R, Morgan LE, Maxwell SM et al (2003) Shifting gears: assessing collateral impacts of fishing methods in the U.S. waters. Front Ecol Environ 10:517–524
Clemen RT (1989) Combining forecasts: a review and annotated bibliography. Int J Forecasting 5:559–583
Clemen RT, Winkler RL (1985) Limits for the precision and value of information from dependent sources. Oper Res 33:427–442
Clemen RT, Winkler RL (1999) Combining probability distributions from experts in risk analysis. Risk Anal 19:187–203
Cohen MJ, Carstenn S, Lane CR (2004) Floristic quality indices for biotic assessment of depressional marsh condition in Florida. Ecol Appl 14:784–794
Collins, HM, Evans R (2007) Rethinking expertise. University of Chicago Press, Chicago
Cooke RM (1991) Experts in uncertainty: opinion and subjective probability in science. Oxford University Press, New York
Cooke RM, Goossens LHJ (2000) Procedures guide for structured expert judgement in accident consequence modelling. Radiat Prot Dosim 90:303–309
Cosmides L, Tooby J (1996) Are humans good intuitive statisticians after all? Rethinking some conclusions from the literature on judgment under uncertainty. Cognition 58:1–73
Cowling RM, Pressey RL (2003) Introduction to systematic conservation planning in the Cape Floristic Region. Biol Conserv 112:1–13

Crance JHBR (1987) Guidelines for using the Delphi technique to develop habitat suitability index curves. U.S. Fish Wildl Serv., Washington. Biological Report#82(10.134)
Crome FHJ, Thomas MR, Moore LA (1996) A novel Bayesian approach to assessing impacts of rain forest logging. Ecol Appl 6:1104–1123
Currie F, Elliott G (1997) Forests and birds: a guide to managing forests for rare birds. Forestry Authority, Cambridge, and Royal Society for the Protection of Birds, Sandy
Czembor CA, Vesk PA (2009) Incorporating between-expert uncertainty into state-and-transition simulation models for forest restoration. For Ecol Manage 259:165–175
Dawes RM, Kagan J (1988) Rational choice in an uncertain world. Harcourt Brace Jovanovich, San Diego
Dickinson JP (1973) Some statistical results in combination of forecasts. Oper Res Q 24:253–260
Dickinson JP (1975) Some comments on combination of forecasts. Oper Res Q 26:205–210
Drescher, MA. Perera AH, Buse LJ et al (2008) Uncertainty in expert knowledge of forest succession: a case study from boreal Ontario. For Chron 84:194–209
Ericsson KA (2004) Deliberate practice and the acquisition and maintenance of expert performance in medicine and related domains. Acad Med 79:S70–S81
Ericsson KA, Charness N (1994) Expert performance: its structure and acquisition. Am Psychol 49:725–747
Ericsson KA, Charness N, Feltovich PJ et al (eds) (2006) The Cambridge handbook of expertise and expert performance. Cambridge University Press, New York
Ericsson KA, Kintsch W (1995) Long-term working memory. Psychol Rev 102:211–245
Ericsson KA, Lehmann AC (1996) Expert and exceptional performance: evidence of maximal adaptation to task constraints. Annu Rev Psychol 47:273–305
Ericsson KA, Towne TJ (2010) Expertise. Wiley Interdisciplinary Reviews: Cognitive Science 1:404–416
Fazey I, Fazey JA, Fazey DMA (2005) Learning more effectively from experience. Ecol Soc 10(2), article 4. Available from http://www.ecologyandsociety.org/vol10/iss2/art4/ (accessed May 2011)
Ferrell WR (1994) Discrete subjective probabilities and decision analysis: elicitation, calibration and combination. In: Wright G, Ayton P (eds) Subjective probability. Wiley, New York
Ferson S, Ginzburg LR (1996) Different methods are needed to propagate ignorance and variability. Reliab Eng Syst Safe 54:133–144
Fischhoff B, Slovic P, Lichtenstein S (1982) Lay foibles and expert fables in judgments about risk. Am Stat 36:240–255
Fisher L (2009) The perfect swarm: the science of complexity in everyday life. Basic Books, New York
Garthwaite PH, Kadane JB, O'Hagan A (2005) Statistical methods for eliciting probability distributions. J Am Stat Assoc 100:680–700
Genest C, McConway KJ (1990) Allocating the weights in the linear opinion pool. J Forecasting 9:53–73
Genest C, Zidek JV (1986) Combining probability distributions: a critique and an annotated bibliography. Stat Sci 1:114–148
Gigerenzer G (1999) Simple heuristics that make us smart. Oxford University Press, New York
Gigerenzer G (2002) Calculated risks: how to know when the numbers deceive you. Simon and Schuster, New York
Gigerenzer G (2008) Rationality for mortals: how people cope with uncertainty. Oxford University Press, New York
Gigerenzer G, Hoffrage U (1995) How to improve Bayesian reasoning without instruction: frequency formats. Psychol Rev 102:684–704
Gilovich T, Griffin D, Kahneman D (eds) (2002) Heuristics and biases: the psychology of intuitive judgement. Cambridge University Press, Cambridge
Grier JW, Elder JB, Gramlich FJ et al (1993) The bald eagle in the northern United States. Bird Conserv 1:41–66

Griffiths SP, Kuhnert PM, Venables WN, Blaber SJM (2007) Estimating abundance of pelagic fishes using gillnet catch data in data-limited fisheries: a Bayesian approach. Can J Fish Aquat Sci 64:1019–1033

Helander B, Marquiss M, Bowerman W (eds) (2003) Sea Eagle 2000. In: Proceedings from an International Conference at Bjökö, Sweden, 13–17 September 2000. Swedish Society for Nature Conservation, Stockholm, pp 129–132

Helander B, Stjernberg,T (2003) Action plan for the conservation of white-tailed Sea Eagle (Haliaeetus albicilla). The Convention on the Conservation of European Wildlife and Natural Habitats, Strasbourg

Hertwig R, Gigerenzer G (1999) The 'conjunction fallacy' revisited: how intelligent inferences look like reasoning errors. J Behav Dec Making 12:275–305

Hofer E (1996) When to separate uncertainties and when not to separate. Reliab Eng Syst Safe 54:113–118

Hoffman FO, Kaplan S (1999) Beyond the domain of direct observation: how to specify a probability distribution that represents the "state of knowledge" about uncertain inputs. Risk Anal 19:131–134

Hoffman RR (1998) How can expertise be defined? Implications of research from cognitive psychology. In: Williams R, Faulkner W, Fleck J (eds) Exploring expertise. Macmillan, New York, pp 81–100

Hogarth RM (1977) Methods for aggregating opinions. In: Jungermann H, DeZeeuw G (eds) Decision making and change in human affairs. Reidel, Dordrecht, pp 231–255

Hogarth RM (1987) Judgment and choice: the psychology of decision. Wiley, New York

Hogarth RM (2001) Educating intuition. The University of Chicago Press, Chicago

Hokstad P, Oien K, Reinertsen R (1998) Recommendations on the use of expert judgment in safety and reliability engineering studies: two offshore case studies. Reliab Eng Syst Safe 61:65–76

Hora SC (1992) Acquisition of expert judgment: examples from risk assessment. J Energy Dev 118:136–148

Hora SC (2004) Probability judgments for continuous quantities: linear combinations and calibration. Manage Sci 50:597–604

Jacobs RA (1995) Methods for combining experts probability assessments. Neural Comput 7:867–888

Johnson CJ, Gillingham MP (2004) Mapping uncertainty: sensitivity of wildlife habitat ratings to expert opinion. J Appl Ecol 41:1032–1041

Kadane JB, Dickey JM, Winkler RL et al (1980) Interactive elicitation of opinion for a normal linear model. J Am Stat Assoc 75:845–854

Kadane JB, Wolfson LJ (1998) Experiences in elicitation. J Roy Stat Soc D-Sta 47:3–19

Kahneman D (1991) Judgment and decision making: a personal view. Psychol Sci 2:142–145

Kahneman D, Tversky A (eds) (1982) Judgment under uncertainty: heuristics and biases. Cambridge University Press, Cambridge

Kangas AS, Kangas J (2004) Probability, possibility and evidence: approaches to consider risk and uncertainty in forestry decision analysis. For Policy Econ 6:169–188

Kaplan S (1992) 'Expert information' versus 'expert opinions'. Another approach to the problem of eliciting/combining/using expert knowledge in PRA. Reliab Eng Syst Safe 35:61–72

Kardes FR (2006) When should consumers and managers trust their intuition? J Consum Psychol 16:20–24

Keeney RL, von Winterfeldt D (1991) Eliciting probabilities from experts in complex technical problems. IEEE Trans Eng Manage 38:191–201

Keith DW (1996) When is it appropriate to combine expert judgments? Climatic Change 33:139–143

Kidd A, Welbank M (1984) Knowledge acquisition. In: Fox J (ed) Infotech state of the art report on expert systems. Pergamon, London

Kuhnert PM, Martin TG, Griffiths SP (2010) A guide to eliciting and using expert knowledge in Bayesian ecological models. Ecol Lett 7:900–914

Kunda Z (1990) The case for motivated reasoning. Psychol Bull 108:480–498
Kynn M (2004) Eliciting expert knowledge for Bayesian logistic regression in species habitat modelling. Department of statistics, Queensland University of Technology, Brisbane
Larkin J, McDermott J, Simon DP, Simon, HA (1980) Expert and novice performance in solving physics problems. Science 208:1335–1342
Lock A (1987) Integrating group judgments in subjective forecasts. In: Wright G, Ayton P (eds) Judgmental forecasting. Wiley, Chichester, pp 109–128
Low-Choy S, O'Leary R, Mengersen K (2009) Elicitation by design in ecology: using expert opinion to inform priors for Bayesian statistical models. Ecology 90:265–277
Ludwig D, Mangel M, Haddad B (2001) Ecology, conservation, and public policy. Annu Rev Ecol Syst 32:481–517
MacMillan DC, Marshall K (2006) The Delphi process: an expert-based approach to ecological modelling in data-poor environments. Anim Conserv 9:11–19
MacNally, R (2007) Consensus weightings of evidence for inferring breeding success in broad-scale bird studies. Austral Ecol 32:479–484
Marsh H, Dennis A, Hines H et al (2007) Optimizing allocation of management resources for wildlife. Conserv Biol 21:387–399
Martin TG, Kuhnert PM, Mengersen K, Possingham, HP (2005) The power of expert opinion in ecological models using Bayesian methods: impact of grazing on birds. Ecol Appl 15:266–280
McCoy ED, Sutton PE, Mushinsky HR (1999) The role of guesswork in conserving the threatened sand skink. Conserv Biol 13:190–194
Meyer M, Booker J (1991) Eliciting and analyzing expert judgment: a practical guide. Academic Press, New York
Morgan MG, Henrion M (1990) Uncertainty: a guide to dealing with uncertainty in quantitative risk and policy analysis. Cambridge University Press, New York
Morris PA (1974) Decision analysis expert use. Manage Sci 20:1233–1241
Morris PA (1977) Combining expert judgments: a Bayesian approach. Manage Sci 23:679–693
Moss R, Schneider, SH (2000) Uncertainties in the IPCC TAR: Recommendations to lead authors for more consistent assessment and reporting. In: Pachauri R, Taniguchi R, Tanaka K (eds) Guidance papers on the cross cutting issues of the third assessment report of the IPCC. World Meteorological Organisation, Geneva, pp 33–51
Murphy AH, Winkler RL (1984) Probability forecasting in meteorology. J Am Stat Assoc 79:489–500
O'Hagan A (1998) Eliciting expert beliefs in substantial practical applications. J Roy Stat Soc D–Statistics 47:21–35
O'Hagan A, Buck CE, Daneshkhah AR et al (2006). Uncertain judgments: eliciting expert probabilities. John Wiley, West Sussex
O'Neill SJ, Osborn TJ, Hulme M et al (2008) Using expert knowledge to assess uncertainties in future polar bear populations under climate change. J Appl Ecol 45:1649–1659
Otway H, von Winterfeldt D (1992) Expert judgment in risk analysis and management: process, context, and pitfalls. Risk Anal 12:83–93
Pate-Cornell ME (1996) Uncertainties in risk analysis: six levels of treatment. Reliab Eng Syst Safe 54:95–111
Payne S (1951) The art of asking questions. Princeton University Press, Princeton
Pearce JL, Cherry K, Drielsma M et al (2001) Incorporating expert opinion and fine-scale vegetation mapping into statistical models of faunal distribution. J Appl Ecol 38:412–424
Pellikka J, Kuikka S, Lindén H, Varis O (2005) The role of game management in wildlife populations: uncertainty analysis of expert knowledge. Eur J Wildlife Res 51:48–59
Peterson CR, Beach LF (1967) Man as an intuitive statistician. Psychol Bull 68:29–46
Price PC (1998) Effects of a relative-frequency elicitation question on likelihood judgment accuracy: the case of external correspondence. Organ Behav Hum Dec 76:277–297
Reading RP, Clark TW, Seebeck JH, Pearce J (1996) Habitat suitability index model for the eastern barred bandicoot, Perameles gunnii. Wildlife Res 23:221–235

Regan HM, Colyvan M, Burgman MA (2002) A taxonomy and treatment of uncertainty for ecology and conservation biology. Ecol Appl 12:618–628
Renooij S (2001) Probability elicitation for belief networks: issues to consider. Knowl Eng Rev 16:255–269
Richman HB, Gobet F, Staszewski JJ, Simon HA (1995) Simulation of expert memory using EPAM IV. Psychol Rev 102:305–333
Roloff GJ, Kernohan BJ (1999) Evaluating reliability of habitat suitability index models. Wildlife Soc Bull 27:973–985
Rosqvist T, Tuominen R (2004) Qualification of formal safety assessment: an exploratory study. Safety Sci 42:99–120
Rothlisberger JD, Lodge DM, Cooke RM, Finnoff DC (2010) Future declines of the binational Laurentian Great Lakes fisheries: the importance of environmental and cultural change. Front Ecol Environ 8:239–244
Saati TL (1980) The analytic hierarchy process. New York, McGraw-Hill
Sanderson EW, Redford KH, Chetkiewicz CLB et al (2002) Planning to save a species: the jaguar as a model. Conserv Biol 16:58–72
Seaver DA (1978) Assessing probability with multiple individuals: group interaction versus mathematical aggregation. Social Science Research Institute, University of Southern California, Los Angeles. Report# SSRI-78-3
Shanteau J (1992) Competence in experts: the role of task characteristics. Organ Behav Hum Dec 53:252–266
Shanteau J, Stewart TR (1992) Why study expert decision-making: some historical perspectives and comments. Organ Behav Hum Dec 53:95–106
Shephard GG, Kirkwood CW (1994) Managing the judgmental probability elicitation process: a case study of analyst/manager interaction. IEEE Trans Eng Manage 41:414–425
Shrader-Frechette K (1996) Value judgments in verifying and validating risk assessment models. In: Cothern CR (ed) Handbook for environmental risk decision making: values, perception and ethics. CRC Lewis Publishers, Boca Raton, pp 291–309
Slottje P, van der Sluijs JP, Knol AB (2008) Expert elicitation: methodological suggestions for its use in environmental health impact assessments. RIVM, Copernicus Institute for Sustainable Development and Innovation., Bilthoven. Report 630004001/2008
Slovic P (1999) Trust, emotion, sex, politics and science: surveying the risk-assessment battlefield. Risk Anal 19:689–701
Slovic P, Finucane ML, Peters E, MacGregor DG (2004) Risk as analysis and risk as feelings: some thoughts about affect, reason, risk, and rationality. Risk Anal 24:311–322
Slovic P, Monahan J, MacGregor DG (2000) Violence risk assessment and risk communication: the effects of using actual cases, providing instruction, and employing probability versus frequency formats. Law Human Behav 24:271–296
Speirs-Bridge A, Fidler F, McBride M et al (2010) Reducing overconfidence in the interval judgments of experts. Risk Anal 30:512–523
Spetzler CS, Stael Von Holstein CAS (1975) Probability encoding in decision analysis. Manage Sci 22:340–358
Stern PC, Fineberg HV (eds) (1996) Understanding risk: informing decisions in a democratic society. National Academies Press, Washington
Sutherland WJ (2006) Predicting the ecological consequences of environmental change: a review of the methods. J Appl Ecol 43:599–616
Sutherland WJ, Bailey MJ, Bainbridge IP et al (2008) Future novel threats and opportunities facing UK biodiversity identified by horizon scanning. J Appl Ecol 45:821–833
Sutherland WJ, Pullin AS, Dolman PM, Knight TM (2004) The need for evidence-based conservation. Trends Ecol Evol 19:305–308
Tallman I, Leik RK, Gray LN, Stafford MC (1993) A theory of problem-solving behavior. Soc Psychol Quart 56:157–177
Tavana M, Kennedy DT, Mohebbi B (1997) An applied study using the analytic hierarchy process to translate common verbal phrases to numerical probabilities. J Behav Dec Making 10:133–150

Teck SJ, Halpern BS, Kappel CV et al (2010) Using expert judgment to estimate marine ecosystem vulnerability in the California Current. Ecol Appl 20:1402–1416
Tversky A, Kahneman D (1974) Judgment under uncertainty: heuristics and biases. Science 185:1124–1131
Tversky A, Kahneman D (1983) Extensional versus intuitive reasoning: the conjunction fallacy in probability judgment. Psychol Rev 90:293–315
Tversky A, Koehler DJ (1994) Support theory: a nonextensional representation of subjective-probability. Psychol Rev 101:547–567
van der Gaag LC, Renooij S, Witteman CLM et al (1999) How to elicit many probabilities. In: Laskey KB, Prade H (eds) Proceedings of the 15th Conference on Uncertainty in Artificial Intelligence, Stockholm, July–August 1999. Morgan Kaufmann, San Francisco
van der Gaag LC, Renooij S, Witteman CLM et al (2002) Probabilities for a probabilistic network: a case study in oesophageal cancer. Artif Intell Med 25:123–148
van Steen JFJ (1992) A perspective on structured expert judgment. J Hazard Mater 29:365–385
von Winterfeldt D, Edwards W (1986) Decision analysis and behavioral research. Cambridge University Press, Cambridge
Walls L, Quigley J (2001) Building prior distributions to support Bayesian reliability growth modelling using expert judgement. Reliab Eng Syst Safe 74:117–128
Wallsten TS, Budescu DV (1995) A review of human linguistic probability processing: general principles and empirical evidence. Knowl Eng Rev 10:43–62
Wallsten TS, Budescu DV, Erev I, Diederich A (1997) Evaluating and combining subjective probability estimates. J Behav Dec Making 10:243–268
Wallsten TS, Budescu DV, Rapoport A et al (1986) Measuring the vague meanings of probability terms. J Exp Psychol Gen 115:348–365
Whitfield DP, Ruddock M, Bullman R (2008) Expert opinion as a tool for quantifying bird tolerance to human disturbance. Biol Conserv 141:2708–2717
Wilson AG (1994) Cognitive factors affecting subjective probability assessment. Duke University, Institute of Statistics and Decision Sciences, Durham. Report #94–02
Windschitl PD, Wells GL (1996) Measuring psychological uncertainty: verbal versus numeric methods. J Exp Psychol-Appl, 2:343–364
Winkler RL, Makridakis S (1983) The combination of forecasts. J Roy Stat Soc A-Sta 146:150–157
Yamada K, Elith J, McCarthy M, Zerger A (2003) Eliciting and integrating expert knowledge for wildlife habitat modelling. Ecol Model 165:251–264

Chapter 3
Elicitator: A User-Friendly, Interactive Tool to Support Scenario-Based Elicitation of Expert Knowledge

Samantha Low-Choy, Allan James, Justine Murray, and Kerrie Mengersen

Contents

S. Low-Choy (✉)
Cooperative Research Centre for National Plant Biosecurity, Canberra, ACT, Australia

Mathematical Sciences Discipline, Queensland University of Technology,
GPO Box 2434, Brisbane, QLD 4001, Australia
e-mail: s.lowchoy@qut.edu.au

A. James
High Performance Computing Unit, Queensland University of Technology,
GPO Box 2434, Brisbane, QLD 4001, Australia

J. Murray
Spatial Ecology Laboratory, The University of Queensland, Brisbane, QLD, Australia

Sustainable Ecosystems, CSIRO Ecosystem Sciences, PO Box 2583,
Brisbane, QLD 4001, Australia

K. Mengersen
Queensland University of Technology, GPO Box 2434, Brisbane, QLD 4001, Australia

A.H. Perera et al. (eds.), *Expert Knowledge and Its Application in Landscape Ecology*,
DOI 10.1007/978-1-4614-1034-8_3, © Springer Science+Business Media, LLC 2012

3.1 Introduction

Expert elicitation is the process of determining what expert knowledge is relevant to support a quantitative analysis and then eliciting this information in a form that supports analysis or decision-making. The credibility of the overall analysis, therefore, relies on the credibility of the elicited knowledge. This, in turn, is determined by the rigor of the design and execution of the elicitation methodology, as well as by its clear communication to ensure transparency and repeatability. It is difficult to establish rigor when the elicitation methods are not documented, as often occurs in ecological research. In this chapter, we describe software that can be combined with a well-structured elicitation process to improve the rigor of expert elicitation and documentation of the results.

When eliciting knowledge, software to automate and manage the mundane computational tasks is helpful, particularly if it provides immediate feedback to let experts review the accuracy of the results. At its most fundamental level, elicitation software can help experts estimate a single quantity in relatively concrete terms, such as the likelihood of species occurrence within a given environment. This is a ubiquitous elicitation need that informs decision-making, risk analysis, parameterization of complex models, and design of data collection or other studies (Sect. 3.2). SHELF is an example of a general-purpose tool that can support this kind of elicitation (Oakley and O'Hagan 2010). *Elicitator* utilizes different statistical encoding algorithms that permit different types of expert knowledge to be captured. Moreover, software such as *Elicitator* helps experts to express and explore their knowledge by providing a platform for feedback and communication that is dynamic, graphical, and interactive (e.g., Kynn 2005; Denham and Mengersen 2007).

Elicitation of a few fundamental quantities may be all that is required when the model is already fully specified. In these cases, direct elicitation simply asks experts to estimate the values of the parameters, which become inputs to the existing model (e.g., Boyce et al. 2005). At a more abstract level of elicitation, experts may be enlisted to help construct a biologically plausible model. Frameworks such as multicriteria decision analysis (e.g., Store and Kangas 2001) and Bayesian belief networks (e.g., Smith et al. 2007) employ a *deductive* approach, in which experts propose the model structure rather than learning it from the data. Alternatively, researchers can use an *inductive* approach to learn the model's structure from experts by revealing patterns underlying the parameters that will be elicited at a more fundamental level (i.e., that will be quantified). This is a form of *indirect* elicitation, since parameters for the model are inferred from the knowledge elicited from the experts. Such an indirect approach is particularly constructive when the expert knowledge is undeveloped at a theoretical level, which is common at the outset of pioneering investigations (e.g., new species or regions). The *Elicitator* software implements a new indirect technique of scenario-based elicitation. It starts at a concrete level, by asking experts to describe the expected ecological responses in a number of well-chosen scenarios, and then, at a more abstract level, lets researchers infer how the input factors defining the scenarios affect the ecological response. In statistical terms, these inputs are called "covariates" because they vary simultaneously and may be "correlated" if they interact to affect each other's values.

Bayesian statistical modeling provides a natural framework for incorporating expert knowledge, which can be treated as a "prior" that describes the initial state of knowledge, using a statistical distribution to reflect the plausibility of each parameter value. Using Bayes' Theorem, the prior is then updated with empirical data to provide a "posterior" that describes the *updated* state of knowledge on the plausible parameter values (Low-Choy et al. 2009b; Chap. 12). In a regression-based habitat model, the posterior can be used to evaluate how likely it is, given the data observed, that a covariate affects the response: "*How likely is it, based on the field data, that each geology type increased, decreased, or had little effect on the probability that sites are occupied by a species*"? By contrast, classical *p*-values focus on the likelihood of data under a null model where the response is *not* affected by the covariate: "*With what chance would we have observed this field data, if geology had no effect on site occupancy*"?

Several elicitation methods have been devised to capture expert knowledge on species–environment relationships within a Bayesian statistical framework using logistic regression (Table 3.1), classification trees (O'Leary et al. 2008), or hierarchical models (Donald et al. 2011). Subtle changes in the wording of the elicitation questions can target quite different expert knowledge, and the results can then be translated into statistical distributions using a variety of statistical encoding methods (Table 1 in Low-Choy et al. 2010) that must be tailored to the particular model (as summarized in Table 3.1 in the present chapter). This diversity in models that incorporate expert knowledge echoes a continuing debate over the most appropriate method for developing data-driven SDMs (e.g., Elith et al. 2006). *Elicitator* currently supports indirect elicitation of parameters in Bayesian generalized linear models. In this chapter, we illustrate the use of *Elicitator* with a case study on the development of an SDM for brush-tailed rock-wallabies (*Petrogale penicillata*) in eastern Australia (Murray et al. 2009).

This chapter is designed to complement existing publications that detail the statistical methodology and software underlying *Elicitator*'s indirect approach (Low-Choy et al. 2009a, 2010; James et al. 2010). Here, we focus on the needs of landscape ecologists by examining the types of research that can benefit from elicitation using *Elicitator*, and compare several methods for eliciting expert knowledge and incorporating the results in a regression analysis for SDMs. To use *Elicitator* effectively, it is also necessary to understand how the tool fits into the overall process of designing scenarios and using the accompanying elicitation approach. We discuss the strengths and weaknesses of the software to help landscape ecologists assess its potential in their own studies.

Table 3.1 Statistical encoding methods that are used for regression analysis using expert knowledge

Sample wording of the questions for habitat modeling, regarding probability of presence (pp)	(In-) Direct (I or D)	Math complexity	Effort	Encoding method	Focus of elicitation in terms of regression parameters	Capturing uncertainty
Does this habitat factor increase, decrease, or have a negligible impact on the pp (compared to a baseline)?	D	M	L	Indicator for sign	Sign of coefficients (Martin et al. 2005)	Not typically done.
What is the probability this habitat factor increases, decreases, or has little impact on the pp (compared to a baseline)?	D	M	L–M	Mixture model on signs	Probability that sign of coefficients is +, – or 0 (O'Leary et al. 2009)	What are the lowest/highest possible probabilities? How large is negligible?
What habitat factors constitute a habitat suitability score? What is the relative weighting of each factor in this score?	D	L	H	Simple direct	Coefficient point estimates (Fleishman et al. 2001)	What is the precision of the coefficient?
As for the simple direct method, then: Classify the score and for each class, describe the range of values for the pp.	D	L	VH*	Multicriteria decision analysis	Coefficients in a score, transformation of a score (Store and Kangas 2001)	Not typically done.
Elicitator: What is the pp at sites with a given habitat profile (e.g., basalt geology, 10% slope)?	I	H	L–M*	scenarios for conditional mean priors	Probability of ecological response given covariates (James et al. 2010)	Lower and upper limits and how sure you are (%) that pp lies within this range.

Plot your estimate of the pp as it changes with one habitat factor, assuming all other factors are held at their optimal value.	I	H	M*	Graphical conditional mean priors	Ecological response given one covariate (Kynn 2005)	Specify value: you are 50% sure the pp lies above or below that value; theoretical lower and upper limits.
For several scenarios, here is the pp. Keeping this in mind, predict the pp for a scenario with a given habitat profile.	I	VH	M*	Predictive	Predicted ecological response given covariates and data (Kadane et al. 1980)	Specify value: you are 50% sure the pp lies above or below that value.
As for the Kadane et al. method, described above.	I	VH	L–M*	Predictive	Predicted ecological response at a site knowing other sites (Denham and Mengersen 2007)	Specify value: you are 50% sure the pp lies above or below that value; limits

Encoding effort: *L* low, *M* medium, *H* high, *VH* very high
Asterisk (*) indicates that appropriate software is used to assist the elicitation

3.2 When to Use *Elicitator*

Elicitator can be used to support a wide range of research purposes that rely on modeling:

- Type 1 – Estimate all parameters required to define an expert-informed *explanatory model.*
- Type 2 – Develop a model (including specification of all parameters) that will be used as the basis for *prediction* based on expert judgments.
- Type 3 – Quantify parameters that describe ecosystem processes or relationships as *inputs to complex models*, such as Bayesian networks and deterministic or simulation models.
- Type 4 – *Explicitly declare the current state of knowledge* to help design an empirical study to guide the formulation of research hypotheses, or to support a Bayesian regression analysis that combines expert knowledge with empirical data.
- Type 5 – Support the elicitation of *evaluation or prediction scenarios* for any type of model.

3.2.1 Formulating an Expert-Informed Explanatory Model

One way of thinking about using *Elicitator* is that it captures expert knowledge on the ecological responses (the response or *Y* variables) for different values of the covariates that define each scenario (the explanatory or *X* variables). In landscape ecology, *Y* variables can include species presence or abundance, ecological impacts or conditions, and other ecological responses. When undertaking explanatory modeling, interest centers on how the ecological response (*Y*) depends on explanatory factors (*X*) such as topography, habitat, climate, pests, predators, and fire.

A distinctive feature of *Elicitator* is that it captures both the expected value of the response *Y* and the range of values that are considered plausible by experts. The knowledge, therefore, represents an expert's balanced estimate (best judgment) among plausible responses. This contrasts with predictive approaches (Table 3.1), in which the expert is presented with a hypothetical set of field data and asked to use this information to predict the ecological response in a new scenario. A statistical distribution with at least two parameters is required to encode this rich information, comprising the expert's best estimate of the ecological response as well as its variability, as depicted in the software's elicitation window (Fig. 3.1).

Elicitator records the encoded expert knowledge about the ecological response for various scenarios (i.e., the provisional set of expert data) in a relational database. To discern the expert's conceptual model underlying all scenarios, this data can be fit to a wide range of statistical models. Users can either implement the model built into *Elicitator* and assess "on the spot" its goodness-of-fit to the expert's beliefs, or can export the expert data so it can be analyzed in another program (e.g., Low-Choy et al. 2010). Options include a wider range of regression models (e.g., nonlinear,

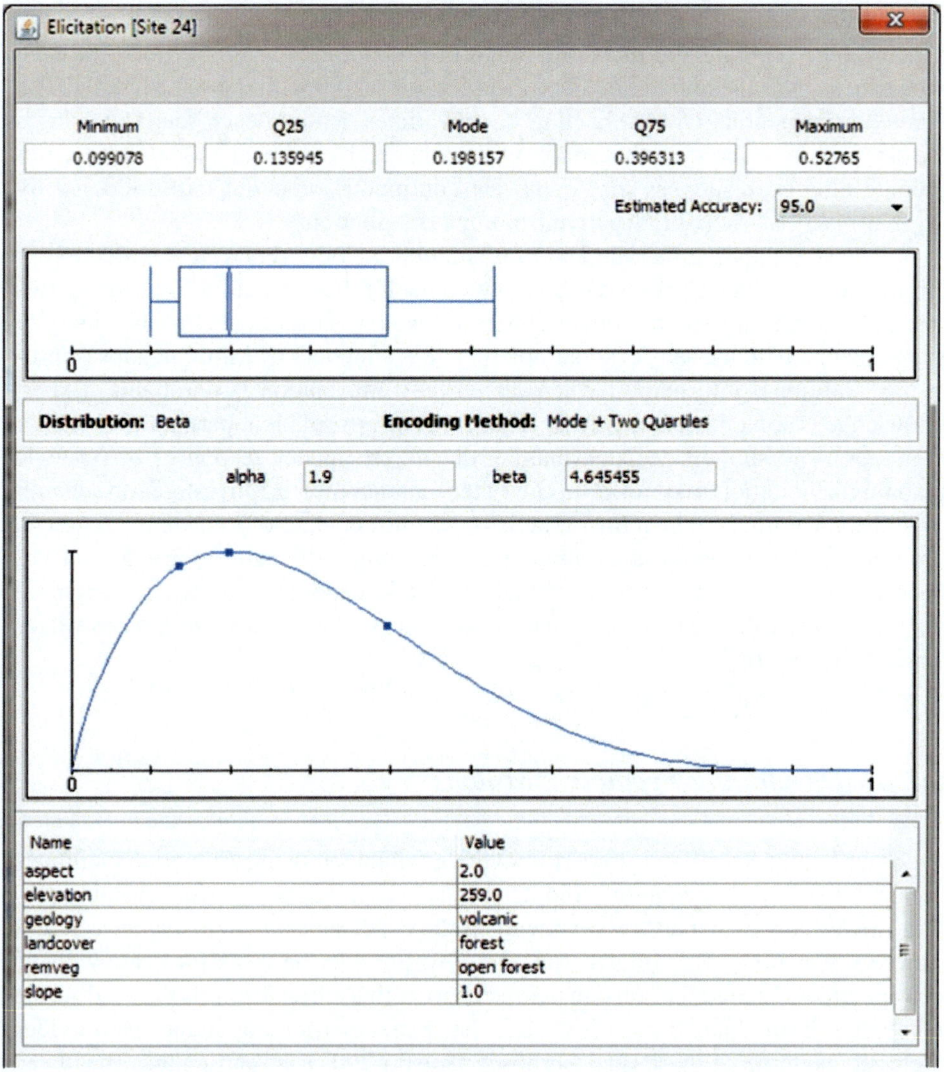

Fig. 3.1 The Elicitation window in the *Elicitator* software. Expert 1 from region Q was asked to estimate the probability of presence of the rock wallaby at site 24. Their estimates were encoded as a Beta (1.9, 4.6) distribution

nonparametric, and mixed effects), classification and regression trees, and Bayesian hierarchical models, including Bayesian networks. For simplicity, *Elicitator* currently uses a regression to fit an expert model to expert knowledge elicited across several scenarios. This provides a straightforward way of relating the output Y to changes in one or more input parameters (X), using a "score" (e.g., of habitat suitability), and regression coefficients that weight the influence of each of the covariates in determining this score.

The first version of *Elicitator* (James et al. 2010) fits a Beta distribution to the expert's best estimate and plausible range of probabilities or proportions for each scenario, such as the number of sites occupied out of 100 with the same habitat. The elicited information is then related to the covariates defining each scenario using a Beta regression. The current version utilizes generalized linear models to accommodate elicited responses that comprise continuous or count data modeled by log-Normal, Gaussian, or Negative Binomial distributions.

We must emphasize that the aim of elicitation is not to replace the collection of empirical data with expert knowledge, which itself relies on fieldwork. Instead, this form of elicitation aims to collate expert knowledge before data are available, to supplement small datasets, or to complement existing data by filling in gaps or providing additional information wherever current information is unreliable. Expert knowledge is sometimes the only timely source of information, particularly when it is necessary to support decision-making during emergency responses or strategic planning. Elicitation may form the first step in a new line of investigation when it provides a rigorous approach to establish the current state of knowledge, and can provide an interim analysis to guide planning until sufficient empirical data has been collated. Elicitation can be valuable in landscape ecology, where empirical data rarely provides a comprehensive view of the ecological response to covariates across a diverse region.

3.2.2 Developing Predictive Models

As described in Sect. 3.2.1, expert knowledge obtained using *Elicitator* facilitates the development of *explanatory* models because it provides tools for revealing patterns underlying an expert's assessments of the expected ecological response across different scenarios. This information can also provide the basis for *prediction*, to extrapolate the ecological response to scenarios that differ from those used in the elicitation. Using habitat models to map the spatial distribution of species provides a classic example of these dual purposes (Austin 2002). In particular, it has been noted that good predictive performance in new situations, such as those that will be induced by climate change, requires emphasis on the ability of the model to adequately explain the underlying ecological processes (Fitzpatrick and Hargrove 2009). However, SDMs with high predictive performance but low explanatory power due to a tenuous link to ecological theory (such as random forests, neural networks and the model fit using the Maxent algorithm) are currently favored over others (such as regression and trees) that have stronger link to ecological theory but lower predictive performance (e.g., Elith et al. 2006). If extrapolation is the main purpose of the expert model, this places greater emphasis on careful choice of the scenarios used for elicitation (Sect. 3.3.2.2) as well as the need to assess the model's predictive performance (Sect. 3.2.5).

3.2.3 Developing Inputs for Other Models

Here, we consider situations in which expert knowledge informs only some of the model parameters, in contrast to Sects. 3.2.1 and 3.2.2, in which expert knowledge is used to develop the entire model (whether explanatory or predictive). Models of complex systems, which are often developed using deterministic or simulation approaches, require specification of the values for input parameters. For example, in conservation planning, the probability of site occupancy by an organism as a function of habitat factors has been elicited using *Elicitator*, and used to predict the organism's cost of traversing the landscape to inform a corridor analysis based on graph theory (Murray 2009). Similarly, models for designing surveillance mechanisms to ensure early detection of exotic pests (e.g., Jarrad et al. 2011) can be based on expert estimates of the likelihood that a pest will become established in a new environment.

Elicitator can also assist in the construction of Bayesian networks from expert knowledge, particularly in complex sub-networks where several covariates (called "parent" nodes) affect an outcome (called a "child" node), as described in Johnson et al. (2010). Using *Elicitator*, the elicitation load can be reduced by asking experts to assess just a few scenarios rather than all scenarios, and can extrapolate the results for other scenarios. *Elicitator* utilizes Beta regression rather than linear regression, so it suits a wide range of probabilities, including those close to zero or one (i.e., below 0.10 or above 0.90). An added benefit is that *Elicitator* not only captures the expert assessments but also quantifies their variability and uncertainty, yielding a plausible range of probabilities for the child nodes under each combination of parent nodes, which may also support uncertainty analysis (Donald et al. 2011).

3.2.4 Explicit Declaration of the Current State of Knowledge

Expert knowledge is valuable because it is sometimes the only source of information, for example, for identifying the habitats of rare and threatened native species or of invasive species. In these situations, knowledge typically starts with opportunistic sampling that yields sparse empirical data that can help the observers to develop their knowledge. As information accumulates, hypotheses are refined, and data collection becomes better targeted and more effective, thereby supporting more quantitative analyses. In this learning cycle, the output from *Elicitator* can be used to take advantage of the Bayesian statistical modeling framework, thereby providing a natural bridge from expert to empirical knowledge. Within the Bayesian learning cycle, expert knowledge can be assigned the status of prior information by representing it in a statistical model called an "informative" prior, and can be subsequently overridden by stronger signals contained in the empirical data. This explicitly treats expert knowledge as a preliminary model that will be revised by empirical data.

A sensitivity analysis can compare the inferences obtained by relying on expert knowledge with those obtained by ignoring prior information (Chap. 12). Within the Bayesian framework, this involves comparing a posterior formed with an informative prior (here, one based on expert knowledge) against a posterior formed with a reference prior that is chosen to be noninformative (McCarthy 2007). By contrast, a validation approach would compare inferences based only on empirical data with those based solely on expert knowledge, and would ignore the cumulative and compensatory effects of using both sources of information. Alternatively, a nondiscriminatory approach may incorporate expert knowledge and empirical data interchangeably, but not distinguish between the two sources. This approach is common with Bayesian networks which frequently draw from multiple information sources (e.g., Donald et al. 2011).

3.2.5 Elicitation of Evaluation and Prediction Scenarios

Scenarios elicited using *Elicitator* at a concrete level may be inherently interesting, but can also be used to infer the underlying expert conceptual model at an abstract level. This meets an emerging need to support the selection of scenarios for evaluating models developed using any methodology. Though it is common to evaluate predictive models using a subset of the training data that was not used in model development or via cross-validation (e.g., Elith et al. 2006), this does not assess the model's ability to predict the response when covariate values in the prediction data differ substantially from those contained in the training data (Fitzpatrick and Hargrove 2009). Integration of scenarios by *Elicitator* at an abstract level can provide predictions of site occupancy, together with their precision, for a range of hypothetical scenarios. This supports decision-making for scenarios that may be far removed from the empirical data, yet within the reach of experts. *Elicitator* can support complex modeling by assessing scenarios in which a model may fail (e.g., Pike 2004).

3.3 How *Elicitator* Supports Expert Elicitation

Substantial preparation is required before using *Elicitator* to capture expert knowledge because the software *supports* the use of a fully developed elicitation process; it does not provide this process. Elicitation can be broken down into six key stages (Low-Choy et al. 2009b): design, preparation, elicitation, encoding, validation, and output.

3.3.1 Design

Elicitator can streamline the elicitation process and manage variability in responses (Low-Choy et al. 2009b) by flexibly tailoring the process to each expert, automating complex calculations, and managing the collected data. Moreover, the software provides ongoing feedback to help experts interactively validate their numerical and graphical assessments, and frees the elicitor to focus on communication rather than calculation. This is important since our experience has shown the importance of constantly reiterating definitions and paraphrasing expert responses to confirm their intended meaning and ensure consistency in the elicited data.

Elicitator automates the encoding process, which speeds up the elicitation process, reduces the chance of arithmetic errors, and provides modelers with different computational choices. The software manages the database of elicited information to ensure accurate recording of expert assessments across multiple experts and scenarios. These efficiencies features let elicitors undertake more elicitations, thereby reducing the risk arising from the consultation of too few experts or the exploration of too few scenarios.

Most of the effort of elicitation occurs *before* the expert and elicitor sit down together. This involves both design and preparation for elicitation. Another benefit of utilizing software is that it helps to standardize the elicitation practice within a project or a field of enquiry, and provides a benchmark elicitation practice that has ideally been developed and peer-reviewed by elicitation designers and practitioners. The design and specification of the structured elicitation method that underlies *Elicitator* comprise six main components (Low-Choy et al. 2010):

1. *Purpose*. Information obtained using *Elicitator* can be used to support scientific investigations or decision-making in the five ways described in Sect. 3.2.
2. *Goal*. Experts are asked for information similar to what they would measure in the field: the expected ecological response under a carefully selected range of scenarios.
3. *Statistical model* (Low-Choy et al. 2010). At a concrete level, a statistical distribution encapsulates the expert's description of the plausible range of expected ecological responses (e.g., probabilities of presence or abundance) for each scenario. At an abstract level, this information is then collated across scenarios by means of regression analysis to infer the underlying conceptual model that experts are implicitly using to predict the relative impact of the covariates on the ecological response.
4. *Encoding* (Sect. 3.3.4). For each scenario, experts are asked to estimate the most likely ecological response (e.g., the number of occupied sites out of 100) together with percentiles (e.g., there is a 2.5% chance that this number lies above the 97.5th percentile).
5. *Managing variability*. To control variability, elicitors can avoid cognitive biases by preparation (training and conditioning) of the experts (Sects. 3.3.2.1–3.3.2.2) and facilitating four forms of feedback (see Sect. 3.3.5). They can also improve accuracy by piloting (pretesting) the elicitation, selecting appropriate experts, and defining appropriate scenarios (Sect. 3.3.2).
6. *Protocol*. Steps 1 to 5 can be organized into an interview script to ensure that the process is transparent and repeatable. The script includes preliminary definitions and precise wording of the questions, which together are designed to elicit the desired response (e.g., the supplementary material in Murray et al. 2009).

3.3.2 Preparation

Preparation involves selection of the experts, determining their background and training, and planning the data management and project management strategies. After defining the kinds of experts that the project will require, they must be selected (Sect. 3.3.2.1), then invited and motivated to participate (e.g., the supplementary material in Murray et al. 2009). Asking experts to methodically enumerate their sources of expertise helps them to recall all relevant information (thereby reducing accessibility bias, Kynn 2008) and helps elicitors to understand the basis (e.g., the research literature versus field experience) for the expert knowledge. This also guides the selection of a representative sample of experts and schools of opinion, and subsequently guides the researcher in combining multiple expert assessments (Albert et al. 2010).

Experts should be trained in any requirements specific to the current elicitation, such as by defining what is to be elicited and the types of information (e.g., probabilities and uncertainties) that will be elicited. The subtleties of interpreting summary statistics (e.g., means, medians, modes, and probabilities) require specific numeracy skills. In particular, if probabilities are to be elicited, then experts must clearly understand what circumstances are to be included and excluded when assessing those probabilities. Training may also involve conditioning (or alerting) experts to common misunderstandings that would affect their ability to quantify their uncertainty or estimate probabilities. Training can also help experts avoid the use of heuristics in an attempt to "shortcut" the process of thinking things through. Extensive practical advice on managing sources of elicitation bias is available (e.g., Appendix C in O'Hagan et al. 2006; Low-Choy and Wilson 2009).

Elicitator requires the modeler to undertake the initial data management by supplying covariate values for each scenario that will be evaluated by the experts (Sect. 3.3.2.2). Where scenarios can be mapped to geographic sites, they must be identified appropriately so that *Elicitator* output can be aligned with a GIS database (see the database window, "Elicitation Sites", in Fig. 3.2a). Then, experts can view the elicitation scenarios in the GIS, to provide better context for the list of covariate values. Due to the difficulty of maintaining ongoing compatibility with GIS packages, *Elicitator* is "loosely coupled" to GIS software to provide a minimal level of communication that facilitates data transfer between these platforms (James et al. 2010). *Elicitator*'s data may be imported or exported in a standard format (comma-delimited text) that is utilized by most GIS packages.

Elicitation results can be organized into projects and phases of a project, with each phase potentially considering several scenarios that correspond to different sets of covariates. Several experts can be consulted during each phase. Using *Elicitator*, researchers select the project, phase, and expert in the Project Properties dialog box at the beginning of each session (Fig. 3.2b). Other settings define the elicitation method to be used throughout the elicitation session. Parameters specific to a method include the data type for the information that will be elicited, the summary statistics to be calculated from this data, the statistical encoding of the distribution, and the encoding method (Fig. 3.2b).

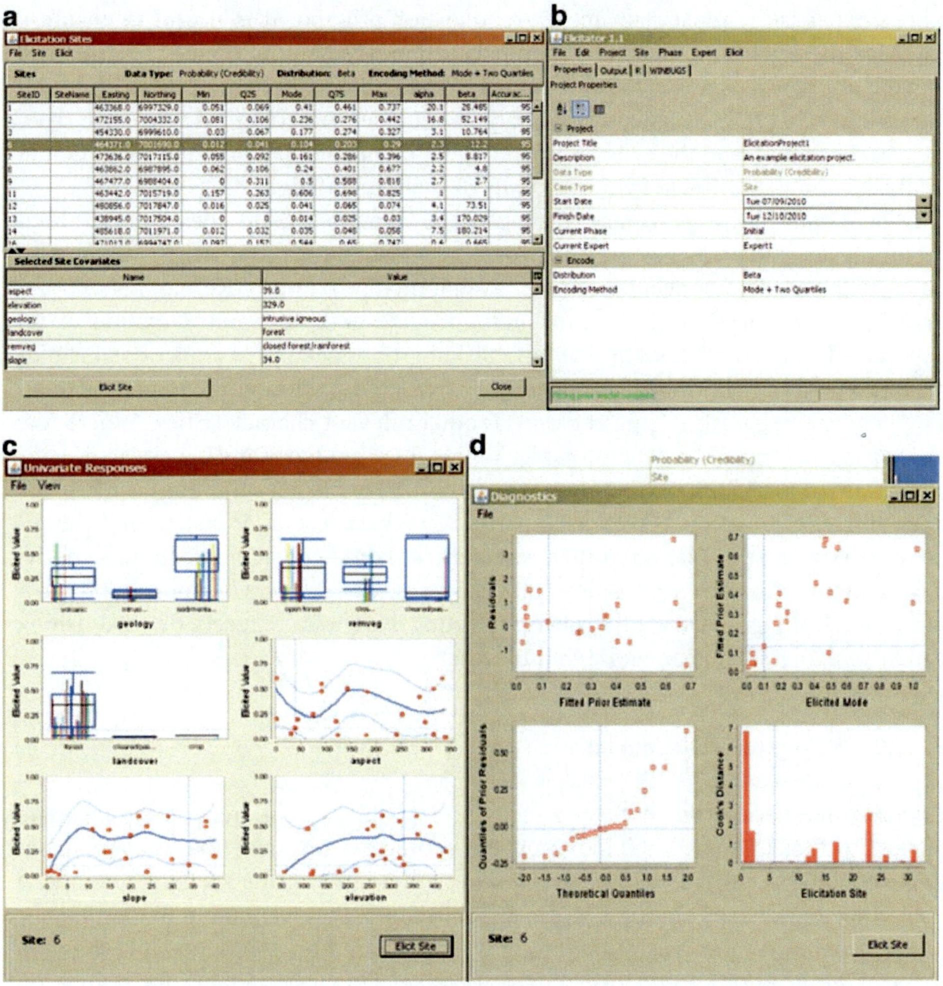

Fig. 3.2 Screenshots from the *Elicitator* software. (**a**) The Elicitation Sites window contains the elicitation database, with one scenario per row, and the columns include the site identifier, covariates, and elicited information. (**b**) The user specifies settings for the elicitation session in the Project Properties dialogue box. After the elicited data has been encoded for a sufficient number of scenarios, users can ask *Elicitator* to fit the expert model to the data they provided and permit feedback. (**c**) The Univariate Responses window provides ecological response curves as a function of each covariate. (**d**) The Diagnostics window provides standard regression diagnostics for the expert model

3.3.2.1 Selection of Experts

To use *Elicitator* and the associated elicitation method effectively, it is necessary to understand how to define and select experts, a subject that has been thoroughly addressed elsewhere (e.g., O'Hagan et al. 2006; Chap. 2). Briefly, the elicitor must first define the characteristics of a suitable expert. It is important to select experts whose expertise is most relevant to the question at hand. It is useful to ensure a balance of specialists, as well as generalists and multidisciplinary investigators who may be able to translate broader knowledge to the context of interest (Keeney and von Winterfeldt 1991). Other considerations include the ecological relevance (species, ecosystems), and the cross-sectional (geographic and habitat) and longitudinal (daily, seasonal or climate cycles) extents of their knowledge.

Where multiple experts are available, the experts should ideally be chosen according to statistical design principles (i.e., to obtain a representative sample) rather than based on convenience (e.g., availability, proximity). Researchers should first determine whether the sample must provide adequate representation of all experts, all schools of thought, identifiable subgroups such as sectors or stakeholders (e.g., government, industry, academia, community), demographic groups (e.g., age, gender), geography (e.g., region), or other relevant characteristics. Where few experts exist, a census may be possible. Where there are too many experts to census, some form of probabilistic sampling (e.g., stratified or cluster sampling) can be used to ensure representation as well as impartiality. Where resources constrain the number of experts that can be consulted, researchers should consider accounting for the most extreme views as well as intermediate ones. Cooke and Goosens (2008) provide useful advice on the criteria for selecting individual experts or establishing panels of experts who can work well together.

3.3.2.2 Selection of Scenarios

Elicitator uses scenarios, each of which corresponds to a different set of covariate values, to focus expert knowledge on the ecological response they expect under those conditions. In the case study presented in Sect. 3.4, the scenarios comprise habitat factors such as geology, slope, and vegetation that were thought to influence site occupancy by the brush-tailed rock wallaby, based on a previous investigation (Murray et al. 2008). Guidance on covariate selection is available for regression (e.g., Austin 2002) and for Bayesian networks (e.g., Uusitalo 2007). The expert model underlying *Elicitator* helps address the choice of covariates by explicitly assessing their impact on the response (relative to the other covariates considered). Elicitation scenarios should be selected carefully to ensure that they are *representative* of the types of scenarios and regions of interest, are sufficiently comprehensive to capture the full range of ecological responses for particular scenarios and subregions, and contain sufficient information to separate expert knowledge (the signal) from elicitation errors (noise).

Scenarios can be hypothetical or can represent real sets of conditions or geographic locations (as described in Sect. 3.3.2.3). This is an important consideration in landscape ecology, where the range of covariate values can contain many gaps (impossible combinations of covariate values) or may otherwise be highly irregular. Whether the scenarios are real or hypothetical, the elicitor and experts should select scenarios that balance a representative set of scenarios with scenarios for which the expert has clear opinions.

Elicitator has no restriction on the number of scenarios for which knowledge can be elicited. Most approaches for selecting scenarios start by categorizing each covariate into categories of relevance to the response. A comprehensive approach would list scenarios formed from every combination of these categories (i.e., a full factorial design). As a rule of thumb, the number of scenarios should be at least twice the number of covariates to provide at least one degree of freedom to assess each "main effect" in the model. However, it is also important to limit the number of scenarios to avoid tiring the experts, which can lead to inaccuracies. In a previous case study, most experts were sufficiently motivated and engaged with the process, so that they had no trouble concentrating for between 1.5 and 6 h to provide 30 elicitations (Murray et al. 2009). Standard statistical approaches can be used to provide a more economical choice of scenarios (Baguley 2004). For instance, Graeco-Latin square, incomplete block, or fractional factorial designs economize by sacrificing information on less important interactions between covariates (e.g., Vikneswaran 2005). Power-based sample size analysis helps set the number of replicates for each scenario when there are few covariates (Sheppard 1999). Bedrick et al. (1996) detail a method to choose scenarios when there is only time to undertake the same number of scenarios as there are covariates.

3.3.2.3 Mapping Scenarios to Geographic Sites

It can be useful to exploit the geographic context of scenarios, particularly when the experts are landscape ecologists (O'Leary et al. 2009). Each scenario is defined by a set of covariates that can correspond to GIS layers. Hence, each scenario can be represented by one or more "real" sites with the desired GIS attributes. A simple method for selecting scenarios is to use a completely randomized sampling protocol for the region of interest. However, this is the least efficient sampling algorithm for achieving full spatial representation. In a previous case study (Murray et al. 2009), scenarios were chosen using a straightforward stratification to ensure representation of the full range of geology, vegetation, and land cover within the study area. Using GIS, one elicitation site was selected randomly within each stratum to ensure that all realistic scenarios were accounted for. The group of scenarios was then checked to ensure adequate representation of fine-scale factors such as elevation, slope, and slope aspect. Experts could instead use the GIS to select sites for assessment (Denham and Mengersen 2007) before importing these scenarios into *Elicitator*.

3.3.3 Elicitation

Using *Elicitator*, assessment of each scenario focuses on the Elicitation window (Fig. 3.1), and begins by asking the expert to inspect the covariates (e.g., GIS habitat attributes) for the scenario. The site's identifying information and the elicited information (columns) for all scenarios (rows) are arranged in the Database window

(Fig. 3.2a), and the covariates are detailed for a selected scenario at the bottom of the Elicitation window (Fig. 3.1). The site identifier in the Database window (column 1, Fig. 3.2a) can be used to link sites to an accompanying GIS.

During elicitation, the expert can interactively modify graphs to specify the expected ecological outcome (e.g., the probability of presence) and the range of plausible values (i.e., the uncertainty in these probabilities) for a scenario with a specific set of covariate values (e.g., a habitat profile). This is achieved by focusing on a few key summary statistics, as detailed in Sect. 3.3.4. *Elicitator* records the theoretical limits, realistic limits (e.g., 95% and 50% credible intervals), and the best estimate (the mode) for the expected ecological outcome. To avoid common cognitive biases (Low-Choy et al. 2010), we recommend that elicitors ask for these quantities in this order, from the "outside in". Figure 3.1 exhibits three alternate methods of entering these assessments: numerically in text boxes (top), by moving the vertical lines in the boxplot (center), or by moving the probability density curve (bottom). Graphs are instantly updated to provide ongoing feedback on the encoded distribution. The expert's uncertainty can also be captured (in the Estimated Accuracy text box, top right, Elicitation dialogue box; Fig. 3.1).

Initially, experts should "walk through" the elicitation process with the elicitor for enough example scenarios to learn the process and the underlying definitions of ecological context and probabilities (Cooke and Goosens 2008). Eliciting opinions for "seed" scenarios lets researchers assess each expert's accuracy against a gold standard if one exists, although this method of calibrating experts is more effective when made transparently (Kynn 2008). To evaluate an expert's consistency, it is helpful to repeat the elicitation of a scenario or to compare the results for similar scenarios. Another approach is to consider alternative "framings" for questions. For example, a series of questions about site occupancy could be supplemented by questions about site nonoccupancy for similar scenarios (Kynn 2008). Throughout the elicitation conversation, the researcher can record the expert's reasoning behind each assessment in the Comments text box of the Elicitation window. Asking an expert to justify their assessment encourages reflection and questions, and thereby improves the quality of the results. This is crucial for very high or very low probabilities (Kynn 2008).

3.3.4 Statistical Encoding

Researchers must first delineate the areas under consideration, perhaps based on administrative or topographic boundaries and based on environmental constraints (e.g., excluding bodies of water and human settlements). It will be necessary to work with the experts to create a shared definition of what a site represents (e.g., a permanent sampling plot versus an ecosystem such as a forest with distinct edges), and to ensure that experts share the same sense of temporal scale (e.g., a single year versus a 100-year successional process). It is also important to recognize preconditions or subpopulations to reduce context dependence. Addressing these issues helps *define* the context, including the window and units in space and time, and the circumstances for including or excluding sites. It, therefore, reduces the biases that may occur through misunderstanding the baseline of probabilities (Low-Choy and Wilson 2009).

When eliciting a probability, more accurate responses are obtained when questions are phrased in terms of whole numbers (e.g., the number of outcomes per 100 with a given result) rather than fractions (Kynn 2008). More generally, elicitation questions should be phrased in a way that it makes it easy for the experts to conceptualize what is required. For this reason, *Elicitator* focuses on "observable" quantities that could be obtained via fieldwork (Low-Choy et al. 2010).

When developing SDMs, the *scenario* that experts should keep in mind during elicitation comprises 100 sites (or some other fixed amount) within the area that meet the scenario's criteria. The problems of inaccurate representativeness and anchoring heuristics (Sect. 3.3.2) can be avoided (Low-Choy et al. 2010) by first asking experts to specify the smallest and largest possible number of sites that would be occupied by an organism *in this scenario*. Hence, they should be 100% sure that the true number lies within these theoretical limits. You can then ask them to supply more realistic bounds, so that there is a 95% (then a 50%) chance that the true number falls within these bounds. Experts can then provide their best estimate of the number of sites occupied for this *scenario*. In the design step (Sect. 3.3.1) or the preparation step (Sect. 3.3.2), it is helpful to ensure that the experts understand the meaning and use of these statistics.

3.3.5 Validation (Feedback)

After elicitation, it is necessary to review the numbers provided by the expert (Low-Choy et al. 2009a; Kuhnert et al. 2010). This is especially important the first time an expert has attempted to address a topic either quantitatively or qualitatively. Four types of feedback include *recording, reflecting, comparing*, and *assessing the implications*. *Recording* elicitations immediately provides evidence of precisely what the expert communicated. The ability to immediately review the recorded elicitation gives the expert an opportunity to *reflect* on each scenario (i.e., to revisit the Elicitation window, via the Database window; Fig. 3.2a) and confirm or revise their original assessment. *Comparing* lets the expert examine several scenarios simultaneously (i.e., revisit the table of elicited information in the Database window; Fig. 3.2a) to confirm that the assessments are consistent. Model summaries and diagnostics (Fig. 3.2c, d) help experts assess the *implications* of their scenario-by-scenario assessments, a crucial step for calibrating expert assessments (Kuhnert et al. 2010).

To illustrate the *implications*, *Elicitator* provides standard regression *diagnostics* (James et al. 2010). Experts can decide whether the implications of their assessments are sensible by inspecting the fitted univariate response curves (Fig. 3.2c): When the elicited results for individual scenarios are combined, does the pattern of ecological response to each covariate still make sense? The diagnostics (Fig. 3.2d) focus on the residuals as a measure of the discrepancy between the elicited expert assessment for each scenario and the value predicted from the expert model constructed to fit all scenarios. The plot of the residuals versus the fitted values (top left of Fig. 3.2d) may highlight outliers that fall beyond an expected "cloud" of residuals centered around the horizontal line through zero. For the quantile–quantile plot of the residuals (bottom left of Fig. 3.2d), deviations from a straight line through the origin with a unit slope (i.e., $y=x$) may indicate difficulties fitting an expert model using this set of covariates, such as a missing covariate or scaling issues (overstating or understating estimated probabilities). Highly influential scenarios will be highlighted in the Cook's residuals plot (bottom right of Fig. 3.2d). Finally, the expert's uncertainty is reflected by the goodness-of-fit, which is reflected in the graph of the elicited best estimate against the fitted value (top right of Fig. 3.2d). The expert may revisit the elicitation for any scenario simply by selecting the corresponding point on any diagnostic plot.

3.3.6 Output (Documenting the Expert Model)

To ensure the repeatability of the elicitation process, it is imperative to document the interview script that was used to structure the process (O'Hagan et al. 2006). Such documentation is rare (Low-Choy et al. 2009b), so *Elicitator* supports the tedious documentation process in several ways. A database of all elicited quantities is maintained, allowing the elicitations to be revisited at any time. This includes revising the choice of covariates, reviewing the elicited quantities for each scenario, and recalculating the statistical distribution to capture the plausible range of expected responses for each scenario (Database window, Fig. 3.2a). This database provides a useful record when the purpose is to elicit scenarios for validation (type 5 in Sect. 3.2).

The expert model parameters are provided in the form of standard statistical output from a regression, including the estimated mean and SD of the regression coefficients. This is sufficient information for modelers to use for explanatory or predictive purposes (types 1 or 2 in Sect. 3.2), or as input to a more complex model (type 3 in Sect. 3.2). The expert model is also reported in the form of a prior to ensure that the elicited information can be readily included in a Bayesian statistical model, using the WinBUGS format (Spiegelhalter et al. 2003). Current developments include provision of the R code (Ihaka and Gentleman 1996) to help fit Bayesian regressions with noninformative and informative priors (as defined in Sect. 3.2.4) using a link to WinBUGs, to support incorporation of the current state of knowledge (type 4 in Sect. 3.2).

3.4 Case Study: An SDM for Brush-Tailed Rock-Wallabies

An SDM was required to support the conservation of a threatened species in eastern Australia, the brush-tailed rock wallaby. In a series of studies, expert knowledge was elicited to support four of the five types of modeling practices described in

Sect. 3.2. First, an expert-derived SDM captured the current state of knowledge on the link between the occurrence of this species and habitat features at a landscape scale (type 1, an explanatory model). When used as a prior in Bayesian regression, this helped to address gaps in the empirical data (Murray et al. 2008) and to link expert knowledge with empirical information (type 4, making the state of knowledge explicit). To enable mapping of suitable habitat and the potential species distribution (type 2, prediction), it was necessary to build a habitat model based entirely on landscape-scale predictors that had been mapped across the habitats of the species. In addition, since the spatial arrangement of suitable habitat is a key consideration for protecting wildlife corridors, posterior predictions of habitat suitability were used as inputs (type 3, inputs for more complex models) for a connectivity analysis (Murray 2009).

This case study is unusual among other SDM approaches due to the substantial effort involved in the design and collection of both expert and empirical data. In Queensland, four experts were interviewed for 2 to 4 h each and fieldwork assessed more than 200 sites. A comparable fieldwork and elicitation effort was expended in an adjoining region in New South Wales, where five experts were consulted, for a total of nine experts across both regions. *Elicitator* was installed on a portable computer used by the elicitor so that each expert could be interviewed in the environment most convenient to them.

Knowledge was elicited from experts who had not been exposed to the recently collected data (Murray et al. 2008) to satisfy a technicality of Bayesian analysis, namely that the prior must be independent from the empirical data it will be combined with. In addition to ensuring consistent definitions (e.g., of site occupancy), the expert and empirical datasets followed a consistent choice of covariates and study area. Expert knowledge was considered complementary to GIS data, with several differences (Murray et al. 2011) including the shorter time-span of the empirical presence–absence observations, the fact that experts extrapolate from local to broader scales whereas GIS layers combine remotely sensed measurements as well as expert interpretations, and the fact that experts "fill in the gaps" using various sources of information.

3.4.1 Developing an Understanding of Habitat Requirements

The aim of the regression analysis was to assess the impact of habitat factors on site occupancy by rock-wallabies. Table 3.2 summarizes the expert-based estimates (combined across experts), data-driven estimates, and combined expert- and data-driven posteriors for these impacts in Queensland, as presented in Murray et al. (2009). In the Bayesian paradigm, regression coefficients are "random" parameters that have a prior and a posterior distribution. The best estimate of a regression coefficient *a posteriori* (after accounting for the empirical and prior information) is its posterior mean, and a measure of plausibility is provided by the posterior standard deviation. These posterior means and SDs are shown in Table 3.2.

Table 3.2 Bayesian posterior mean and standard deviation for coefficients that reflect the impact of habitat factors (data definitions and sources detailed in Murray et al. 2008) on the probability of site occupancy by rock wallabies (Murray et al. 2009)

	Increasing reliance on data (from left to right) → Increasing reliance on expert opinion (from right to left) ←					
	Expert model not informed by data		Bayesian posterior model estimated using empirical data, with different priors			
	Prior informed by information elicited from Queensland experts 1 to 4				Noninformative Priors	
Habitat	Mean	SD	Mean	SD	Mean	SD
Intercept (baseline scenario)	−1.089	0.958	−0.441	0.649	−1.620	1.305
Intrusive igneous rock	−1.416	0.684	−1.570	0.475	−1.682	0.733
Sedimentary/metamorphic rock	−0.508	0.529	−1.170	0.370	−1.596	0.564
Closed forest	−0.358	0.406	−0.369	0.390	−0.682	1.058
Cleared forest	−0.856	0.465	−0.960	0.373	−1.437	0.667
All vegetation removed	−1.229	0.537	−1.131	0.474	−0.548	1.021
Agricultural crop	−0.742	0.723	−0.777	0.727	−0.915	2.644
Elevation	0.001	0.003	0.000	0.001	NA	NA
Slope	0.023	0.018	0.049	0.016	0.104	0.039

Positive and negative values indicate an increase and decrease (respectively) in probability of site occupancy; values near zero indicate little or no impact on this probability. The table summarizes the progression of models built from purely expert opinion (left) to purely empirical data (right), with a model that bridges between the two types of data in the center. NA indicates the effect was not estimable from the data

In *Elicitator*, the intercept represents the response (here, probability of site occupancy) for baseline covariates (here, habitat), where all categorical variables are set to their baseline value. In *Elicitator*, this baseline is determined by two rules. For each categorical variable, the baseline category has the smallest numeric code. For each continuous variable, the baseline category has a value of zero. These interpretations of the baseline result directly from the implementation of the Beta regression underlying *Elicitator* (specifically, the use of treatment contrasts).

In landscape ecology, habitat factors are often coded so that increasing values correspond to increasingly suitable habitat. Then the regression coefficients can be interpreted so that increasingly larger and more positive values lead to increasingly more suitable habitat, and higher probability of site occupancy; and the underlying baselines correspond to the least suitable habitat. However, in the rock wallaby case study, habitat covariates had different effects in the two regions so that no single coding would intuitively link the sign of the regression coefficient and habitat suitability. Here, codings used for categorical variables – geology, forest, and land cover – led to a *baseline habitat in Queensland that comprised optimal habitat categories of volcanic rock, open forest, and forested land cover*. Thus, all other categories decreased the probability of site occupancy in comparison to the baseline, as reflected by their *negative* effects in Table 3.2. However, for the two continuous variables (slope and elevation), the baseline value is zero. Thus, larger values of their regression coefficients *increase* the probability of site occupancy in comparison to the baseline, as reflected by *positive* effects in Table 3.2.

The choice of codings (and contrasts) does not influence the predicted univariate responses (Fig. 3.2c), but greatly influence the interpretation of the regression coefficients reported in Table 3.2. These nuances are avoided in applications (similar to Elith et al. 2006) where the emphasis is on reporting predictive performance (type 2 in Sect. 3.2) rather than on the explanatory ability of the SDM. These nuances are also avoided during elicitation with *Elicitator*, since the expert focuses on the predicted univariate response curves (Fig. 3.2c) rather than the estimated effects. However, these nuances will be encountered if the elicitor wishes to understand the implications of the model discerned from the expert assessments (type 1 in Sect. 3.2) or use the expert assessments to inform a Bayesian regression (type 4 in Sect. 3.2), as in this case study.

This study followed a smaller study that elicited knowledge from just two experts, who were inconclusive about the effect of slope, with credible intervals that included a value of zero (Method A of Table 5 in O'Leary et al. 2009). The change in the expert-informed models can be attributed to the different data collection methods for empirical and expert data. The elicitation tools used similar interfaces, although *Elicitator* (used by Murray et al. 2009) also included a boxplot, which was preferred by ecologists. *Elicitator* also targeted each expert's best estimate (viewed statistically as a mode) of the number of sites occupied (out of 100) and the plausible range of values for this expectation (absolute then realistic 95% and 50% intervals). By contrast, the tool used by O'Leary et al. targeted the predicted ecological response (viewed statistically as a median) and prediction interval (quartiles).

3.4.2 Predicting (Mapping) Habitat

As shown in Fig. 3.3, the potential species distribution is focused in ribbons (dark areas) within the granite belt in both the expert model (Fig. 3.3a) and the posterior model that combines expert knowledge with empirical data (Fig. 3.3b). However, the posterior model shows areas with a moderately high predicted probability of presence surrounding the narrower areas identified by the experts. Discrete locations with high habitat values (i.e., high probabilities of presence) are clearly identifiable, with one or two large patches located to the far south, and more than four large patches located in the north. The posterior analysis (Fig. 3.3b) softens the differentiation between the areas predicted to have low and high probability of occupancy deduced purely from expert knowledge (Fig. 3.3a). The uncertainty in the posterior predictions is highest in the areas mapped as highly probable site occupancy (Fig. 3.3c), which is a characteristic of logistic regression when predicted probabilities fall under 0.5.

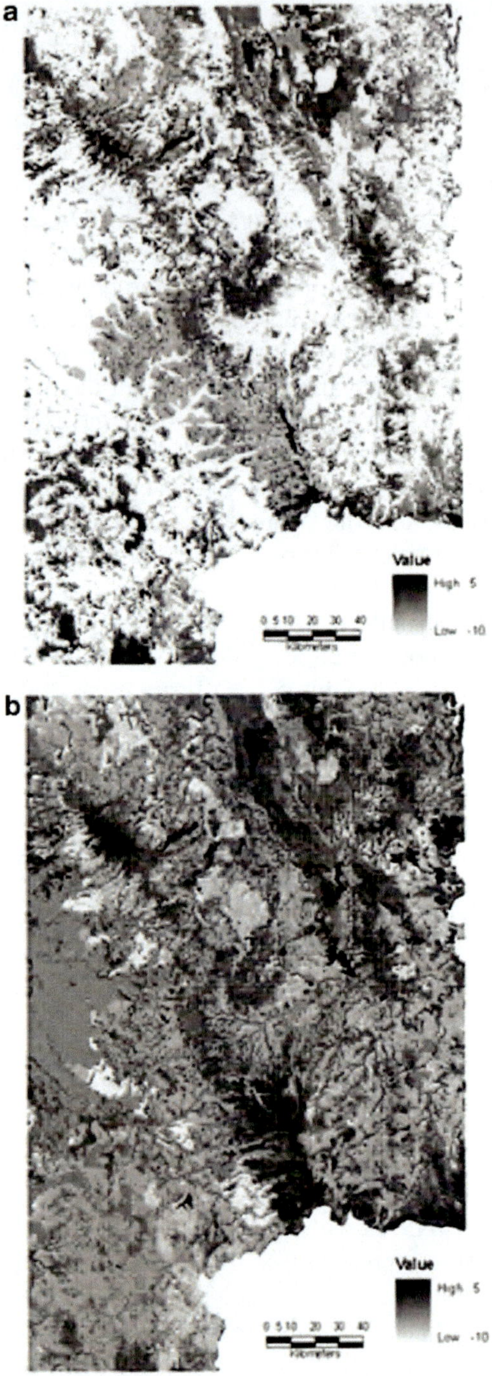

Fig. 3.3 Map of the predicted spatial distribution of the brush-tailed rock wallaby (Murray et al. 2009). Maps (**a**) and (**b**) illustrate the logarithm of the odds of presence (i.e., a habitat suitability score) with zero indicating a 50–50 chance of presence. The three maps correspond to the following: (**a**) predicted presence based on the expert model fit using *Elicitator* based solely on elicited information (prior predictions by the experts), (**b**) predicted presence from the posterior model that balances input from empirical and expert data (posterior predictions), and (**c**) the standard deviation of (uncertainty in) the posterior predictions shown in map (**b**)

Fig. 3.3 (continued)

It is also possible to visually discern potential corridors, most of which comprise areas with moderately high habitat quality (mid-gray tones) that connect the highest-quality habitat patches. Graph theory has been used to delineate these corridors more precisely (Murray 2009), based on an analysis in which the cost of traversing the landscape was expressed as a function of the predicted probability of occurrence using these habitat quality models.

3.4.3 Demonstrating the Benefits of Structured Elicitation Supported by Elicitator

Experts found the *Elicitator* interface to be user-friendly; all nine experts became comfortable with the software and familiar with its use after the initial walkthrough. Comments made during the elicitation process indicated that the experts found the tool helped them to "voice" their knowledge. They appreciated the flexible graphical interface, and initially focused on the interactive boxplot as the most familiar display. The elicitor noted that all experts utilized a combination of the software's features to provide the four types of instantaneous feedback, and used this feedback to enhance their accuracy and consistency (Sect. 3.3.5): they reflected on their assessments, compared their assessments across scenarios, considered the implications for the inferred impact of habitat factors on rock wallaby presence, and communicated with the elicitor.

By supporting the statistical encoding of expert knowledge, *Elicitator* allowed different aspects of uncertainty about the rock wallaby's habitat requirements to be easily captured:

1. For each scenario, experts were asked to consider 100 such sites, then specify the lower and upper bounds for the number of sites that would be occupied by at least one brush-tailed rock wallaby. This translated into a plausible range for the probability of site occupancy, which varied among experts and scenarios; Low-Choy et al. (2010) provide examples in their Fig. 1. Experts tended to give a narrower range of likely probabilities of site occupancy for scenarios considered unattractive to rock-wallabies.
2. Being able to assign different levels of confidence across scenarios provided a useful way for experts to express their uncertainty.
3. When information was combined across scenarios (using a Beta regression within *Elicitator*), the error bars in univariate response curves showed the strength of their convictions regarding the impacts of habitat factors on the probability of site occupancy.
4. Combining expert knowledge as informative priors within a Bayesian regression with field observations lowered the standard deviation, and therefore the uncertainty (compared to the data-driven model with noninformative priors) of the estimated impact of each habitat factor.
5. The software streamlined the collation of different sources of expert knowledge by managing the data obtained from the different experts (Fig. 3.2a), at different phases of the elicitation process, for 21 different scenarios. The software could also be used in interview environments that were convenient to the expert, including remote locations.

3.4.4 Improving the Feasibility of Habitat Modeling

Field sampling over spatially extensive and often complex landscapes and at broad scales (i.e., landscape rather than site scales) can be costly in terms of time, money, and other necessary resources. It can significantly affect budgets that might be better spent on management. In such cases, expert knowledge can provide an alternative. As illustrated in our case study, Bayesian analyses that supplement small datasets with expert-informed priors can lead to more balanced interim results, with accurate estimation of regression coefficients (in the sense that the standard deviations are small). Out-of-sample predictive performance has been assessed by comparing models developed using the same set of predictors with field data and experts sourced from two different regions (Murray et al. 2011).

In this study, we have an intercept term (reflecting the baseline scenario) and five variables (Table 3.2): three categorical measures (land cover, remnant vegetation, and geology) that only have three categories, so they were each represented by two dummy variables, and two continuous measures (elevation and slope). Here, the number of field sites was comparable to the number of unique combinations of the covariates (five covariates simplified to three levels) in a fully factorial design ($3^5 = 243$, versus more than 200 sites). Although some of these combinations did not occur in the study area, the inherent ecological variability in the occurrence of rock-wallabies in each habitat scenario means that the field data alone was not sufficiently informative about some covariates (e.g., land cover). In some cases, expert knowledge changed conclusions on whether factors increased or decreased occupancy (e.g., sedimentary rock type). Threatened species such as the brush-tailed rock wallaby can benefit from the approach described in this chapter by taking advantage of expert knowledge when empirical data is in limited supply or is only available at local scales rather than the landscape scale that is required for environmental and biodiversity management and planning.

3.5 Strengths and Weaknesses of Encoding via *Elicitator*

Several aspects of how *Elicitator* encodes expert knowledge require further consideration. We address each of these separately, noting their strengths and weaknesses in comparison to alternatives.

The interface of *Elicitator*, inspired by Denham and Mengersen (2007), provides feedback using graphical visualization techniques. As noted by those authors, many ecologists gain their expertise during field trips to particular locations, and can easily recall their knowledge of these sites. Allowing experts to visualize scenarios that correspond to a particular location within GIS software provides contextual information that improves their recall and therefore the accuracy of their predictions. *Elicitator* is designed to operate in parallel with a GIS, rather than being embedded within a GIS (Denham and Mengersen 2007). This loose coupling avoids the need to redundantly provide GIS functionality that is available elsewhere. *Elicitator* can, therefore, be used when scenarios do not correspond to mapped locations; for example, scenarios could correspond to patients, and the predicted changes in their health could be in response to various risk factors (covariates). Further development could provide more integrated GIS functionality, such as mapping elicited expert assessments.

With *Elicitator*, experts can interact with the statistical graphs and record their assessments of the ecological response (here, the probability of presence) for each scenario. *Elicitator* provides three representations of the expert assessments: a boxplot, a probability density function, and text boxes. The graphs offer handles that experts can use to intuitively reshape the statistical distribution by shifting the position of key summary statistics. Other software packages provide only one type of graph (e.g., Kynn 2005; Denham and Mengersen 2007; Oakley and O'Hagan 2010). In our case study, we found that most ecologists preferred to use the boxplot for initial specification of their assessments, but found the probability density function and text boxes useful for refining them, particularly as their statistical understanding improved. The tool, therefore, accommodates different ways of thinking (numerical, graphical, or – after exporting the data for use in a GIS – geographical) that might be employed by a diverse group of experts. Other graphical tools could be developed to help address common logical fallacies and biases.

Unlike most other elicitation tools and approaches (e.g., see Chapter 2 of O'Hagan et al. 2006), *Elicitator* asks experts for their estimate of the *mode* of the ecological response – their *best* estimate or the value they consider most likely. It is more common to ask for the median estimate (e.g., Kynn 2005; Denham and Mengersen 2007), where experts choose a value for which there is a 50% chance that the true value lies above or below this value. *Elicitator* captures quantiles to provide information on the expected range of values and requires experts to specify particular credible intervals.

Elicitator is unusual among elicitation methods since it captures more than two summary statistics, which is the bare minimum required by a deterministic encoding approach (e.g., Bedrick et al. 1996; Chapter 2 of O'Hagan et al. 2006). By "deterministic," we mean that the elicited quantities are related to the parameters by equations that do not incorporate a measure of uncertainty. The advantage of collecting additional information beyond the minimum is that the expert model can then account for and quantify the expert uncertainty. This additional information can help researchers to calibrate the opinions of multiple experts, which can contribute to comparing or combining their assessments (Albert et al. 2010). It also reduces the pressure on experts to stipulate their assessments exactly.

The *Elicitator* prototype (v1.0) used in our case study implemented only one method of capturing expert knowledge about each site; it required experts to follow an "outside-in" approach by first assessing the minimum and maximum values, and then quantifying the mode and two quantiles. The next release will include additional methods that permit encoding based on up to seven summary statistics. These methods are sufficiently detailed to capture expert uncertainty yet rapid enough to permit elicitations of multiple scenarios within a single session 2–3 h long. In some cases, only a few fundamental quantities may need to be quantified to quantify input parameters for a larger model (type 3 in Sect. 3.2). For these situations, the SHELF tool (Oakley and O'Hagan 2010) may also be appropriate, since it provides four encoding techniques and some graphical feedback: the P or Q methods are based on cumulative probabilities or quantiles (respectively), whereas *Elicitator* uses a hybrid PQ method, and the R and T methods are based on (respectively) roulette (elicitation of a histogram, a discretized version of the hybrid PQ method) and tertiles (i.e., dividing the distribution into equal thirds). Both packages implement statistical encoding via least-squares regression.

One major motivation for developing *Elicitator* was to streamline elicitation and therefore facilitate experimentation on various aspects of the elicitation procedure, on the encoding algorithms, and on expert characteristics. This functionality is yet to be fully developed. For instance, a reviewer suggested inclusion of a module that helps researchers develop and test the elicitation script. This could guide the researcher through the process of pilot testing of the questions on a small number of experts to assess their accuracy; this could be done by recording the script, linking it to some test questions or seed scenarios (Sect. 3.3.3), and supporting an analysis to assess consistency and bias.

3.6 Conclusions

Elicitator contributes to a more robust, transparent, and repeatable elicitation of expert knowledge. This is an important contribution because, despite the large investment of resources in other elements of modeling, little effort is sometimes devoted to capturing expert knowledge. Presenting a less-intimidating statistical methodology for encoding expert knowledge makes it easier to incorporate expert knowledge in a structured way.

Nevertheless, considerable effort is required to utilize an elicitation tool effectively; researchers must carefully design and test the script, select and train experts, conduct the elicitation, and review and validate the outputs before they can be used to support management decisions, and they must document the process and its results. Moreover, when knowledge is elicited from multiple experts, it is necessary to consider how to calibrate then combine the resulting information. An important issue that requires further exploration within the landscape ecology context is the complementary nature of expert knowledge and empirical data that has been measured at different scales.

Elicitator is a work in progress, and researchers are extending the range of data types allowed for the ecological response and supporting the corresponding calculations, extending the range of feedback mechanisms, and improving error handling and integration across the software's four components (statistical computation, relational database, GIS links, and interactive graphics). To learn more about the status of the software, please contact us (Elicitator@qut.edu.au). To our knowledge, no other tools support an indirect technique based on eliciting the expected ecological response under a range of scenarios and are suitable for use in inferring an expert's conceptual model.

References

Albert I, Donnet S, Guihenneuc C et al (2010) Combining expert opinions in expert elicitation. Available from http://eprints.qut.edu.au/40812 (accessed April 2011)

Austin M (2002) Spatial prediction of species distribution: an interface between ecological theory and statistical modelling. Ecol Modell 157:101–118

Baguley T (2004) Understanding statistical power in the context of applied research. Appl Ergon 35:73–80

Bedrick EJ, Christensen R, Johnson W (1996) A new perspective on priors for generalized linear models. J Am Stat Assoc 91(436):1450–1460

Boyce MS, Irwin LL, Barker R (2005) Demographic meta-analysis: synthesizing vital rates for spotted owls. J Appl Ecol 42:38–49

Cooke RM, Goosens LLJH (2008) TU Delft expert judgment data base. Reliab Engin Syst Safe 93:657–674

Denham R, Mengersen K (2007) Geographically assisted elicitation of expert opinion for regression models. Bayesian Anal 2(1):99–136

Donald M, Mengersen K, Toze S et al (2011) Incorporating parameter uncertainty into Quantitative Microbial Risk Assessments (QMRA). J Water Health 9(1):10–26

Elith J, Graham C, Anderson R et al (2006) Novel methods improve prediction of species' distributions from occurrence data. Ecography 29:129–151

Fitzpatrick MC, Hargrove WW (2009) The projection of species distribution models and the problem of non-analog climate. Biodivers Conserv 18:2255–2261

Fleishman E, MacNally R, Fay JP, Murphy DD (2001) Modelling and predicting species occurrences using broad-scale environmental variables: an example with butterflies of the great basin. Conserv Biol 15:1674–1685

Ihaka R, Gentleman R (1996) R: A Language for Data Analysis and Graphics, J Comput Graph Statist 5:299–314

James A, Low-Choy S, Murray J, Mengersen K (2010) *Elicitator*: An expert elicitation tool for regression in ecology. Environ Modell Softw 25(1):129–145

Jarrad F, Barrett S, Murray J et al (2011 Improved design method for biosecurity surveillance and early detection of non-indigenous rats. New Zeal J Ecol 35(2), Available from http://www.nzes.org.nz/nzje/new_issues/NZJEcol_JarradIP.pdf (accessed April 2011)

Johnson S, Fielding F, Hamilton G, Mengersen K (2010) An Integrated Bayesian Network approach to Lyngbya majuscula bloom initiation. Marine Environ Res 69:27–37

Kadane JB, Dickey JM, Winkler RL et al (1980) Interactive elicitation of opinion for a normal linear model. J Am Stat Assoc 75:845–854

Keeney RL, von Winterfeldt D (1991) Eliciting probabilities from experts in complex technical problems. IEEE Trans Eng Manage 38(3):191–201

Kuhnert P, Martin TG, Griffiths SP (2010) A guide to eliciting and using expert knowledge in Bayesian ecological models. Ecol Lett 13:900–914

Kynn M (2005) Eliciting expert opinion for a Bayesian logistic regression model in natural resources. PhD thesis, School of Mathematical Sciences, Queensland University of Technology, Brisbane. Available from http://eprints.qut.edu.au/16041/ (accessed April 2011)

Kynn M (2008) The "heuristics and biases" bias in expert elicitation. J Roy Stat Soc A Sta 171(1):239–264

Low-Choy S, James A, Mengersen K (2009a) Expert elicitation and its interface with technology: a review with a view to designing *Elicitator*. In: Anderssen RS, Braddock RD, Newnham LTH (eds) 18th World IMACS Congress and MODSIM09 International Congress on Modelling and Simulation. Modelling and Simulation Society of Australia and New Zealand and International Association for Mathematics and Computers in Simulation, pp 4269–4275. Available from http://www.mssanz.org.au/modsim09/J2/lowchoy.pdf (accessed April 2011)

Low-Choy S, O'Leary R, Mengersen K (2009b) Elicitation by design for ecology: using expert opinion to inform priors for Bayesian statistical models. Ecology 90:265–277

Low-Choy S, Wilson T (2009) How do experts think about statistics? Hints for improving undergraduate and postgraduate training. In: Murphy P (ed) Next Steps in Statistics Education, Durban, South Africa 2009. International Association for Statistics Education, Auckland. Available from http://www.stat.auckland.ac.nz/~iase/publications/sat09/4_3.pdf (accessed April 2011)

Low-Choy S, Murray J, James A, Mengersen K (2010) Indirect elicitation from ecological experts: from methods and software to habitat modelling and rock-wallabies. In: O'Hagan A, West M (eds) Oxford Handbook of Applied Bayesian Analysis. Oxford University Press, Oxford

Martin TG, Kuhnert PM, Mengersen K, Possingham HP (2005) The power of expert opinion in ecological models: A Bayesian approach examining the impact of livestock grazing on birds. Ecol Appl 15(1):266–280

McCarthy MA (2007) Bayesian Methods for Ecology, Cambridge University Press, Cambridge

Murray JV (2009) Spatial modeling for the conservation of threatened species: distributions, habitats and landscape connectivity of the brush-tailed rock wallaby (Petrogale penicillata). PhD thesis, School of Biological Sciences, The University of Queensland, Brisbane, Available from: http://espace.library.uq.edu.au/view/UQ:188140 (accessed April 2011)

Murray JV, Goldizen AW, O'Leary RA et al (2009) How useful is expert opinion for predicting the distribution of a species within and beyond the region of expertise? A case study using brush-tailed rock-wallabies Petrogale penicillata. J Appl Ecol 46:842–851

Murray JV, Low-Choy S, McAlpine CA et al (2011) Evaluating model transferability for a threatened species to adjacent areas: implications for rock-wallaby conservation. Austral Ecol 36(1):76–89

Murray J, Low-Choy S, Possingham H, Goldizen A (2008) The importance of ecological scale for wildlife conservation in naturally fragmented environments: A case study of the brush-tailed rock-wallaby (Petrogale penicillata). Biol Conserv 141:7–22

Oakley JE, O'Hagan, A (2010) SHELF: the Sheffield Elicitation Framework (version 2.0). School of Mathematics and Statistics, University of Sheffield, Sheffield. Available from http://tonyohagan.co.uk/shelf (accessed April 2011)

O'Hagan A, Buck CE, Daneshkhah A et al (2006) Uncertain judgements: eliciting experts' probabilities. John Wiley & Sons Inc, Chichester

O'Leary R, Low-Choy S, Murray J et al (2009) Comparison of three elicitation methods for logistic regression on predicting the presence of the threatened brush-tailed rock-wallaby Petrogale penicillata. Environmetrics 20:379–398

O'Leary RA, Murray JV, Low-Choy S, Mengersen K (2008) Expert elicitation for Bayesian classification trees. J Appl Prob Stat 3:95–106

Pike WA (2004) Modeling drinking water quality violations with Bayesian networks. J Am Water Resour Assoc 40:1563–1578

Sheppard CRC (1999) How large should my sample be? Some quick guides to sample size and the power of tests. Mar Pollut Bull 38(6):439–447

Smith CS, Howes AL, Price B, McAlpine CA (2007) Using a Bayesian belief network to predict suitable habitat of an endangered mammal—the Julia Creek dunnart (Sminthopsis douglasi). Biol Conserv 8:333–347

Spiegelhalter DJ, Thomas A, Best NG, Lunn D (2003) WinBUGS version 1.4 user manual. MRC Biostatistics Unit, Cambridge. Technical report

Store R, Kangas J (2001) Integrating spatial multi-criteria evaluation and expert knowledge for GIS-based habitat suitability modelling. Landsc Urban Plan 55:79–93

Uusitalo L (2007) Advantages and challenges of Bayesian networks in environmental modelling. Ecol Modell 203:312–318

Vikneswaran (2005) An R companion to "experimental design". Available from http://cran.r-project.org/doc/contrib/Vikneswaran-ED_companion.pdf (accessed April 2011)

Chapter 4
Eliciting Expert Knowledge of Forest Succession Using an Innovative Software Tool

Michael Drescher, Lisa J. Buse, Ajith H. Perera, and Marc R. Ouellette

Contents

4.1 Introduction

Professionals add a wealth of experiential knowledge to the application of scientific data and the implementation of procedures in many fields of work. In the forestry sector, expert knowledge is used in developing strategic plans for forest management, including large-scale land-use planning to manage the timber supply (OMNR 2010),

M. Drescher (✉)
School of Planning, University of Waterloo, 200 University Avenue West, Waterloo, ON, Canada N2L 3G1
e-mail: mdresche@uwaterloo.ca

L.J. Buse • A.H. Perera • M.R. Ouellette
Ontario Forest Research Institute, 1235 Queen Street East, Sault Ste. Marie, ON, Canada P6A 2E5

A.H. Perera et al. (eds.), *Expert Knowledge and Its Application in Landscape Ecology*,
DOI 10.1007/978-1-4614-1034-8_4, © Springer Science+Business Media, LLC 2012

or for conservation planning (McNay et al. 2005, 2006). Tactical applications of expert knowledge in forestry include landscape mapping (Walton and Meidinger 2006) and forest management operations (Willoughby and Thomson 2004; Bone et al. 2007). Despite the common use of expert knowledge in forestry applications, this use is neither formal nor rigorous, and the knowledge is often implicit and latent, with unknown characteristics and reliability.

Predicting future forest composition is a common example of the use of expert knowledge in applications of forest management planning, and depends on experiential knowledge of forest succession (e.g., Czembor and Vesk 2009). Frequently, however, its use is neither rigorous nor explicit (Davis and Ruddle 2010). Often, the rationale for the use of expert knowledge is not provided, and the extent of its use relative to other sources of information is unclear (e.g., Chap. 3). Furthermore, even if provided, the rationale rarely describes how experts were selected, how expert knowledge was elicited, or whether further analysis and evaluation of expert knowledge occurred. Consequently, the reliability of expert knowledge of forest succession (EKFS) is unknown, and the reliability of applications of that knowledge becomes uncertain. Lack of formality in the elicitation of expert knowledge may lead to the use of inadequate and inconsistent methods and to difficulties in subsequent assessment of results and subsequent comparisons among studies (Chap. 2). In addition, a lack of documentation often makes it impossible to repeat a study, which contravenes one of the fundamental principles of scientific research (i.e., replication) and increases uncertainty in the subsequent application of the knowledge that results from a study. Given the importance of EKFS in long-term planning of forest resources, it is imperative that its characteristics and levels of reliability be better known. Only then can we assess the likelihood of whether forest management plans will produce the desired forest states and whether the plans will be robust as a result of lowering the degree of uncertainty.

Most forest succession models (e.g., simulation models and decision-support systems) that are used in forest management are quantitative and require numerical input (Taylor et al. 2009). Although such models often rely on expert knowledge as a primary input, it is not usually clear how this knowledge is elicited, formalized, and quantified, and no evidence of thoroughness in following these steps or of repeatability is provided (e.g., Forbis et al. 2006; Hemstrom et al. 2007; Koniak and Noy-Meier 2009). The process of knowledge elicitation – including the selection of experts, the elicitation situation, and the elicitation technique – determines the quality and reliability of the knowledge. Therefore, rigorous methods of elicitation are essential for successful use of expert knowledge.

Our specific goals in this chapter are threefold: First, we present a case study of the elicitation of EKFS. We demonstrate how expert knowledge was elicited, formalized, and quantified, and how we captured ancillary information that may reveal additional aspects of the expert knowledge. Second, we describe the development and use of a software tool that facilitated the elicitation and formalization of EKFS. Specifically, we highlight how the elicitation process was supported by using the customized software tool and the advantages of such an approach for researchers and experts. Third, based on our experiences with the elicitation and use of EKFS, we describe general challenges for eliciting knowledge that we encountered and how we overcame them (i.e., lessons learned). Ultimately, we suggest improvements to the practice of eliciting expert knowledge in the hope that our findings will benefit future studies of knowledge elicitation.

4.2 Case Study

To illustrate our approach for formalizing and characterizing expert knowledge, we present a case study of eliciting EKFS in the boreal forest of Ontario, Canada. This study was motivated by the need to bring more rigor to the extensive use of expert knowledge in Ontario's planning process for forest landscape management. A suite of simulation models that aid in the design of future forest landscapes, such as Patchworks (Lockwood and Moore 1993) and BFOLDS (Perera et al. 2008), rely on EKFS. Attempts have been made to document EKFS (e.g., Ride et al. 2004; Vasiliauskas et al. 2004), but details of the methodology and of the characteristics (e.g., uncertainty, variability) of the expert knowledge remain ambiguous. In our project, we elicited expert knowledge to develop rules of forest succession that could then be used to parameterize the BFOLDS model for simulating forest landscape dynamics. We paid particular attention to developing an elicitation process that was both explicit and repeatable and to transparent methods of assessing characteristics of the knowledge. Drescher et al. (2006, 2008a, b), Drescher and Perera (2010a, b) and Ouellette and Drescher (2010) provide details of this study. Here, we describe the three key phases of this exercise: selecting the experts, developing a software tool to facilitate the elicitation of their knowledge, and eliciting and formalizing the expert knowledge. Figure 4.1 illustrates the major steps and their sequence.

4.2.1 Selecting the Experts

A key step in the use of expert knowledge is selecting appropriate experts. For this study, our intent was to elicit judgments from experts with experience in developing and using their knowledge of forest succession obtained from across the diverse geography of boreal Ontario. First, we contacted an array of forest management professionals in Ontario to identify those who are considered by their peers to be experts in forest succession based on their extensive local or regional experience. This was a form of *sampling by referral* (Welch 1975), in which initial contacts led to subsequent referrals to prospective experts. The resulting list of prospective experts consisted of individuals from a variety of educational, professional, and geographical backgrounds.

From this list, we identified *primary* experts (i.e., individuals who are leaders in their field and are well-respected). We believe that these primary experts, in addition to possessing expert knowledge, are well-connected and positioned to encourage other potential experts to participate. *Secondary* experts were individuals who may

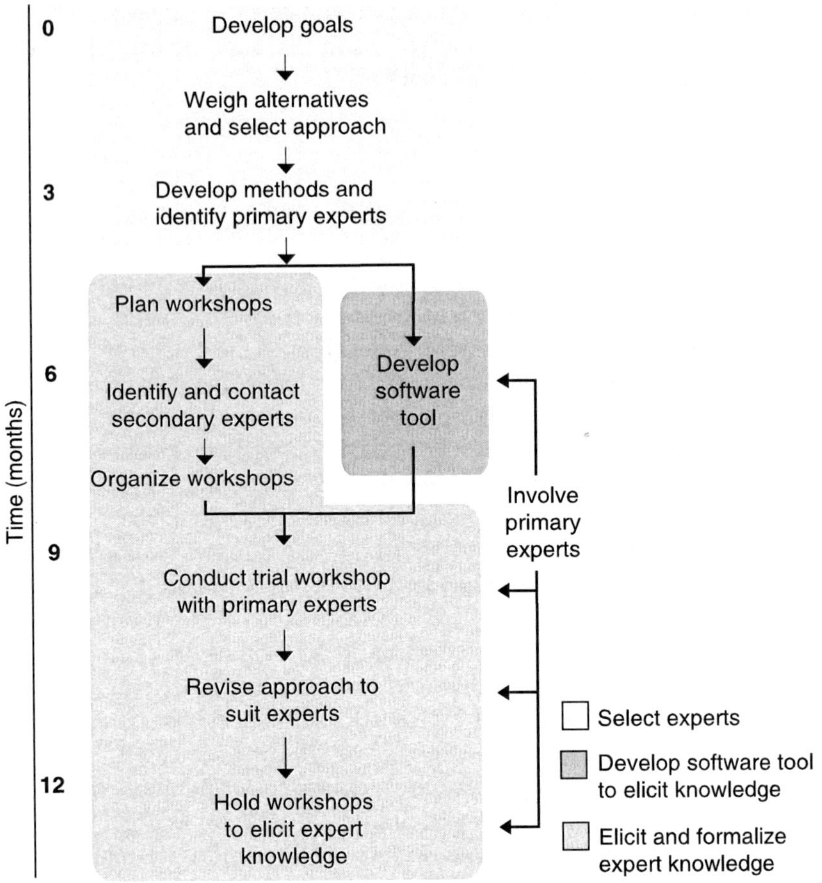

Fig. 4.1 Flowchart of the process used to elicit expert knowledge of forest succession (EKFS) with assistance from a software tool and from workshops that brought researchers together with primary and secondary experts

not be leaders in their field and may not be as well-connected as primary experts, but who nonetheless regularly use their knowledge of forest succession in their day-to-day work.

Our initial contact with the primary experts occurred during the planning stages of the study so we could ensure that any primary experts who agreed to participate were also actively engaged in the design phase of the study and in the construction of the software tool that would be used to support the knowledge elicitation process. Unfortunately, despite the efforts of the primary experts to recruit secondary experts, expert participation was limited because of time and resource constraints on the part of both the participants and the researchers. This resulted in a smaller pool of experts and a lower diversity of backgrounds than we expected. The nine experts (two primary and seven secondary) who ultimately participated in our study included foresters, ecologists, and planning analysts. Their degree of expertise is indicated by

Table 4.1 Characteristics of the experts who participated in our study to elicit expert knowledge of forest succession

Background and training	Professional experience (years)
Forestry	29
Ecology and biology	27
Forestry and ecology	25
Forestry	22
Biology	20
Forestry	19
Forestry	13
Forestry	10
Forestry	6

their years of experience in using their knowledge of forest succession in government or private management applications (Table 4.1). Their collective years of experience exceeded 170 years.

4.2.2 *Developing a Software Tool to Facilitate the Elicitation of Knowledge*

To elicit EKFS, we preferred an approach that allowed experts to express their knowledge in a quantitative and standardized manner. We also wanted to offer experts the opportunity to self-assess their knowledge in terms of its consistency with forest dynamics and to report their own levels of uncertainty about that knowledge, as suggested by Cleaves (1994). A secondary goal was to provide a simple and convenient way to store and manage the expert knowledge so as to facilitate subsequent applications, analyses, and transfer of the knowledge. For this purpose, we developed a customized software tool, the "Succession Pathway Tool" (SPT) that enabled experts to interactively express their knowledge and visually explore the resulting succession pathways. Details about the tool are provided by Ouellette and Drescher (2010). Throughout our development of the tool, primary experts were involved to ensure that the resulting product met their needs. For example, they provided input about the information that had to be entered into the tool and the information that was offered by the tool, as well as its structural logic and display options.

Forest succession is a process that is difficult to comprehend and to express unequivocally. For example, though individual processes such as seed dispersal, germination, and establishment occur over short periods, the entire process of forest succession occurs over long periods, often at time intervals that exceed the human lifespan, and is influenced by a multitude of environmental factors. Consequently, forest succession is a complex process that varies both spatially and temporally and that arguably can best be expressed in stochastic terms.

SPT was designed to simplify this view of forest succession by collapsing it into two hierarchical levels and three stochastic aspects. SPT allowed experts to compartmentalize their knowledge at two hierarchical levels (1) a lower level that described the interactions between a few individual forest types in detail and (2) a higher level that provided a general overview of the interactions among all forest types simultaneously. Although the lower level allowed experts to express their detailed knowledge of forest succession pathways, the higher level enabled them to explore long-term interactions among all forest types in the form of networks. We asked the experts to express their knowledge of three stochastic aspects of forest succession: the probability of forest succession, as well as its direction and timing. In other words, the tool was designed to help experts express their knowledge parsimoniously, without overwhelming them with complex and stochastic networks of interactions that would be difficult to understand. SPT also offered them a means to assess whether the expressed knowledge was logically consistent with their mental models and led to long-term dynamics that conformed with their experience.

The experts used a graphical user interface (GUI; Fig. 4.2) instead of complex tabular data matrices to enter their knowledge of forest succession at two levels: a detailed forest succession pathway and a broad forest succession network. The use of a GUI and the ability to switch between low and high levels of detail for the forest succession knowledge simplified the expression and exchange of expert knowledge. The GUI also minimized the level of computing and modeling knowledge required for individuals to participate in our study.

SPT also let the experts enter ancillary information that characterized their own knowledge. Our aim was to gather information related to each expert's level of confidence about their knowledge of forest succession and their perception of the level of complexity of forest succession. Both characteristics contribute to uncertainty (Chap. 2). Confidence is mainly determined by the amount of evidence supporting an expert's knowledge, and relates to "epistemic" uncertainty. Epistemic uncertainty is determined by the amount of available information about a system, and can be reduced by increasing the number of observations (e.g., Hora 1996). Complexity mainly refers to variability in forest succession that goes beyond the stochastic forest succession processes that can be quantified by SPT. This is a stochastic component that relates to "aleatory" uncertainty, which is the degree of uncertainty inherent to a system that cannot be reduced by obtaining more observations (e.g., Hora 1996).

SPT prompts experts to enter their self-assessed levels of confidence in their knowledge and to define the complexity at the level of individual forest succession pathways. Here, experts are free to classify their relative confidence and complexity as low, medium, or high. The ancillary information about uncertainty is linked to the elicited EKFS. SPT stores both types of information together so that they can easily be retrieved for further analysis. For example, patterns in the levels of confidence or complexity can be investigated and their relationships to other knowledge attributes can be assessed.

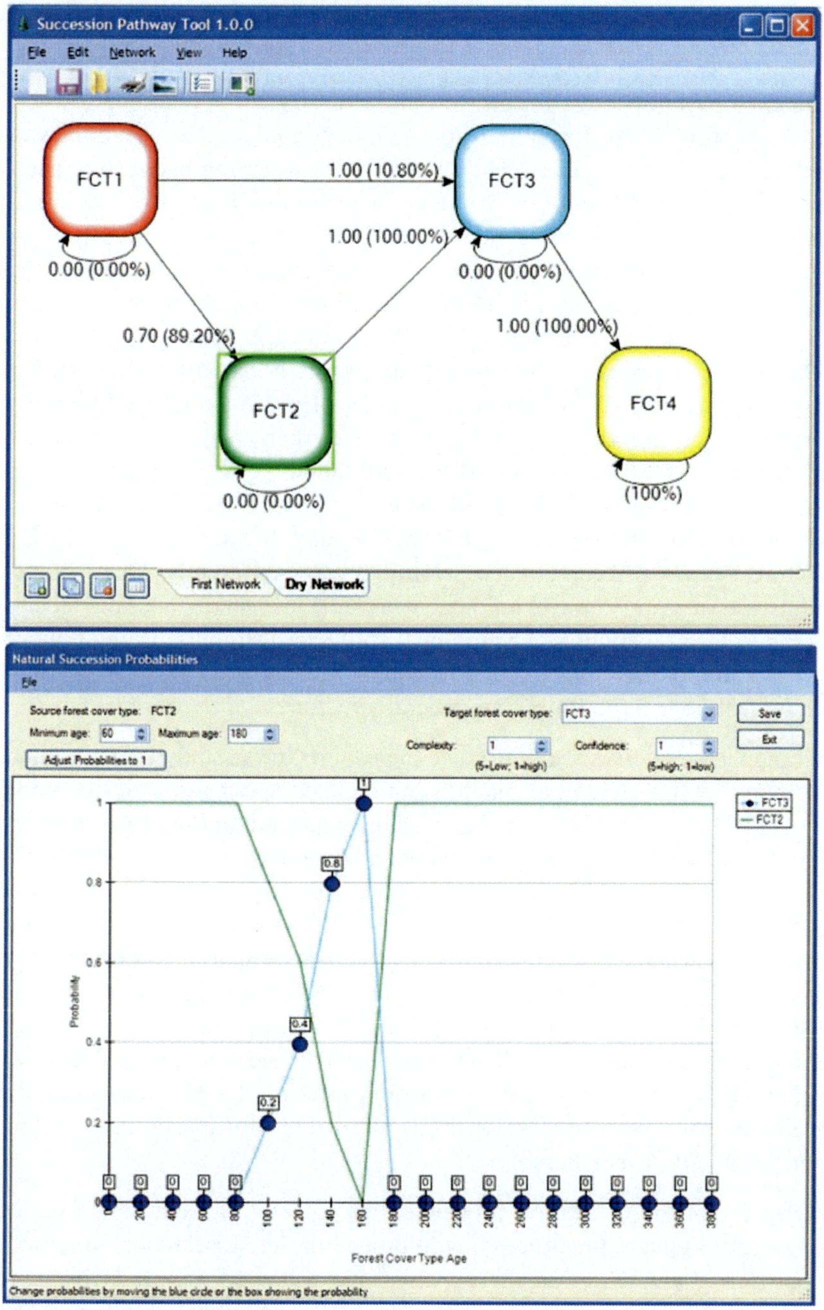

Fig. 4.2 Illustration of the general user interface of the Succession Pathway Tool (SPT): The general forest succession network level shows connections among all forest types (*top*). At the detailed successional pathway level, the display shows the temporal course and probability of succession between individual forest types (*bottom*)

4.2.3 Eliciting and Formalizing Expert Knowledge

When an expert's knowledge is complex and extensive, it is common to use face-to-face workshops as the principal communication mode for the elicitation of knowledge (Meyer and Booker 1991). The direct interaction between experts and researchers during workshops facilitates clear communication of concepts and terminology, provides an environment free of distractions from the expert's daily routine, and allows efficient exchange of a large volume of information (Meyer and Booker 1991). Moreover, the interaction allows researchers to pursue discoveries that might be missed through more automated approaches.

We originally planned a series of one-day workshops with the goal of maximizing participation from experts with a range of professional backgrounds and from a range of locations. A series of such small workshops would have let us tailor each workshop to the preferences of the individual experts (e.g., time commitments, topics of interest). We grouped the experts based on Ontario's administrative regions (i.e., northwest and northeast), because foresters in the two regions classify forest types and forest succession pathways differently. Due to logistical constraints, we were only able to hold one workshop in each region, timed to accommodate the experts' schedules. The regional split led to slightly different elicitation approaches and modifications to our subsequent analyses.

Each workshop had the following general structure:

1. We provided an overview to introduce the experts to the study's background, the workshop timelines and goals, and the anticipated applications of the workshop results. (This full disclosure of the study background and purposes was important to build trust among the researchers and experts.)
2. We explained the research questions and defined the terms and concepts that are used in mental models of forest succession. This allowed the research team and participating experts to reach a common understanding of the processes, to communicate effectively about those processes, and to use the same measures or units when addressing their knowledge of forest succession.
3. We introduced the knowledge elicitation tool (SPT) and demonstrated its functions for entering, reviewing, and editing an expert's knowledge of forest succession.
4. We elicited the knowledge using SPT, augmented by discussions among the experts and with researchers.

To ensure that the elicitation proceeded as expected, a facilitator (one of the primary experts) guided the process, with input from the researchers upon request. This approach guaranteed that the expert group followed the planned elicitation steps and stayed focused on the workshop goals. It also helped to prevent the most dominant or vocal members of the group from overly influencing the expression of other experts. The experts discussed their knowledge in a group, then subsequently entered their knowledge individually and anonymously using SPT (Fig. 4.3). Nevertheless, some experts felt uncomfortable expressing their individual knowledge and preferred a group-consensus approach. Although seeking group consensus

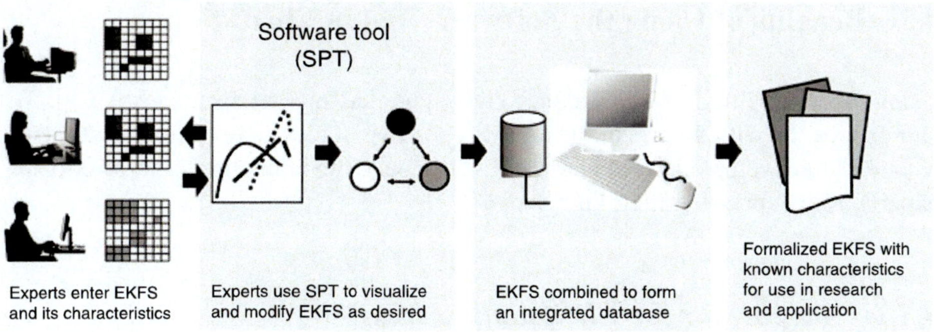

Fig. 4.3 Experts used SPT to enter, visualize, and explore their EKFS, which was then integrated into a database for use in further research and application of the knowledge

is a legitimate approach for expressing expert knowledge, it does not allow an assessment of individual knowledge characteristics, which was our original intent. However, we modified the elicitation approach to accommodate the preferences of these experts and focused on the characteristics of the expert group's knowledge. Further details of the workshop proceedings and time frame were documented by Drescher et al. (2006, 2008a).

Our chosen elicitation situation, and specifically the format of the meeting in which expert knowledge was elicited (Meyer and Booker 1991), had elements of a focus group in which all participants freely discussed each topic, and also included aspects of a "Delphi" process, in which participants were also allowed to provide anonymous responses. The elicitation technique we chose to prompt experts to reveal their knowledge was a mixture of structured interviews and knowledge modeling (Meyer and Booker 1991). This combination of elements was chosen to facilitate information exchange and critical reflection and to counteract any biasing processes such as "groupthink" (Janis 1972).

Effective communication is crucial to the success of eliciting expert knowledge. Primary experts were involved in the study design and execution as early as possible, thereby making them partners in our study. We mainly communicated formally with secondary experts through e-mail updates. However, the familiarity between the primary and secondary experts provided another communication channel that enabled more personal exchanges that strengthened everyone's motivation. Other important contributions of the primary experts were the selection of workshop participants, providing advice on how to refine SPT, and designing the workshops, which they reviewed by participating in a trial workshop before we began the actual data collection. Furthermore, the primary experts functioned as workshop facilitators. As they were intimately familiar with the study's goals and methods and understood the language and thinking patterns of both the researchers and the experts, the primary experts acted as translators between the other experts and the researchers. Though most interactions with secondary experts occurred via the primary experts, researchers occasionally communicated directly with the secondary experts.

4.3 Benefits of Using the Software Tool in Elicitation

Using a software tool to elicit expert knowledge has many advantages (Chap. 3). Our experience with the use of the tool was positive: it did not require much time or resources to develop and test, and it did not impede the elicitation process or deter experts from expressing their knowledge.

4.3.1 Support for the Elicitation Process

Given the complexity of forest succession, individual experts may perceive its details, organize their observations, and derive mental models of forest succession differently. Using SPT, which presented a platform with a common *succession language* (i.e., a set of unambiguous definitions, concepts, and terms), minimized any variability in the mental models introduced by ambiguity in terminology or in the scale of the succession process. Furthermore, SPT assisted experts in extending their knowledge beyond qualitative and deterministic perceptions of succession toward more quantitative and stochastic statements. Thus, SPT guided the experts to be more precise in their expressions and to consider the full range of possible successional pathways, including rare ones. In the field of knowledge elicitation, SPT appears to be nearly unique. To our knowledge, only one similar tool exists: Elicitator (James et al. 2010; Chap. 3). However, Elicitator is more complex in its application of Bayesian statistics for updating insights from empirical data using expert knowledge.

Our design of SPT was aimed at capturing a rich, multifaceted dataset that could be analyzed from many perspectives. In addition to capturing expert knowledge, SPT helped us to collect important ancillary information *about* expert knowledge, such as its uncertainty. Using SPT, the experts self-assessed their uncertainty in two dimensions: (1) *confidence,* which relates to the strength of the evidence supporting a belief and (2) *complexity*, which relates to the perceived stochasticity of the system about which a belief is expressed. This information was used to generate "meta-knowledge" (i.e., knowledge about the knowledge of the experts). The efficient design of this tool not only increased the effectiveness of the elicitation process, but also provided a rich dataset that could be applied to answer a variety of questions, such as potential sources of uncertainty about the expert knowledge.

4.3.2 Benefits to Experts

SPT helped the experts simplify the complex process of forest succession, especially when addressed at larger spatial and temporal scales, by allowing them to articulate their knowledge of succession in smaller, better-defined steps, and subsequently helping them to assemble that knowledge into a coherent whole. Because SPT required the experts to formulate their knowledge quantitatively – and thus move it from approximate to exact – and as individual statements, SPT helped them to articulate their knowledge unequivocally. As well, SPT's visualization functions helped experts who lacked modeling experience to examine and express their knowledge of forest succession in a model-compatible format. In addition, by prompting experts to self-assess the characteristics of their knowledge, SPT increased their awareness of meta-traits of their knowledge, such as variability, confidence, and uncertainty.

Because experts entered their knowledge within an organized framework, the tool also helped them to identify gaps in their knowledge. The visualization aspect of SPT helped experts to discover emergent properties of their knowledge (Neilsen and Dane-Nielsen 2010), such as the effects of interactions among their individual beliefs and the emergence of broad succession networks that they designed using SPT. Experts could also iteratively revise their statements to obtain outcomes that better reflected their mental models. This helped to increase confidence in their knowledge and assured consistency in their statements, as well as in the subsequent application of that knowledge in their respective professions. SPT also ensured long-term storage of the expert knowledge, which is becoming a stated goal in knowledge-based organizations that face the loss of expertise and erosion of professional knowledge as expert employees retire or move to other jobs (Kozlowski and Salas 2009). Documenting expert knowledge in SPT ensures that it will remain available for future use, including data sharing and further analyses.

4.3.3 Benefits to Researchers

The visual expression of succession by SPT simplified communication between experts and researchers and allowed us to gather the experts' knowledge at various hierarchical levels, from individual succession pathways to larger succession networks. The increased consistency of the expert knowledge gained by using SPT is tremendously beneficial for both future research and application of the knowledge. SPT ensured that the expert knowledge was captured in a quantitative and explicitly stochastic framework and stored in an electronic database. The standardized database format ensured that the knowledge will remain available for future use, without a need to repeatedly gain access to the experts. The database format enables further statistical analyses and characterization of the data, thereby expanding its potential applications. The capture and storage of the expert knowledge in an electronic format also make it readily accessible for updates and additions.

4.4 Lessons Learned from Interacting with Experts to Elicit Knowledge

Our interaction with the experts was not without challenges. Although we did not encounter problems with their use of the software tool, we did have some problems interacting with the experts. These challenges were highly circumstantial (i.e., they depended on the local culture and on individual personalities), and are not directly relevant to other situations. Instead of describing the specifics of these problems, we offer some generalized lessons learned from our experiences that may be of value to other researchers. Ecologists who are not formally trained to interact with people as research subjects may need to acquire new skills to interact with experts – to identify, communicate with, engage, and build trust, and finally to elicit knowledge. Our experience suggests that for maximum effectiveness, researchers should partner with social scientists who are familiar with the relevant elicitation approaches.

4.4.1 Motivating Experts to Participate

Professionals who possess the level of expertise required for a specific study are rare. For various reasons, those who are willing to participate in the elicitation of expert knowledge are even rarer. However, researchers can motivate experts to share their knowledge in several ways (e.g., King et al. 2002):

- Articulating the significance of each expert's contribution to the advancement of science and the benefits for future applications in their field of expertise.
- Engaging experts in all phases of the study, from the initial design to final application of the knowledge.
- Encouraging experts to collaborate *after* the knowledge elicitation process, such as by providing an opportunity to coauthor publications and gain first access to any tools and applications developed from the process.
- Enlisting primary experts to persuade and recruit other professionals who can contribute their knowledge.
- Accounting for local knowledge priorities and the needs of experts in the elicitation process.

These strategies not only help to initially motivate experts to share their experiences, but also minimize the attrition rate (i.e., the loss of experts before finalization of the study). In this study, only 2 of the 11 experts initially selected withdrew from the study.

4.4.2 Communicating with Experts

Effective dialogue with experts is crucial to the success of knowledge elicitation, but it requires concerted effort by researchers to select appropriate language, modes of discussion, timing, and topics. We found the following factors to be important in this context:

- Reach a common understanding of the terminology, concepts, and goals of the study to avoid ambiguities and misunderstanding of the scope, scale, and resolution of the knowledge exchange.
- Use multiple means to communicate directly with experts as well as indirectly through primary experts to help translate and reinforce messages.

- Use a customized software tool that provides an effective visual platform that helps the experts to illustrate, enunciate, and clarify their ideas, and thereby facilitates the communication process.
- Maintain continuous communication with the experts, from first contact through publication of the study results, to build their trust and ensure that you receive critical feedback.

4.4.3 *Eliciting Knowledge Adaptively*

Many approaches are available to elicit knowledge. The choice of the most appropriate method depends on the specific circumstances of the study. In most instances, a combination of approaches is more effective than a single approach (Meyer and Booker 1991), and we followed that advice. Selecting the exact combination of approaches may be a challenge, but this can be minimized by being adaptable (i.e., being willing to change the elicitation approaches, timing, and scope of the knowledge elicitation based on feedback from the experts). Pilot testing of the elicitation process with selected experts helps to confirm the efficacy of specific techniques, language, and terminology, and allows researchers to refine their approach before beginning to gather the actual knowledge. Pilot tests also provide an opportunity for the primary experts to become more familiar with the tools, such as the SPT software used in the present study.

4.4.4 *Minimizing Biases*

Using a small and interconnected pool of experts may introduce sampling biases, and innovative statistical techniques may be necessary to minimize their effect on the knowledge that is elicited (Heckathorn 1997). Other biases associated with experts are behavioral (Meyer and Booker 1991) and are induced by social pressures such as groupthink (Janis 1972), or cognitive (Meyer and Booker 1991) and occur due to errors in perceiving, processing, and storing information. A key aspect of the knowledge elicitation process is to detect such biases and minimize their effects though appropriate countermeasures, such as those outlined in Table 4.2.

4.5 Conclusions

This chapter describes a case study of eliciting expert knowledge about forest succession with the aid of a software tool, and using the results to create a body of knowledge that could then be characterized, analyzed, and compared among experts (Fig. 4.4). Drescher and Perera (Chap. 9) provide more details of this process.

Table 4.2 Important types of expert bias that should be considered when designing elicitation approaches, their causes, and countermeasures that can minimize their impact

Bias type	Bias	Causes	Countermeasures
Motivational and behavioral	Groupthink	Excessive influence by a dominant individual or by the group	Avoid leading questions and encourage individual and anonymous expression of knowledge
	Impression management	Conforming to societal or corporate expectations	Avoid value-laden questions and encourage individual and anonymous expression of knowledge
	Wishful thinking	Conflict of interest	Choose experts and tasks to avoid the possibility of personal gain
	Misinterpretation	Frame of reference differs between expert and researcher	Record all knowledge statements literally and independent of their apparent value; analyze statements using an objective coding scheme
	Misrepresentation	Knowledge encoding does not accurately capture the knowledge statement and its characteristics	Involve the experts in designing the measures for encoding knowledge
Cognitive	Anchoring	Sequential processing of information and a lack of re-evaluation	Avoid unjustified extrapolation from previous experience
	Inconsistency	Limited processing ability leads to insufficient discrimination	Provide real-time feedback about consistency of statements and keep the level of difficulty low
	Availability	The ability to recall information depends on the noteworthiness of the event	Use several information sources or involve multiple experts to provide a broader inference base
	Overconfidence	Limited examination of evidence	Provide complete evidence and feedback about consistency of responses with expectations

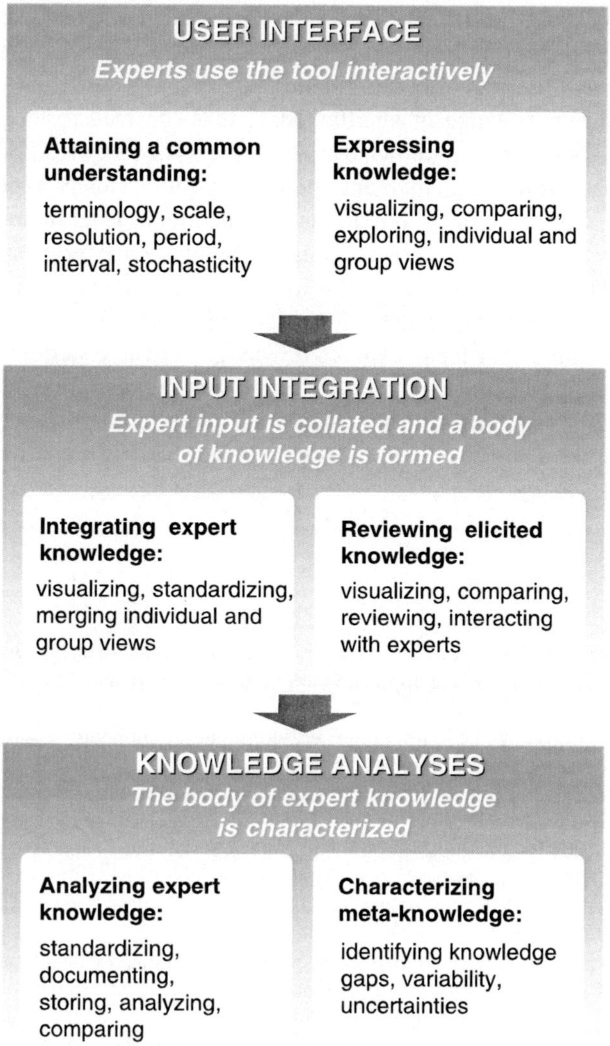

Fig. 4.4 Overview of the functions and characteristics of SPT and its role in the broader knowledge integration process. The process for eliciting expert knowledge is explained in this chapter, and the subsequent knowledge integration and analysis steps are presented by Drescher and Perera (Chap. 9)

We found that the software tool, SPT, helped the experts to visualize and address the process of boreal forest succession not just as isolated steps, but as complete networks, and to explore their own knowledge in stochastic terms. Thus, we were able to elicit what was previously fragmented and *ad hoc* knowledge of boreal forest succession that resided with individual experts and integrate this knowledge into a cohesive database. This database has properties similar to a typical empirical dataset used in ecological experimentation because it is quantitative, and was produced using explicit and repeatable data collection methods.

We learned the importance of ancillary data that can lead to the generation of valuable meta-knowledge (i.e., knowledge about the knowledge). Although EKFS is being used in many forest management applications, it is mostly applied in a fragmented manner, with much detail but no view to the higher level interactions among successional processes. Expert knowledge is not typically viewed at synoptic levels, where the details would coalesce and be integrated into succession networks that reveal the stochasticity of outcomes and geographic variability. SPT enabled such integration of detailed and local expert knowledge into a generalized knowledge base with broader applicability. Finally, we found that the definition and selection of appropriate experts was crucial to the success of the elicitation process and to the characteristics of the body of expert knowledge assembled through this process.

References

Bone C, Dragicevic S, Roberts A (2007) Evaluating forest management practices using a GIS-based cellular automata modeling approach with multispectral imagery. Environ Model Assess 12:105–118

Cleaves DA (1994) Assessing uncertainty in expert judgments about natural resources. USDA For Serv, South For Exp Stn, New Orleans, Gen Tech Rep SO-110

Czembor CA, Vesk PA (2009) Incorporating between-expert uncertainty into state-and-transition simulation models for forest restoration. For Ecol Manage 259:165–175

Davis A, Ruddle K (2010) Constructing confidence: rational skepticism and systematic enquiry in local ecological knowledge research. Ecol Appl 20:880–894

Drescher MD, Perera AH (2010a) A network approach for evaluating and communicating forest change models. J Appl Ecol 47:57–66

Drescher MD, Perera AH (2010b) Comparing two sets of forest cover change knowledge used in forest landscape management planning. J Environ Plan Manage 53:591–613

Drescher M, Perera AH, Buse LJ, et al (2006) Identifying uncertainty in practitioner knowledge of boreal forest succession in Ontario through a workshop approach. Ont Min Nat Resour, Ont For Res Inst, Sault Ste. Marie, For Res Rep No 165

Drescher M, Perera AH, Buse LJ, et al (2008a) Boreal forest succession in Ontario: An analysis of the knowledge space. Ont Min Nat Resour, Ont For Res Inst, Sault Ste. Marie, For Res Rep No 171

Drescher M, Perera AH, Buse LJ, et al (2008b) Uncertainty in expert knowledge of forest succession: A case study from boreal Ontario. For Chron 84:194–209

Forbis TA, Provencher L, Frid L, Medlyn G (2006) Great Basin land management planning using ecological modeling. Environ Manage 38:62–83

Heckathorn DD (1997) Respondent-driven sampling: A new approach to the study of hidden populations. Soc Probl 44:174–199

Hemstrom MA, Merzenich J, Reger A, Wales B (2007) Integrated analysis of landscape management scenarios using state and transition models in the upper Grande Ronde River Subbasin, Oregon, USA. Landsc Urban Plan 80:198–211

Hora SC (1996) Aleatory and epistemic uncertainty in probability elicitation with an example from hazardous waste management. Reliabil Engineer Syst Saf 54:217–223

James A, Low-Choy S, Mengersen K (2010) Elicitator: An expert elicitation tool for regression in ecology. Environ Mod Softw 25:129–145

Janis IL (1972) Victims of groupthink. Houghton Mifflin Company, Boston
King, W, Mark PV Jr, McCoy S (2002) The most important issues in knowledge management. Commun ACM 45:93–97
Koniak G, Noy-Meier I (2009) A hierarchical, multi-scale, management-responsive model of Mediterranean vegetation dynamics. Ecol Mod 220:1148–1158
Kozlowski SW, Salas E (eds) (2009) Learning, training, and development in organizations. Routledge, Taylor and Francis Group, New York
Lockwood C, Moore T (1993) Harvest scheduling with spatial constraints: a simulated annealing approach. Can J For Res 23:468–478
McNay RS, Apps C, Wilson S, et al (2006) Use of habitat supply models to establish herd based recovery targets for threatened mountain caribou in British Columbia: Year 2 Progress Report. Wildlife Infometrics Inc., Mackenzie, Rep No 180
McNay RS, Caldwell JL, Sulyma R (2005) Impact of riparian management on timber supply in the Mackenzie TSA, British Columbia. Wildlife Infometrics Inc., Mackenzie, Rep No 167
Meyer, MA, Booker JM (1991) Eliciting and analyzing expert judgement: A practical guide. Knowledge-based Systems Vol 5. Academic Press, New York
Neilsen, C, Dane-Nielsen, HC (2010) The emergent properties of intellectual capital; a conceptual offering. J Human Resour Cost Account 14:6–27
OMNR (2010) Ontario's forest management guides: an introduction. Ont Min Nat Resour, Sault Ste. Marie, Ontario. Available from http://www.mnr.gov.on.ca/stdprodconsume/groups/lr/@mnr/@forests/documents/document/mnr_e000257.pdf (Accessed November 2010)
Ouellette M, Drescher, MD (2010) Succession Pathway Tool user's guide. Ont. Min Nat Resour, Ont For Res Inst, Sault Ste. Marie, For Res Inf Pap No 173
Perera AH, Ouellette MR, Cui W, et al (2008) BFOLDS 1.0: A spatial simulation model for exploring large scale fire regimes and succession in boreal forest landscapes. Ont. Min Nat Resour, Ont For Res Inst, Sault Ste. Marie, For Res Rep No 152
Ride KR, Longpre TWF, Racey GD, Elkie PC (2004) Estimates of ecoregional forest composition derived using modelled bounds of natural variation in Northwestern Ontario. Ont Min Nat Resour, Northwest Sci Inf, Thunder Bay, Tech Rep TR-136
Taylor AR, Chen HYH, Van Damme L (2009) A review of forest succession models and their suitability for forest management planning. For Sci 55:23–36
Vasiliauskas S, Chen HYH, Parton J, et al (2004) Successional pathways: Proposed SFMM rules for northeast regional standard forest units, Version 2. Ont Min Nat Resour, Northeast Sci Inf, South Porcupine, Tech Note TN-018
Walton A, Meidinger D (2006) Capturing expert knowledge for ecosystem mapping using Bayesian networks. Can J For Res 36:3087–3103
Welch S (1975) Sampling by referral in a dispersed population. Publ Opinion Quart 39:237–245
Willoughby I, Thomson JA (2004) Expert help on herbicide selection. For Brit Timber 33:30

Chapter 5
Expert Knowledge as a Foundation for the Management of Secretive Species and Their Habitat

C. Ashton Drew and Jaime A. Collazo

Contents

5.1 Introduction

In this chapter, we share lessons learned during the elicitation and application of expert knowledge in the form of a belief network model for the habitat of a waterbird, the King Rail (*Rallus elegans*). A belief network is a statistical framework

C.A. Drew (✉)
Department of Biology, North Carolina Cooperative Fish and Wildlife Research Unit, North Carolina State University, David Clark Labs, Box 7617, Raleigh, NC 27695, USA
e-mail: cadrew@ncsu.edu

J.A. Collazo
United States Geological Survey, North Carolina Cooperative Fish and Wildlife Research Unit, North Carolina State University, David Clark Labs, Box 7617, Raleigh, NC 27695, USA

A.H. Perera et al. (eds.), *Expert Knowledge and Its Application in Landscape Ecology*, DOI 10.1007/978-1-4614-1034-8_5, used to graphically represent and evaluate hypothesized cause and effect relationships among variables. Our model was a pilot project to explore the value of such a model as a tool to help the US Fish and Wildlife Service (USFWS) conserve species that lack sufficient empirical data to guide management decisions. Many factors limit the availability of empirical data that can support landscape-scale conservation planning. Globally, most species simply have not yet been subject to empirical study (Wilson 2000). Even for well-studied species, data are often restricted to specific geographic extents, to particular seasons, or to specific segments of a species' life history. The USFWS mandates that the agency's conservation actions (1) be coordinated across regional landscapes, (2) be founded on the best available science (with testable assumptions), and (3) support adaptive management through monitoring and assessment of action outcomes. Given limits on the available data, the concept of "best available science" in the context of conservation planning generally includes a mix of empirical data and expert knowledge (Sullivan et al. 2006).

The King Rail served as the focus of our pilot study because it presented data challenges common to many nongame species. Our study area was the USFWS Eastern North Carolina and Southeastern Virginia Ecoregion (hereafter, "the ecoregion"). Within the ecoregion, the King Rail (1) had never been studied locally (local systems were ecologically distinct from those regions that had been studied), (2) had only recently been prioritized (so local biologists and managers had not previously paid close attention to the population), and (3) is secretive (so the bird is difficult to detect even if it is present). In the Southeast Coastal Plain, the ecoregion offers a large amount of the fresh and brackish water emergent marsh habitat preferred by the King Rail (USGS Southeast Gap Analysis Program [GAP], http://www.basic.ncsu.edu/segap/index.html). However, GAP models only identify *potential* habitat, and provide no indication of the quality or occupancy of that habitat. Given that the ecoregion may be a critical management unit for achieving national and regional population objectives (Cooper 2008), the USFWS desired a model with finer spatial resolution of the distribution and abundance of breeding King Rail within the potential habitats. Ideally, the model would estimate the number of King Rail currently supported by the ecoregion and indicate where management of existing conservation lands or acquisition of new conservation lands would protect the greatest numbers of this waterbird. Expert-based modeling provided a means to address these questions despite the presence of significant empirical data gaps for the study species.

5.2 Case Study Approach

5.2.1 Belief Network Models of Habitat Occupancy

Expert-based belief network models that support conservation efforts are constructed through elicitation of the assumptions and logic of experts regarding how an ecological system is structured and how it will respond to management actions

(Marcot et al. 2006; Kuhnert and Hayes 2009). The belief network (also referred to as an "influence diagram" or "causal diagram") represents the hypothesized cause and effect relationships elicited from experts. Relationships between variables and their expected effects are defined using conditional-probability tables, which can be populated using available data, expert knowledge, or a combination of the two. In an expert-based *Bayesian* belief network, the initial expert model is updated through the addition of new data, which incrementally decrease the weight given to the expert knowledge and increase the weight given to the empirical data as monitoring or research continues. Uncertainty is explicitly incorporated, displayed, and propagated throughout the network.

We elected to use belief network models for the same reasons they have been promoted to the conservation community (McCann et al. 2006; Nyberg et al. 2006; Uusitalo 2007; Howes et al. 2010; Chap. 7): they distill complex systems into easily visualized and communicated diagrams, accommodate data from a diversity of sources, provide opportunities for filling data gaps using professional judgment, provide quantitative output in the form of probabilities of various outcomes, and easily integrate new monitoring data to support adaptive management. Primary constraints include challenges associated with rigorously eliciting probabilities, the common necessity of representing data in discrete formats (i.e., as categorical rather than continuous variables), and the acyclical nature of the model structure (e.g., the inability to incorporate feedback loops, but see Bashari et al. 2009). Kuhnert and Hayes (2009) provide an introduction to some of the potential pitfalls, which relate to data scaling (e.g., from local to landscape scales), discretization (i.e., categorizing data that would be better represented by continuous variables), the network structure (e.g., defining the cause and effect relationships among variables), and the complexity of the models. An additional challenge is how to elicit probability estimates that accurately and precisely reflect the experts' knowledge (Renooij 2001; Chap. 3). Most people, even those with science training, provide poor estimates of probability values (O'Hagan et al. 2006).

Our research objective was to predict the probability of occupancy of a given site by a breeding population of King Rails within areas modeled as potential habitat by the Southeast GAP models (Drew et al. 2006). In occupancy modeling, repeated presence–absence data or encounter histories allow the calculation of a detection probability. Occupancy estimates can then be adjusted for differences in detection probabilities, allowing for more accurate inferences about species–habitat associations. Occupancy estimation and similar techniques (MacKenzie et al. 2006) grew out of population surveys in which observers noted that species are not always detected even when they were present (Royle and Nichols 2003). We considered processes influencing occupancy at two spatial scales: marsh patches within the ecoregion, and sites within the marsh patches. We then conducted field surveys and compared the field-derived occupancy estimates, calculated using the program PRESENCE (version 3.1, http://www.mbr-pwrc.usgs.gov/software/presence.html), to the predictions of an expert-based belief network model.

5.2.2 *Expert Selection and Elicitation Procedures*

No published studies or data existed to define the ecology of King Rails in the ecoregion. Incidental observations provided the sole source of knowledge and hypotheses concerning the waterbird's response to various landscape and microhabitat variables locally. The four participating experts were USFWS biologists serving at local National Wildlife Refuges (Table 5.1). These experts offered strong, local knowledge of marsh habitats in the refuges where they work, but not necessarily regional knowledge (e.g., marshes in neighboring public or private land). Although none had previously researched or monitored secretive marsh bird populations, all had observed the King Rail and all were aware of the bird's basic ecology through research literature from other regions and through professional meetings. We elicited information using two techniques (defined below): discussion interviews and image-based interviews. For both elicitation techniques, experts were interviewed privately at their office or an alternate site of their choice.

5.2.2.1 Discussion Interviews

Discussion interviews followed a semistructured approach, presented in two parts over a period of approximately 3 h. The interviews followed a script to ensure that each expert received the same introductory material (e.g., orientation to the objectives and terminology) and the same wording for each question and task. Part One sought to characterize the domain of the expert's experience (Table 5.1), and Part Two captured their knowledge of King Rail ecology. Within the boundaries of this structured design, the conversation was free to evolve around an expert's requests for clarification of terminology and the modeling process, discussion of the proposed variables of interest, and justifications of hypotheses.

Our questions in Part One addressed the spatial and temporal extent and resolution of an expert's relevant work experience, the contribution of different resources (e.g., observations, literature and colleagues) to the elicited knowledge, the expert's familiarity with interpreting digital orthophotos (both printed and shown on a laptop computer), the expert's self-confidence in their local knowledge of the King Rail, and the expert's overall belief that the King Rail responded to certain landscape features. This provided information to evaluate or weight an expert's knowledge and also allowed time to orient the expert to the interview method and vocabulary.

In Part Two, experts first identified landscape and microhabitat variables of potential importance and then directly quantified the relationships between occupancy of a site and each variable. If an expert failed to independently identify any variables mentioned in the literature or by another expert, we asked them to consider these variables too. If later experts identified variables not elicited from an earlier expert, the earlier experts were contacted by phone or e-mail and given the opportunity to comment on the additional variable. Experts provided probability

Table 5.1 Examples of questions posed to characterize the experts during Part One of the discussion interviews

Background information gathered during elicitation	Expert 1[a]	2	3	4	Statistics		Primary objectives	Example insights gained
Field experience (temporal and spatial)					Mean	SD		
1. How long have you worked in marsh habitats?	20	12	32	25	22	8	Characterize where and how experts gained their knowledge	Years of experience is a poor indicator of the knowledge's value to project objectives (most observations may have occurred in a small area of the refuge, at the wrong time of year, or at sites with very different past conditions)
2. How long in Atlantic Coastal Plain marsh habitats?	15	12	25	25	19	7		
3. How long in marsh habitats within the Ecoregion?	13	12	16	25	17	6		
4. How long at this refuge?	8	6	16	13	11	5		
5. On the aerial imagery provided, please indicate the areas that you visit, the seasons when you visit, and the frequency of your visits. Distinguish between the locations visited for surveys of secretive marsh birds, other bird surveys, or activities unrelated to birds								
Use of knowledge resources					Percent			
6. Primary literature/professional meetings	Yes	No	No	Yes	50		Characterize whether and how an expert's knowledge was shaped by their professional community	Personal interest (effort beyond work requirements) seems to be an important factor contributing to expertise
7. Gray literature/field guides/Internet	Yes	Yes	Yes	Yes	100			
8. Informal field observations	Yes	Yes	Yes	Yes	100			
9. Formal field research or monitoring	No	Yes	Yes	No	50			
Confidence (0 = strongly disagree, 5 = neutral, and 10 = strongly agree)					Mean	SD		

(continued)

Table 5.1 (continued)

Background information gathered during elicitation	Expert				Statistics		Primary objectives	Example insights gained
	1[a]	2	3	4				
10. King Rail distribution reflects selection for or against visible, mappable habitat characteristics	8	10	8	5	8	2	Orientation to landscape-scale data and metrics; identify experts skeptical of the project approach to understand their concerns and improve their buy-in; clarify the role of the experts and the methods of knowledge analysis and application	Though the experts were comfortable working with maps and DOQQ imagery, this did not necessarily translate into an awareness of landscape-scale patterns and processes. Experts used maps to identify sites with known unique microhabitat characteristics
11. I can identify at least some of these important characteristics accurately	6	7	10	7	8	2		
12. I can identify familiar features in these GIS data	8	9	9	9	9	1		
13. I can identify familiar features in these DOQQ[b] data	9	8	8	9	9	1		
14. If I went to the best King Rail habitat on my refuge at the peak of the breeding season, I would detect a KIRA the first place I surveyed using the call-back method	6	4	5	1	4	2		
Detectability (all questions refer to call-back surveys conducted at the peak of the breeding season under ideal conditions in the best available habitat)					Mean	SD		
15. At how many of ten survey sites would you detect a King Rail?	3; 1	4	8	3	6	2	Orientation to probability estimation; check for internal consistency; test for different baseline detectability assumptions	Experts have different baseline assumptions about the probability of occupancy and the probability of detection
16. If surveys were repeated for ten visits at each of these ten sites, what would be the maximum number of visits when a single site produced at least one detection?	6; 2	4	9	3	5	3		

These questions provided metadata that could be used to compare the experience, confidence, and baseline assumptions of the experts. These questions also helped the experts to understand the project's purpose and demonstrated our methods for probability elicitation

The full interview script and the responses of the experts will be published as an appendix in the final project report (expected in 2011)

[a]Expert 1 answered based on experience at two refuges, with the combined response reported except for questions 15 and 16

[b]*DOQQ* digital orthophoto quarter quadrangle images

estimates of response rates for call-back surveys in hypothetical landscapes. Call-back surveys play recorded bird songs to prompt responses from local birds (Conway 2009). By imagining ten sites for a given landscape scenario (e.g., a marsh patch size of 20 acres), experts stated the number of sites where they would expect to detect at least one King Rail during a call-back survey. In answering, we instructed the experts to assume that these hypothetical surveys were conducted at the peak of the breeding season, under ideal sampling conditions, and in otherwise suitable habitat to maximize the probability of occupancy. For each variable (e.g., patch size) experts also identified, if relevant, the value beyond which detection would drop to zero or reach a minimum threshold and the value or range of values that would offer the maximum number of detections. Finally, to facilitate our comparison of the responses, we asked the experts to identify what they considered to be the highest possible proportion of the population detected relative to the proportion actually present under ideal conditions at their refuge.

In all steps, the experts explained the reasoning behind their hypotheses. After full discussion of all variables, experts identified the top five and then the top two variables that they expected to most strongly influence King Rail occupancy throughout their refuge. Audio recordings of the interviews let us review this information during model construction.

5.2.2.2 Image-Based Interviews

The image-based interviews required experts to assign an occupancy probability class [low (1–33%), moderate (34–66%), or high (67–100%)] to preselected potential habitat sites visualized using an aerial image (digital orthophoto quarter quadrangle, DOQQ; 1998 imagery obtained from US Geological Survey: http://eros.usgs.gov/#/Home). The selected sites represented a random stratified sample, with two strata representing landscape variables that were highly ranked by the experts during the discussion interviews (distance to open water and marsh patch size), and a third that they considered to be potentially important but about which they were highly uncertain (dominant land cover, the dominant land cover class within 1 km surrounding a potential habitat site). Some sites were also randomly set in nonmarsh habitats to ensure that experts correctly distinguished marsh from nonmarsh habitat in the DOQQ images. The images were provided in paper format, but were also available on a laptop computer to allow the experts to zoom in or out around features of interest. Experts classified sites for three landscapes: their own refuge, an unfamiliar refuge with similar habitats, and an unfamiliar refuge with dissimilar habitats. For each landscape, they viewed and classified a total of 100 sites (in sets of ten) to reduce visual confusion among the sites. Experts were asked to verbally describe the habitat traits they were considering as they classified each site. These comments were captured in our notes and using an audio recorder.

5.2.3 *Model Construction*

We used version 4.1.6 of the Netica software (http://www.norsys.com/) to construct the model, and followed guidelines for the development of belief networks for ecological and conservation applications (Marcot et al. 2006). We first generated an influence diagram to describe the system structure (i.e., the relationships between causes and effects), then defined the categorical state of each model node (Kuhnert and Hayes 2009), and finally defined probabilistic relationships among the variables. The model structure and the underlying conditional probability tables drew upon experts' combined knowledge synthesized from their personal observations, review of the literature from other regions, and interaction with colleagues involved in marsh bird management and research.

Construction of the influence diagram and conditional probability tables required us to reconcile the predictions of different experts. We compared each expert's stated maximum detection probability because experts could disagree on an absolute scale while still agreeing on a relative scale up to that maximum probability; the range of landscape values represented within an expert's refuge so that we could distinguish between responses based on local experience versus informed conjecture; and the expert's relative experience (e.g., years at their refuge, time spent in marshes during the breeding season, and number of King Rail sightings). Where we could not reasonably explain differences among the experts, we conducted follow-up conversations or allowed evidence from the literature to guide our development of the belief network. Through such follow-up discussions, we learned that differences among experts' rankings of marsh patch size as an informative variable reflected the presence (high rank) or absence (low rank) of a strong patch-size gradient in the landscapes of their experience. For example, we confirmed that given the strong gradient in patch size across the regional landscape, all experts agreed that marsh patch size was a potentially informative predictor variable for King Rail habitat occupancy. When this phase of the analysis was complete, experts reviewed and accepted the model as a reasonable representation of their hypotheses before we applied it to the ecoregional landscape. As model output, we reported the probability of occupancy both as an expected value (the statistical expectation with a standard deviation) and as the probability of a site ranking low, moderate, or high for occupancy.

5.2.4 *Validating and Updating the Expert Model with Field Data*

In 2008 and 2009, we surveyed 89 marsh sites using the National Marsh Bird Survey Protocol (Conway 2009). The sites represented a stratified random sample, with the strata selected to represent the belief network variables that contributed most to variability in the model predictions (patch size, distance to open water, and dominant land cover). After estimating the probability of occupancy using the PRESENCE software, we validated the model by comparing the occupancy predicted by the belief network with the detection-adjusted occupancy estimates.

We used the test-with-cases function in Netica, which evaluates how well the belief network's predictions match the observed data. We also assessed whether the predictions of the expert-based belief network model consistently over- or under-estimated the detection-adjusted occupancy estimates.

After comparing the expert-based belief network model against the detection-adjusted occupancy estimates calculated from empirical data, we tested the belief network model's ability to learn (i.e., be refined) based on subsequent monitoring data. A primary justification for using these models in conservation and resource management settings is that they can be updated through the incorporation of field monitoring data, thereby gradually offering more accurate and precise predictions (Marcot et al. 2006; Uusitalo 2007; Howes et al. 2010). We entered one-third of the field data, randomly drawn from the full dataset, as observed case studies to enable the belief network to "learn." During the learning process, the prior model elicited from the experts is updated based on the data (likelihood) to produce the Bayesian posterior probabilities. We compared the abilities of the expert-only (prior) versus the expert+data (posterior) models to accurately predict the remaining two-thirds of the observed data. We repeated this process 100 times to compare the distributions of the accuracy statistics for the two models.

5.2.5 *Experts and Their Knowledge*

5.2.5.1 Expert Participation and Feedback

Our experts were "experienced wildlife professionals" rather than species specialists and, in general, all experts were initially nervous regarding how their limited knowledge would be applied. Some had experienced the misuse or misrepresentation of expert information. Though we did not directly evaluate their discomfort or concerns, several observations emerged based on the level of effort required to reassure and encourage experts at different stages of the process.

All experts asserted that their knowledge was suitable only for formulating hypotheses but not as a direct substitute for empirical data (Bunnell 1989). Although they were confident of their personal observations, they were often uncertain how typical or atypical their observations might be in the context of the full study region or even their own refuge. Also, they feared that their hypotheses could be misrepresented as observation and result in the development of regulations and management guidelines without further testing. Their discomfort was greatest for quantitative tasks; they were most comfortable identifying potentially important habitat variables, less comfortable ranking these variables, and least comfortable providing probability values. This increasing degree of discomfort likely reflects their degree of exposure to these types of information. Although each had detected the King Rail in their respective refuges, most of their knowledge of the bird came from the published literature, professional meetings, and interactions with colleagues. They also drew from experience with more common, taxonomically related species. These resources provided support for their hypotheses about associated environmental variables, but would not have provided support for the most quantitative hypotheses.

The experts expressed greater discomfort when providing information at landscape scales than at microhabitat scales. This is unsurprising because when experts observe species in the field they have immediate access to microhabitat data (e.g., vegetation species and height, water depth), but would need to plot these observations on a map to draw associations with landscape characteristics. Also unsurprising was the increased level of discomfort when, during the image-based interviews, we asked experts to classify sites in unfamiliar landscapes. In familiar landscapes, they drew heavily on their knowledge of local microhabitat characteristics to assign probability classes. In unfamiliar landscapes, they could not draw upon such fine-scale site-level knowledge and were forced to consider landscape features and infer microhabitat conditions.

5.2.5.2 Variance Among Experts

Individuals' habitat hypotheses differed in multiple aspects, including variable identification, ranking, and probability estimates. Although the variance in responses contributed to the overall uncertainty, it also revealed information about the bird's ecology (Murray et al. 2009; Aspinall 2010). We observed two interesting trends.

First, experts tended to identify and rank highest those variables that best described the variability in the landscapes that they manage. Drawing experience from unique spatial and temporal domains, their responses differed accordingly. For example, experts with experience in fringing coastal marshes with strong salinity gradients ranked the potential influence of salinity above that of patch size or distance to open water, whereas those with experience in fresh-oligohaline marshes of highly variable size hypothesized a stronger role for patch size and distance to open water (Fig. 5.1a).

Second, we found that some differences among experts could be attributed to differences in their baseline expectations: the probability of detecting a King Rail in ideal habitat under ideal conditions using the National Secretive Marshbird Survey Protocol. If their responses were standardized as a function of their personal baseline (i.e., maximum probability) prior to comparison, we obtained a very different perspective on the extent of expert (dis)agreement (Fig. 5.1b).

5.2.5.3 Variance Between the Two Elicitation Methods

The two elicitation methods led to different results and had different strengths and weaknesses from the perspectives of both the elicitator and the expert. The color infrared DOQQ imagery provided the most current aerial image data for the full extent of our project area at the time of the interviews. These images were captured in the spring of 1998, 8 years prior to our interviews. Given the significant and rapid landscape changes that have occurred in our study region, this led to immediate confusion. We had not anticipated the question of whether experts should answer

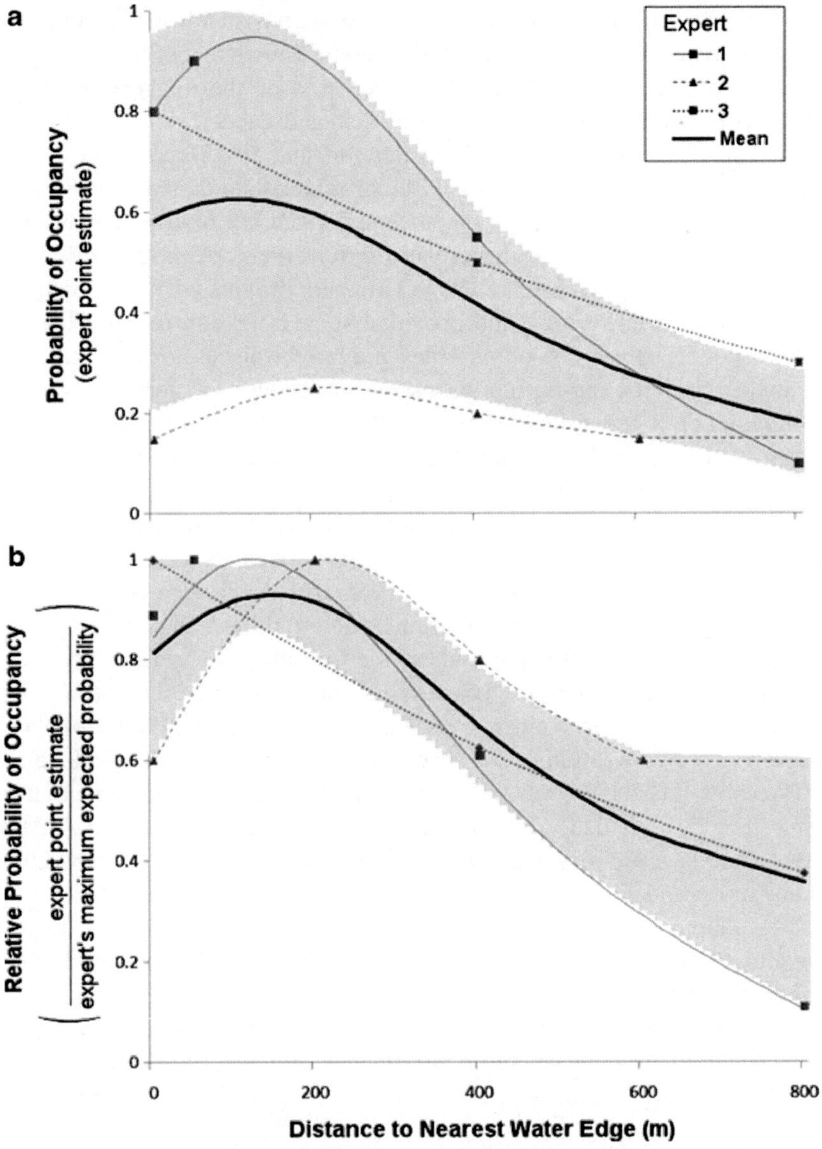

Fig. 5.1 An illustrative example of the differences observed among experts' predictions, showing responses elicited from three experts (*lines*) and the standard deviation (*gray region*). The degree of uncertainty depended on the question posed. (**a**) When we asked the experts to estimate the probability associated with a specific distance value, disagreement was greatest at short distances. (**b**) Using a relative scale (based on the proportion of each expert's maximum value) changed the nature of the disagreement among the experts. The experts agreed more strongly that shorter distances were best, but disagreed about how greater distances would reduce the probability of occupancy

according to the information they perceived in the imagery, what they believed the habitat to be like at present, or what they remembered the habitat to be like when they had observed King Rail (their observations ranged from the 1970s through 2006). We directed the experts to answer based on the visual cues provided by the imagery (the 1998 information) to ensure temporal consistency among the experts, but it is possible that their responses were biased by knowledge of the subsequent or current status of the landscape. Interviews centered on the imagery did have the desired effect of forcing the experts to consider landscape-scale variables such as marsh patch size, marsh–open water interspersion, and proximity to forest, agriculture, and urban areas. If an expert was familiar with a site, they always referenced known microhabitat conditions or management histories first and ignored landscape-scale variables until we directly prompted them. Although eliciting the probability scores for sites using the DOQQ imagery provided a method with strong repeatability for a single expert and provided standardization across experts, this method generated a shorter list of variables and briefer discussions.

Information gathered during the discussion interview was more informative for constructing the draft belief network models than the information gathered from the image-based interviews. The issue of temporal mismatches between an expert's experience, the imagery, and the model application was just one of several challenges with the image-based results. As noted by others who have used map data in elicitation (Yamada et al. 2003; Murray et al. 2009), experts varied in their expertise in interpreting landscape features from aerial imagery, and were cued by different features that they used as their primary basis for interpreting a landscape's value. For example, one expert primarily focused on the proximity to and density of shrub cover, whereas another focused on features that indicated the presence of an invasive marsh plant. In this manner, their responses evidenced a bias similar to "anchoring," which involves reaching a judgment based on prior conceptions (Kahneman et al. 1982). Typically, this anchor was a variable that they identified as a potential factor at the first site they classified. Also, similar to the results of Yamada et al. (2003), we found that experts made variable use of the information provided; some chose to simply classify points on the map that we had printed (ignoring the digital version provided on a laptop computer), whereas others wanted to zoom in on each point in the image to obtain clues to the microhabitat features. Finally, we found that their attention and interest faltered sooner during the image-based interviews, such that the quality of the elicited information may have declined as time passed. As they tired, they began to classify sites more rapidly and verbally considered fewer potential variables than they had considered for sites early in the interview process.

5.2.5.4 Poor Relationship Between Expert Knowledge and Landscape-Scale Information

The value of expert information elicited for spatially explicit species-habitat modeling is debated, particularly in terms of the value of empirical data (Chap. 8; Chap. 11; Burgman et al. 2011). However, most discussions have focused on individual and group biases or the fallacies common in probabilistic thinking (e.g., Kahneman et al. 1982; Renooij 2001; Kynn 2008; Low-Choy et al. 2009). Fewer authors have explored the potential sources of expert uncertainty and error that may be unique to landscape ecological applications (Elith et al. 2002; Yamada et al. 2003; Murray et al. 2009).

We found no published guidance on best practices to match expert knowledge elicited in discussions with map or image data, yet we recognized that many choices made in the processing of such spatial data within a geographic information system (hereafter, "GIS data") would affect the outcome of applying elicited judgments to the landscape. Three particular problems stand out: (1) eliciting expert knowledge prior to gathering available GIS data, (2) processing GIS data without a formal review of the ecological implications of the methodological choices, and (3) failing to account for GIS data inaccuracies when assessing the value of expert knowledge.

If the characteristics of the GIS data (e.g., resolution and range of values) are unknown at the time of the elicitation, serious problems may be encountered later in the project. The most common of these would be discovering a scale mismatch between the elicited knowledge and the GIS data. In our case, experts indicated that the King Rail was strongly associated with marsh–open water edges. However, this perceived correlation follows from observations that link the King Rail to fine-scale features (1–10 m in scale) such as ditches, muskrat runs, and small ponds, which are typically not visible in remotely sensed or interpolated spatial data. Only by knowing the proposed habitat modeling scale (e.g., the 30-m resolution of land cover data) could we explore the relevance of this correlation to our project objectives. In fact, after collecting field data, we documented a significant negative relationship between King Rail occupancy of a site and the marsh–open water edges depicted in the GIS data. Thus, at some sites where the expert-based models performed poorly, we found that the errors may have reflected a mismatch in spatial scale between the experts' observations and the GIS data, rather than simply expert error.

Careful consideration of the available GIS data clarified the relationships between the experts' knowledge and the many closely related landscape metrics that might be used to represent that knowledge spatially. For example, our experts required direct assistance to communicate their personal definition of "lots of edge" in terms of the landscape metrics available to spatially quantify "edge" within a GIS. We used cartoon illustrations of common metrics (e.g., interspersion, edge density, and distance to edge) to clarify how different metrics represented a given landscape to ensure we selected the metric that best matched the expert's meaning.

Despite taking great care to elicit accurate expert knowledge, we found that this information was easily distorted through GIS processing decisions. For example, our experts indicated that marsh patch size was a critical variable. To identify patches and calculate their area, we had to select a suitable land cover classification system and then choose a neighborhood rule to aggregate raster cells (pixels) into distinct patches. Although our land cover data defined six marsh habitat classes within our study region, the experts argued that some distinctions were not ecologically meaningful because they were based on geographic rather than ecological boundaries (e.g., Mid-Atlantic salt marsh was equivalent to Southeast Coastal Plain salt marsh and these classes could be merged). Had we analyzed the landscape using the original data rather than grouping closely related types of marsh, very few marsh patches would have appeared large enough to support the King Rail. Similarly, given that movement of this bird is not restricted to the four cardinal compass directions, we elected to use an eight-neighbor rule (versus a four-neighbor rule, with one neighbor per cardinal direction) when calculating patch size. This meant that two raster cells would be aggregated into a patch if they were directly adjacent, including diagonal alignment, with the patches touching only at their corners. A four-neighbor rule would have counted two diagonal cells as distinct patches. Our decision further increased the amount of marsh identified as belonging to a large patch without any change in the number of cells mapped as marsh.

In some cases, experts are correct and their knowledge has been well modeled, but the GIS data are inaccurate. Inaccurate data are likely if the spatial datasets used to spatially project the expert-based model predictions are very old (e.g., land cover maps in rapidly developing regions) or if interpolated surfaces are based on few points (e.g., the case for many soil maps). When we reviewed the areas of our map where predictive performance was low, we often found that the GIS data were erroneous. A large region of our study area had been incorrectly mapped as salt marsh (with a very low probability of occupancy by the King Rail), when in fact it was found to be oligohaline marsh (with a very high probability of occupancy). Model prediction errors in this region at least partially reflected problems with the GIS data rather than expert error or model error. It is unclear whether past studies comparing expert-based model predictions to field data made an effort to first assess and exclude inaccuracies of the spatial data layers prior to attributing errors to the expert knowledge.

5.2.6 Expert-Based Versus Data-Based Model Performance

Predictions from the expert-only belief network model differed significantly from those of the detection-adjusted occupancy estimates primarily due to two factors (Fig. 5.2a; mean difference in a paired t-test = −0.11, $P < 0.001$). First, the experts hypothesized a strong positive association between occupancy and marsh–open water edges, when in fact the relationship between the mapped edge and the detection-adjusted occupancy was negative. Second, experts exhibited two biases in their elicited probability estimates. The first bias was one of conservatism; experts resisted assigning any variable a probability value of 0 or 1. That is, they were uncomfortable saying that a King Rail would never or would always be present under certain conditions. This reflected their uncertainty regarding the strength of their knowledge of this species. The second bias was one of detection; experts provided predictions of occupancy based on their recollected presence or absence observations. Yet naïve estimates of occupancy (unadjusted for detection) typically underestimate the true occupancy (MacKenzie et al. 2002). By failing to account for the probability of detection, experts systematically underestimated occupancy in all conditional probability tables.

Fortunately, the performance of the expert-based model markedly improved through the addition of even a small amount of data for Bayesian updating (Fig. 5.2b). Initially, we planned to update the model with the first-year data so we could then compare the expert-only versus expert + data models' predictions of the

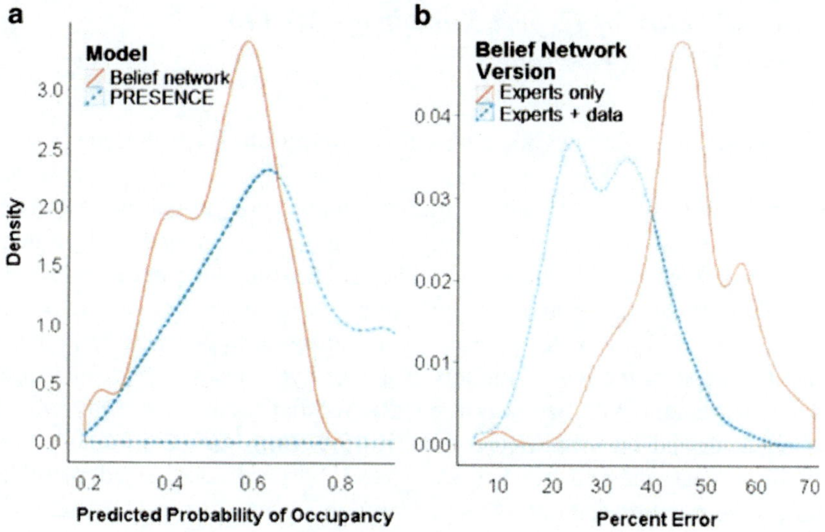

Fig. 5.2 (**a**) Comparison of the predictions made by the expert-based belief network model in Netica with predictions by the data-based model in PRESENCE. Predictions based on expert knowledge were lower than those based on the empirical data. Experts never assigned a probability of 0 or 1 to any level of any variable. In the two seasons of field data, however, the King Rail were consistently detected in certain kinds of habitat (and not detected in others), so the PRESENCE occupancy data included the full range of predicted values from 0 to 1. (**b**) Although the original expert-based belief network model initially showed high error rates (a mean value near 50% error in the contingency table), the addition of even a small amount of field data greatly reduced the error (to a mean of around 25%). The original, expert-only model is the Bayesian prior probability of occupancy; the updated expert+data model is the posterior probability of occupancy

second-year observations. However, we encountered a serious, yet possibly common, problem with this plan: our first field season was one of the driest years on record, whereas the second season was one of the wettest. Our expert elicitations had focused on average or best conditions, not extreme conditions. Updating an average-based model with data collected under one set of extreme conditions and then testing under the opposing set of extreme conditions seemed an unreasonable measure of success. Thus, we chose the randomized approach described in Sect. 5.2.4. However, this situation highlights concerns regarding how best to assess expert models. Even when validation data have been collected from the same study region as the region covered by the modeling exercise, it is not always clear whether they were collected under similar conditions (e.g., El Niño versus La Niña cycles; before or after the impacts of invasive exotic species). When developing and testing spatial models, it is equally important to reflect on the temporal dimensions of the expert knowledge and the associated validation data. Differences between the two might point to valuable information rather than simple inaccuracies or biases in the expert knowledge.

5.2.7 *Application of Expert Knowledge Within Belief Network Models*

5.2.7.1 Expert Knowledge Can Be Used to Formulate Hypotheses

Our experts were hesitant to identify variables and provide probability estimates in the absence of a structured, empirical study of King Rail breeding ecology and habitat associations. We found it helpful to emphasize the role of the belief network as a means to formulate, structure, and visually communicate assumptions and hypotheses. In our case, we could motivate experts by reminding them that USFWS conservation practices support the use of models to depict hypotheses to be tested through adaptive management. We also reassured experts that their uncertainty would be directly incorporated into the model and reported along with the model's predictions. In this manner, their elicited knowledge and "best professional judgment" were less likely to be misrepresented as "empirically observed and experimentally tested fact." Furthermore, their knowledge would serve as the foundation of an adaptive management and monitoring program, rather than functioning as a final product that would constrain all future decisions.

As a second aid to eliciting the expert knowledge, we structured the belief network model around two distinct ecological processes: habitat access and habitat selection. After first eliciting variables that influence the probability that a King Rail will encounter a given marsh patch, we asked what factors would influence its decision to remain in the patch and establish a breeding territory. We elicited their responses regarding site selection in three categories: the probability of finding a suitable nesting habitat and of finding a suitable forage habitat, and the probability of avoiding anthropogenic disturbance. Interestingly, although the experts initially struggled to name the landscape attributes that would influence the King Rail's habitat occupancy patterns, they quickly suggested patch characteristics that might inhibit access, most of which were landscape attributes (e.g., too distant from water and a too small or too isolated marsh).

5.2.7.2 Belief Networks Can Guide Adaptive Monitoring and Management

Belief network models can guide experimental design in support of adaptive monitoring. In adaptive monitoring programs, the sampling design evolves iteratively over time as new information emerges and as the research questions change (Lindenmayer and Likens 2009). In an adaptive management setting, multiple variables can potentially influence population dynamics or species–habitat associations. Resources allocated to monitoring are often inadequate to permit sampling across all levels of all variables every year. Adaptive monitoring can allocate the available resources most efficiently to support learning by the model and subsequent adaptive management. Belief network software includes tools to analyze model sensitivity to initial assumptions and uncertainty (Marcot et al. 2006), and this information facilitates designing a monitoring strategy that will maximize opportunities to refine the model and the hypotheses upon which it is based.

For a belief network model to be most useful in an adaptive management *and monitoring* setting, the empirical research methods and measurement units should be clearly defined throughout the elicitation process and should directly match those that will be used in the monitoring program. Different methods of empirical data collection have very different sampling efficiencies and detection probabilities. If experts offer information based on different empirical methods, then differences in opinion could be due to methodological differences rather than due to true disagreement about habitat value. When we asked experts about the probability of detecting the King Rail, we clarified that their response should assume that the detection method would be the National Marsh Bird Survey Protocol. By also using this empirical method in our validation surveys, we obtained a more direct understanding of errors within the model.

5.3 Recommendations for Good Practice and Further Research

Although several past projects required us to use expert knowledge to supplement or interpret available empirical data, this study was our first to elicit quantitative estimates of probability values from experts as the sole source of information to construct a belief network. The USFWS envisioned and requested a modeling approach that, although initiated with expert knowledge, could guide management decisions and monitoring efforts and gradually move them toward data-based decisions. We therefore spent significant time evaluating potential sources of uncertainty and error, some of which have been highlighted in this chapter. A fuller discussion of these sources will appear in the final project report to USFWS, which is expected to be produced by the end of 2011. Based on our elicitation experiences, we offer some recommendations to landscape ecologists who will develop or critique expert-based models.

5.3.1 Consider the Source of the Experts' Knowledge of Landscape Ecology

Though all of our experts were local biologists, their experience was gained in different refuges, each with unique landscape and microhabitat features. Not surprisingly, individual experience defined the temporal and spatial domain of their expertise and bounded their responses. We observed that if, for example, a refuge offered no interior marsh habitat, the local expert discounted the importance of edge versus interior habitat. Prior consideration of the landscapes where the experts gained their expertise will improve the elicitation process (e.g., identify regions of overconfidence) and assist with weighting of the experts' responses (if applicable). By reviewing the domains of expertise, it would also be possible to evaluate whether the experts' combined knowledge provides a full representation of the variability present in the landscape. Knowing the extent of their knowledge might also provide a means to determine the optimal number of experts. Although sociology provides insights into optimal sample sizes based on group dynamics (e.g., 3–8 experts; Clemen and Winkler 1985), landscape ecological applications must also ensure that the experts fully represent the degree of variability present within the study area.

Recommendations:

- Review the landscape characteristics of each expert's domain before and after elicitation.
- Define both the spatial and the temporal characteristics of an expert's knowledge and use this to distinguish responses based on direct experience from those based on plausible extrapolation.
- Do not combine expert estimates or resolve differences without first verifying whether these differences reflect differences in their domain of expertise.

5.3.2 Consider an Expert's Knowledge of Landscape Ecology

Most field biologists, wildlife ecologists, and other potential experts are not landscape ecologists. The local-scale experience and perspectives of these experts more closely resembles the species view from the ground (e.g., fine-scale variations in local microhabitat) than the view provided by a satellite (e.g., for a broad-scale regional landscape). It is therefore important to consider how well land cover classifications and metrics reflect an expert's experience of their landscape. Although an expert has the opportunity to directly observe and then recall local habitat features, most landscape metrics require calculations using mapping software, a process that is removed both spatially and temporally from the field observations. Furthermore, the experts have been trained to be attentive to microhabitat differences because, typically, these are the habitat characteristics that they have traditionally manipulated to manage wildlife populations.

Recommendations:

- Prior to eliciting the key variables and their associated probabilities, assess the expert's knowledge of landscape ecological patterns and processes. Provide orientation to help them understand key concepts and terms, and supplement this discussion with visual illustrations or examples.
- If you are using GIS data as a tool for the experts to view sites, ensure that the experts are appropriately oriented to the software and offered spatial information at a standardized scale (i.e., grain size and spatial extent). Also determine the

reliability of the GIS data, as errors in model outputs could be errors in the GIS data rather than in expert knowledge.

- Elicit the field methods and measurement units that underlie an expert's knowledge (e.g., different research methods and the difference between occupancy and detection) to ascertain whether and how these factors inform models of landscape patterns and processes.
- Consider the time frame of an expert's knowledge acquisition and whether this period captures the full range of temporal variability (e.g., 50-year flood events, El Niño and La Niña cycles) relevant to the period over which the model will be applied or tested.
- Distinguish between quantitative uncertainty (e.g., how much) and qualitative uncertainty (e.g., more or less).
- GIS processing decisions should be informed by the expert's knowledge of ecological processes and species behavior, otherwise the spatial data layers will not accurately represent the elicited information.
- Elicitation is fatiguing, especially when it involves repetitive tasks, new terminology, or unfamiliar technology. Allow ample time for mental and physical breaks to ensure consistent levels of expert attention throughout the elicitation process.

5.3.3 Develop Better Elicitation Tools

A primary lesson from this modeling experience is that extra effort applied to design and test a rigorous elicitation is time and money well spent. Obtaining a precise and accurate summary of an expert's knowledge requires the same thoughtful attention to detail as the design of any other empirical study. Indeed, for sociologists and others who work with human subjects, eliciting knowledge is empirical research. Formal elicitation methods (e.g., the Elicitator software; James et al. 2009) and associated statistical analysis can help researchers manage potential bias, quantify and track uncertainty, explore sources of conflicting knowledge, and support complex decisions in the face of competing interests and risks (O'Hagan et al. 2006). However, the application of these elicitation methods within a spatially explicit context remains an area of ongoing development (Chap. 3; DBL Interactive, http://www.decisionbasedlearning.org/). Currently, however, the elicitation of landscape ecological knowledge remains hampered by the inability to interactively display an expert's elicited knowledge within a GIS. Such visualization support would facilitate the identification of potential scale mismatches, characterization of an expert's domains of experience, and assessment of GIS data accuracy, among other benefits. These improvements would greatly facilitate project planning and implementation.

5.4 Conclusions

Controversy still surrounds the use of expert-based models to support natural resource management and conservation decisions because such models often lack rigor, and poor methods can easily lead to biased results (e.g., Chap. 8 and Chap. 11). In our models, expert knowledge did exhibit some biases and errors (e.g., underestimating true occupancy), and it is critical to point out that the original expert-only (prior probability) belief network model offered little predictive value. However, the expert-only belief network model guided the development of the field sampling design by targeting limited monitoring resources toward the greatest sources of model uncertainty. The predictive performance (percent accuracy) of the resulting (posterior) belief network improved greatly through the process of Bayesian updating through the addition of even a relatively small sample of monitoring data. However, we first had to address the effects of GIS data errors and differences in climate or other monitoring conditions prior to validating or updating the models. We concluded that expert-based models can support the USFWS vision for adaptive monitoring and management that is promoted in their Strategic Habitat Conservation Plan, so long as the models are implemented with an effective monitoring and data management program.

References

Aspinall W (2010) A route to more tractable expert advice. Nature 463:294–295

Bashari H, Smith C, Bosch OJH (2009) Developing decision support tools for rangeland management by combining state and transition models and Bayesian belief networks. Agric Syst 99:23–34

Bunnell F (1989) Alchemy and uncertainty: what good are models? USDA Forest Service, Northwest Research Station, Portland, General Technical Report PNW-GTR-232

Burgman M, Carr A, Godden L, et al (2011) Redefining expertise and improving ecological judgement. Conserv Lett 4:81–87

Clemen RT, Winkler RL (1985) Limits for the precision and value of information from independent sources. Operat Res 33:427–442

Conway CJ (2009) Standardized North American marsh bird monitoring protocols, version 2009–2. U.S. Geological Survey, Arizona Cooperative Fish and Wildlife Research Unit, Tucson, Wildlife Research Report #2009–02

Cooper TR (2008) King Rail conservation plan, version 1. U.S. Fish and Wildlife Service, Fort Snelling

Drew CA, McKerrow A, Earsome S (2006) Stepping-down regional habitat and population objectives to individual National Wildlife Refuges: A pilot project in the Roanoke-Tar-Neuse-Cape Fear Ecosystem. U.S. Geological Service, Washington, Gap Analysis Bulletin 14:53–61. Available from http://pubs.er.usgs.gov/djvu/GAP/Bulletin14.pdf (accessed May 2011)

Elith J, Burgman MA, Regan HM (2002) Mapping epistemic uncertainties and vague concepts in predictions of species distribution. Ecol Modell 157(2–3):313–329

Howes AL, Maron M, McAlpine CA (2010) Bayesian networks and adaptive management of wildlife habitat. Conserv Biol 24(4):974–983

James A, Low-Choy S, Mengersen K (2009) Elicitator: a software-based expert knowledge elicitation tool for regression in ecology. Environ Model Softw 25:129–145

Kahneman D, Slovic P, Tversky, A (eds.) (1982) Judgment under uncertainty: heuristics and biases. Cambridge University Press, Cambridge

Kuhnert PM, Hayes KR (2009) How believable is your BBN? In: Anderssen RS, Braddock RD, Newham LTH (eds) 18th World IMACS/MODSIM Congress, Cairns, Australia, 13–17 July 2009, Modelling and Simulation Society of Australia and New Zealand, Cairns, pp 4319–4325. Available from http://mssanz.org.au/modsim09 (accessed May 2011)

Kynn M (2008) The 'heuristics and biases' bias in expert elicitation. J Roy Stat Soc Series A 71(1):239–264

Lindenmayer DB, Likens GE (2009) Adaptive monitoring: a new paradigm for long-term research and monitoring. Trends Ecol Evol 24:482–486

Low-Choy S, O'Leary R, Mengersen K (2009) Elicitation by design in ecology: using expert opinion to inform priors for Bayesian statistical models. Ecology 90:265–277

MacKenzie DI, Nichols JD, Lachman GB, et al (2002) Estimating site occupancy when detection probabilities are less than one. Ecology 83:2248–2255

MacKenzie DI, Nichols JD, Royle JA, et al (2006) Occupancy estimation and modeling: inferring pattern and dynamics of species occurrence. Associated Press, Burlington

Marcot BG, Steventon JD, Sutherland GD, McCann RK (2006) Guidelines for developing and updating Bayesian belief networks applied to ecological modeling and conservation. Can J For Res 36:3063–3074

McCann RK, Marcot BG, Ellis R (2006) Bayesian belief networks: applications in natural resource management. Can J For Res 36:3053–3062

Murray JV, Goldizen AW, O'Leary RA, et al (2009) How useful is expert opinion for predicting the distribution of species within and beyond the region of expertise? A case study using brush-tailed rock-wallabies *Petrogale penicillata*. J Appl Ecol 46:842–851

Nyberg JB, Marcot BG, Sulyma R (2006) Using Bayesian belief networks in adaptive management. Can J For Res 36:3104–3116

O'Hagan A, Buck CE, Daneshkhan A, et al (2006) Uncertain judgements: eliciting experts' probabilities. Wiley, Chichester

Renooij S (2001) Probability elicitation for belief networks: issues to consider. Knowl Eng Rev 16(3):255–269

Royle JA, Nichols JD (2003). Estimating abundance from repeated presence–absence data or point counts. Ecology 84:777–790

Sullivan PJ, Acheson JM, Angermeier PL, et al (2006) Defining and implementing best available science for fisheries and environmental science, policy, and management. Fisheries 31(9):460–465

Uusitalo L (2007) Advantages and challenges of Bayesian networks in environmental modeling. Ecol Modell 203(3–4):312–318

Wilson EO (2000) On the future of conservation biology. Conserv Biol 14(1):1–3

Yamada K, Elith J, McCarthy M, Zerger A (2003) Eliciting and integrating expert knowledge for wildlife habitat modeling. Ecol Modell 165(2–3):251–264

Chapter 6
Incorporating Expert Knowledge in Decision-Support Models for Avian Conservation

Allison T. Moody and James B. Grand

Contents

A.T. Moody (✉)
School of Forestry and Wildlife Sciences, Auburn University, 3301 Forestry and Wildlife Science Building, 602 Duncan Drive, Auburn, AL 36849, USA
e-mail: atm0005@auburn.edu

J.B. Grand
USGS, Alabama Cooperative Fish and Wildlife Research Unit, 3236 School of Forestry Wildlife Sciences, Auburn University, 602 Duncan Drive, Auburn, AL 36849, USA

A.H. Perera et al. (eds.), *Expert Knowledge and Its Application in Landscape Ecology*, DOI 10.1007/978-1-4614-1034-8_6,

6.1 Introduction

Bird abundance in the United States has been declining for more than half a century, likely as a result of habitat changes (Valiela and Martinetto 2007; North American Bird Conservation, U.S. Committee 2009). In the Southeastern United States, habitat and management changes including deforestation, reforestation, urban growth, and fire suppression have reduced the availability of high-quality habitats and have increased habitat fragmentation (Wear 2002; Griffith et al. 2003; Van Lear et al. 2005). Short-term projections suggest that urbanization will continue to reduce forest areas and increase their fragmentation (Wear et al. 2004). In the long term, climate change will alter precipitation and temperature patterns, and rising sea levels will reduce coastal habitat (IPCC 2007). It is therefore important to conserve what is currently available (species, habitats, and ecosystems) and plan for future conservation. To effectively protect or increase bird populations in this context, conservation must maintain or increase habitat quality and quantity. However, given limited resources, it is important to focus efforts where they have the greatest benefit rather than where land is economically unimportant (Pressey et al. 1996). Furthermore, complex systems with multiple species and habitats may require trade-offs among conflicting conservation objectives.

The Southeastern United States has high conservation importance because of the region's habitat and species diversity, ecological processes, and evolutionary potential; however, it also warrants strong concern because of historical habitat loss, future threats, and inadequate protection (Olson and Dinerstein 1998). Based on data from USGS (2010), we estimate that only 12% of Southern Coastal Plain ecoregion is under permanent protection. Conservation planning for this large and varied ecoregion is complicated by variable data availability across regions, habitats, and species. Therefore, to support conservation planning, experts may be the best source of information (Low-Choy et al. 2009).

In conservation planning, experts are often used to evaluate potential threats (Coppolillo et al. 2004; Teck et al. 2010; Chap. 12), select high-priority areas (Noss et al. 2002; Cipollini et al. 2005; MacNally et al. 2008), define initial values for Bayesian modeling (Low-Choy et al. 2009; Kuhnert et al. 2010; Chap. 5), and propose conservation targets (Hess and King 2002; Didier et al. 2009; Amici et al. 2010). Experts can provide critical insights when there are multiple conflicting objectives, when empirical data about species and habitats is lacking, threats are uncertain, and it is necessary to focus on a few key species. This is common when developing large-scale, long-term plans. However, when expert knowledge supports conservation decisions, little information may be provided or recorded about the experts or how their knowledge was collected and used (Kuhnert et al. 2010; Chap. 8).

To support avian conservation in the Southeastern United States by ensuring adequate current and future habitat, we elicited expert knowledge of species–habitat associations, habitat management needs, and threats. This chapter highlights one aspect of our project: the use of expert knowledge to define a suite of focal species for species–habitat modeling that would subsequently support the development of a decision-support tool. These species represent the present and future habitat needs of other species that cannot be modeled given time and resource constraints. The final tool will be a series of spatially explicit landscape models based on the habitat needs of focal species that indicate where to focus conservation efforts.

6.2 Case Study Context

The elicitation exercises reported here formed the foundation for a much broader study that had three major objectives: to assess the current ability of habitats to sustain avian populations; to model future conditions based on projected urban growth, conservation programs, and climate change and predict the response of avian populations; and to enhance coordination among stakeholders during all planning stages. Stakeholders provide access to information that may be unavailable elsewhere, and help us to address the concerns of those who will enact conservation actions, thereby leading to better outcomes (Pressey and Bottrill 2009).

The project covered the South Atlantic Migratory Bird Initiative (SAMBI) area (Fig. 6.1; Watson and Malloy 2006). The area extends from the Atlantic Coast in the east to the boundary between the Coastal Plain and the western Piedmont. Historically, this area was dominated by fire-maintained longleaf pine (*Pinus palustris*) savanna (Outcalt and Sheffield 1996), but only 2% of this habitat remains after conversion to agriculture, pine plantations, and urban areas (Van Lear et al. 2005). Frequent fires created high biodiversity (Van Lear et al. 2005), including a high proportion (40%) of endemic plant species (Walker 1998) and 30 threatened or endangered vertebrates (Van Lear et al. 2005). Other important habitats include bottomland hardwood forest dominated by flood-tolerant species such as cypress (*Taxodium distichum*) and tupelo (*Nyssa aquatica*; Hodges 1997). Unique non-alluvial forested wetlands include rainfall-driven pocosins, Carolina bays, and pitcher plant (*Sarracenia* spp.) bogs (Richardson 2003). SAMBI's coastal area has extensive barrier islands and highly productive estuarine wetlands (Dame et al. 2000).

6.3 Focal Species Approach

When ecosystem management targets focal species, the goal is to protect many other species (Margules and Pressey 2000). In contrast to conservation based on ecosystems or ecosystem functions, focal species indicate the quantity and arrangement of conservation areas and allow planning at a finer scale (Roberge and Angelstam 2004). Among focal species, sub-categories include indicator, keystone, flagship, umbrella, and landscape species (Caro and O'Doherty 1999). Indicator species reflect ecosystem health or biodiversity (Landres et al. 1988). We did not explicitly select biodiversity indicators because the low spatial resolution of

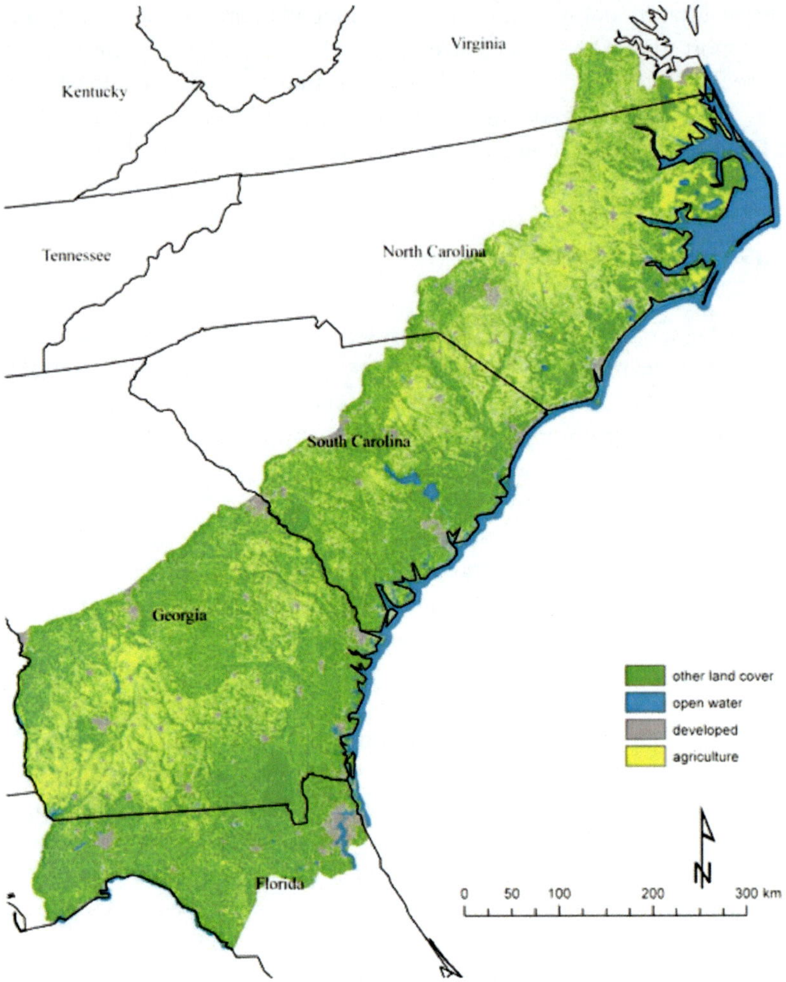

Fig. 6.1 The study region in the Southeastern United States included coastal plain regions of Virginia, North Carolina, South Carolina, Georgia, and Florida

remote-sensing data leads to the apparent co-occurrence of many species (Favreau et al. 2006). In comparison, keystone species are more influential than their abundance suggests (Power et al. 1996); in the SAMBI area, they include gopher tortoises (*Gopherus polyphemus*), which excavate burrows used by many other species (Guyer and Bailey 1993). Flagship species are species that attract public support and may promote conservation of associated species, even though this may not be an explicit conservation goal in flagship species management (Simberloff 1998). Umbrella species require large habitat patches, so their conservation explicitly protects many other species in those large areas (Noss 1990). Landscape species resemble umbrella species in requiring large areas, but also require a specific habitat composition (Sanderson et al. 2002). Because the SAMBI area has so many different conservation goals (e.g., restoring rare species, increasing populations of hunted species, preserving common species), we did not want to restrict experts to any one type of focal species. Using multiple focal species reduces the risk of missing endemic or range-restricted species when planning reserves (Lambeck 1997; Hess and King 2002) and explicitly includes species with substantial public interest or conservation resources.

Using focal species to guide conservation efforts has been criticized. Andelman and Fagan (2000) showed that selecting focal species using a range of criteria did not improve protection of the greatest number of species at a minimum number of sites than randomly chosen species. The effectiveness of focal species also varies with the taxa that are selected (Roberge and Angelstam 2004). For instance, basing conservation areas on birds did not protect butterflies (Fleishman et al. 2001), nor did protecting large mammals protect smaller mammals (Caro 2001). However, focal species can be effective in more limited situations; for example, protecting focal butterflies protected other butterflies and protecting focal birds protected other birds (Fleishman et al. 2001). Since our objective was to use avian focal species to represent other birds, rather than overall biodiversity, the focal species approach was appropriate for our purposes.

Although focal species are commonly used for conservation planning, selecting them based on expert knowledge is less common. For the Bolivian Andes and the Republic of the Congo, Coppolillo et al. (2004) selected landscape species for conservation planning. They selected 4–6 large vertebrate species at each site to represent the habitat requirements, threat sensitivity, and ecological function of other species, and their importance to humans. At both locations, experts identified potential focal species, scored each species using the above-mentioned criteria, and selected the final suite of focal species. These experts were field biologists, managers, and people who knew the species or area; they scored species using (in order) published and unpublished literature and their own knowledge. Although Coppolillo et al. (2004) used experts to select these species, they reported insufficient detail to guide other researchers interested in using experts to support conservation efforts. For example, they did not discuss the extent to which the experts resorted to non-literature information sources nor did they detail how they elicited information from the experts.

6.4 Elicitation of Focal Species

For our purposes, we wanted a species suite that would represent all habitat types defined by the SAMBI Plan (Watson and Malloy 2006), including species with large area requirements and species requiring management. Our initial list of potential focal species comprised 65 key species identified in the SAMBI Plan. We subsequently used the two processes described in Sect. 6.4.1 to develop lists based on expert knowledge using two selection methods. Finally, we validated the two subsets of the overall list against the original list of 65 species.

6.4.1 Two Approaches

To select focal species, we used Lambeck's (1997) selection process and a method rooted in structured decision making (SDM). The former method has been used to select focal species (e.g., Roberge and Angelstam 2004); the latter was a modification of Gregory and Keeney's (2002) decision-making process. We designed our elicitation process to work with the SDM method but added the Lambeck method because it refines the species selection by focusing on landscape design and management rather than expert elicitation.

Lambeck (1997) modified the umbrella species concept by systematically selecting species based on their threat category, with an emphasis on protecting the most sensitive species (Roberge and Angelstam 2004). For example, connectivity should support species with restricted dispersal ability, and patch size should sustain species with large area requirements. This method used empirical data from published literature and field research rather than expert opinion. Rather than eliciting quantitative data from experts during the Lambeck analysis, we modified the method to accept qualitative expert data. We believed this would help experts reach a consensus more quickly and maximize participation by experts who lacked confidence in their ability to provide precise data.

Gregory and Keeney (2002) have broad experience in decision analysis and have used their SDM methods to define and solve resource management issues. SDM, unlike the Lambeck method, helps stakeholders to make decisions, and we modified the process to use expert opinion. SDM comprises a five-step procedure for solving problems: state the problem, establish objectives that can be evaluated, design alternative solutions, evaluate each alternative's consequences, and assess trade-offs before reaching a decision. Although SDM can help individuals to reach a decision, it is especially useful for groups.

In our modified SDM process, we used habitat characteristics as objectives and potential focal species as alternatives based on their association with each habitat characteristic. Experts prepared an alternatives table that rated each species according to the strength of its linkage with each characteristic. As experts characterized habitat needs, similarities emerged among species. High similarity between the habitat requirements of two species justified removal of the species of lower conservation or management concern from the species list. The species list was reduced using criteria that will be used to manage the conservation system (Wiens et al. 2008); in the SAMBI project, this will be through habitat acquisition (coarse scale) and enhancement (medium scale), so we emphasized similarities among coarse- and medium-scale habitat characteristics. The level of spatial detail is an important aspect of the present exercise, since habitat planning and management will be based on remote-sensing data (satellite photos used to provide land-use and vegetation type data) stored in the geographical information system software that will be used in a subsequent stage of this project to develop landscape models.

6.4.2 Expert Selection

The Atlantic Coast Joint Venture, a partnership of governmental and non-governmental organizations that strives to provide healthy ecosystems to support healthy avian populations across jurisdictions, organized the SAMBI project. Working through the Joint Venture gave us access to many experts. We limited participation to experts associated with SAMBI but did not limit their number. We wanted the largest group possible because no individual understands all potential focal species (Teck et al. 2010), and broad participation reduces the bias caused by extreme views (Low-Choy et al. 2009; Chap. 2). We invited all SAMBI members, including biologists and managers, from the Joint Venture team. Of 278 invitees, 53 attended elicitation meetings. During follow-up surveys, we again invited all SAMBI members; of those who attended the elicitation meetings, 16 participated in a conference call and 15 completed at least part of the survey.

Experts included representatives from state and federal government agencies in Virginia, North Carolina, South Carolina, Georgia, and Florida; non-governmental organizations included The Nature Conservancy, Ducks Unlimited, Audubon Society state chapters, the Tall Timbers Research Station, the North Carolina Museum of Natural Sciences, the University of Florida, and the University of Georgia.

Our initial list included 65 potential focal species identified in the SAMBI Conservation Plan (Watson and Malloy 2006). We wanted experts to consider the species associated with particular habitat characteristics (Table 6.1). Large scale species–habitat associations were found in the literature (Hamel 1992), but medium-scale details of habitat preferences were difficult to determine. We felt that experts who study or work with a species would know this information, even if they did not publish it. Our preliminary work with the experts suggested that certain habitat characteristics extended across habitats and could be considered apart from the larger habitat types. For instance, bare ground is found in both grasslands and wetlands, and closed canopies are found in both deciduous and mixed forests. We presented the species list alphabetically to avoid biasing expert responses.

6.4.3 Focal Species Identification Meetings

From August to November 2008 we held 2-day meetings in each state. The first day introduced the SDM process and summarized the project; during the afternoon, we began species selection. During the selection process, we divided experts into four groups based on their stated area of knowledge or comfort: waterfowl (e.g., ducks, geese, and swans), land birds, waterbirds (e.g., herons, rails, gulls, and terns), and shorebirds (e.g., sandpipers and plovers). We generally had more waterfowl experts than other types, but we also had several land bird experts. We usually combined

Table 6.1 Characteristics of the habitats used to inform the selection of focal species and to define key functional characteristics

Habitat class	Characteristics	Comments
Hydrological	coastal	Use areas adjacent to coast, not necessarily marine habitat
	Water type	
	Water depth	
	Salinity	
	Presence of submerged aquatic vegetation	
	Aquatic macroinvertebrates	
	Turbidity	
	Flooding	Includes both seasonal and tidal flooding
	High-energy shore	
	Low-energy shore	
Disturbance	Any	
	High fire frequency	Every 3–5 years
	Growing season fires	
Vegetation	Canopy cover	
	Mid-story	
	Understory	
	Low basal area	Basal area <50 ft^2 $acre^{-1}$
	Old or mature trees	Individual trees, required for nesting or foraging
	Mature forest	
	Bare ground	
Other	Patch size	
	Social aggregation	Individuals or pairs associate with others with or without overlap
	Large forest patch	Requires a large patch of contiguous forest
	Elevation	
	Urban avoidance	
	Edges	Between habitat types or between land and water
	Large home range	
	Invasive species	

shorebird and waterbird experts because so few were present. At the Georgia meeting, only one individual had shorebird and waterbird expertise, but North Carolina and Florida had numerous experts in this category.

On the second day, we reviewed the previous day's work and discussed landscape design issues. To encourage discussion, we started with simple examples. For example, we picked a bird with well-known, well-defined habitat preferences and asked experts to review that example. We knew some experts personally and could direct questions to an appropriate expert. Facilitators answered questions and clarified characteristics during meetings to reduce bias due to imprecise language (Kuhnert et al. 2010). To elicit information, we asked experts to identify important habitat characteristics for the SAMBI priority species (Watson and Malloy 2006). At the first meeting, we did not initially present the species–habitat association tables because creating an alternatives table without preconceptions is a key step in the SDM process (Gregory and Keeney 2002). However, this made the process unworkably slow because experts wanted to assign the species to habitats rather than to habitat characteristics, so we subsequently presented our prepared tables and were much more successful at focusing experts on the process. For subsequent meetings, we started with matrices of potential focal species and habitat characteristics. We asked each expert to identify and score the habitat characteristics required for each potential focal species in their group (Table 6.2). Their scoring choices ranged from 1 (beneficial or preferred) to 5 (detrimental or avoided). Experts could also report insufficient information or that the relationship was neutral by not scoring the species. We also let experts answer in more detail, for example, to describe a relationship where the species preferred a moderate level of a habitat characteristic but avoided either extreme. For each species, we also asked the experts to note whether the species were umbrella, flagship, biodiversity indicator, keystone, and habitat or dietary specialist or generalist. We did not provide access to published data (e.g., field guides, species accounts and Internet searches), so they answered based on their own knowledge or experience.

Experts were comfortable with the scoring system except when we did not clarify the direction of the scoring. For example, the "depth of water" characteristic was confusing because we did not specify whether this meant shallow or deep water. When opinions differed about species preferences for deep versus shallow water, the results were ambiguous. When experts revealed this problem, we asked them to add a brief description after their score to indicate how they interpreted the scale so we understood their intent when we compiled our data. Subsequently, we provided definitions so that all experts used the same scoring criteria.

During our elicitation meetings, experts were given equal weight and group members worked to achieve consensus. Because we held meetings in each state, experts tended to know each other. This made it possible that professional relationships influenced their answers (Gregory and Keeney 2002), such as when someone deferred to a superior in their organization, so the answers may have been biased towards the opinions of the most senior experts. We did not address this source of bias because we assumed that the most senior experts had the most experience and knowledge and that this therefore provided an acceptable, if unmeasured, weighting.

After experts completed the exercise, we compiled their answers and presented them to the whole group the next day. We did not prevent them from commenting on the results from other groups. During this stage of the process, we did not need a high level of individual participation because the smaller groups had already

Table 6.2 Example of a table used to select the focal species using SDM in which experts ranked species preference for or avoidance of specific habitat characteristics. Thirteen additional habitat types and 42 associated species were similarly assessed. Scores ranged from 1 (prefer) to 5 (avoid). If experts disagreed, the range of answers was reported

			Hydrology			Structure	
Habitat	Species	Selected as focal species	Coastal	Water type	Depth	Forest type	Canopy
Beach[a]	**American Oystercatcher**	**Both**	**1**	**1 – Ocean**	**5**		
	Piping Plover	**L**	**1**	**1 – Ocean**	**5**		
	Wilson's Plover	**L**	**1**	**1 – Ocean**	**5**		
	Whimbrel	**L**	**1**	**1 – Ocean**	**5**		
	Least Tern	NS	2	1 – Ocean	4		
	Black Skimmer	NS	2	1 – Ocean	4		
	Royal Tern	NS	1	1 – Ocean			
Beach[b]	**Red Knot**	**Both**					
Deep-water coastal[c]	**Redhead**	**Both**					
	Black Scoter	**L**	**1**	**1 – Ocean**	**1**		
	Canvasback	**L**	**1–4**	**1–2 Open**			
	Lesser Scaup	**L**	**1**	**1 – Open**	**1 – Medium**		
Inland[d]	**Sandhill Crane**	**SDM**	**4**				
	American Woodcock	NS	4				
Riparian[d]	**Louisiana Waterthrush**	**SDM**		**1 – Riparian**	**5**	**Deciduous**	**2**
	Acadian Flycatcher	**Online**		**1 – Riparian**		**Deciduous**	**1**
	Cerulean Warbler	**L**		**1 – Riparian**		**Deciduous**	**1**
Non-riparian[d]	**Black-throated Green Warbler**	**SDM**		**2 – Not rivers**		**Mixed**	**1**
	Prothonotary warbler	**Online**		**2**		**Deciduous**	

Species in bold are the focal species selected based on Lambeck's (1997) method (*L*), using SDM (*SDM*), using both methods (*both*), or using an online survey (*online*). Other species were not selected as focal species (NS). Common names conform to those in the American Ornithologists' Union (1998) standard

[a]Breeding and non-breeding habitat

[b]Migration habitat

[c]Non-breeding habitat

[d]Breeding habitat

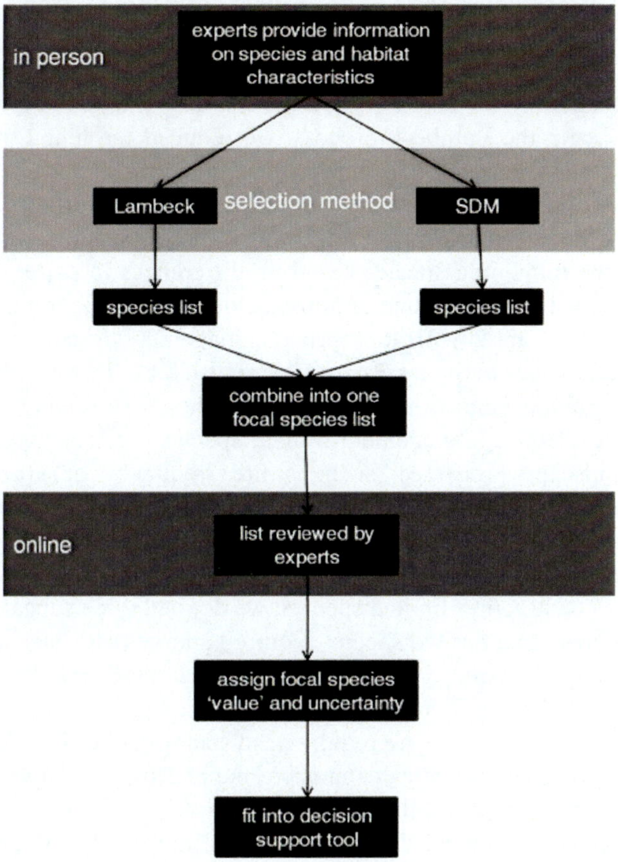

Fig. 6.2 The process we used to create the list of focal species based on an elicitation of expert knowledge using two selection methods

reached a consensus before sharing their results with the full group. A few experts sometimes monopolized the discussion, which may have introduced bias (Kuhnert et al. 2010). When this happened, we asked the original group to confirm whether their results should be modified. When group members had different opinions, we recorded all answers rather than forcing an artificial consensus.

The initial set of meetings provided a framework for selecting potential focal species and modeling the habitat configuration. For both selection methods (Lambeck and SDM), we used the same data tables, but we used different processes to create the focal species lists (Fig. 6.2). For each species and each corresponding habitat characteristic, we created an overall score by combining the scores from all states. We used the majority score unless there was a disagreement (a characteristic was said to be both avoided and preferred by a species), in which case we kept the range of scores.

6.4.4 Analysis of the Elicited Data Using the Lambeck Method

We used Lambeck's (1997) process to create a list of the potential focal species (Fig. 6.3a). To begin the Lambeck process, we reduced the length of the expert-elicited list by excluding species that had secure populations, abundant game species, and species identified as being of moderate concern (thus, low priority) in the SAMBI Plan. If there was any uncertainty, we retained a species. In the next step, we subdivided the remaining species based on differences in pattern and process; that is, we distinguished species that required habitat reconstruction from those that could live in existing habitat with appropriate management. Reconstruction-limited species required changes to the landscape pattern, such as creating additional habitat patches, improving connectivity between patches, or creating larger habitat patches (Lambeck 1997). Management-limited species are sensitive to the rate or intensity of landscape processes, such as fire frequency or grazing intensity (Lambeck 1997). We did not make these categories mutually exclusive because some species may currently lack adequate habitat and their habitat may require management once it has been established.

Among the reconstruction-limited species, we defined three subcategories based on the expert scores: area-limited species required a large patch size, and resource-limited species had preferred or required habitat characteristics that could not be detected by remote sensing. For example, several duck species rely on submerged aquatic vegetation and other species require dead standing trees for nesting, but currently these habitat characteristics cannot be mapped using remote sensing. The third subdivision was dispersal-limited species. However, experts concurred that this was not a limiting factor for the priority species in the SAMBI area.

We defined management-limited species as any species that scored "beneficial" or "preferred" for any of the disturbance categories, as well as any species that required human-based management, including several duck species that relied on managed wetlands for their winter habitat.

In the final step, Lambeck (1997) suggests that for each habitat, one should select the *most* limited species for each pattern and process and design the landscape based on the needs of those species. For example, the species with the largest area requirement would define the minimum patch size, the species with the shortest dispersal distance would define the maximum distance between patches, and the species most sensitive to disturbance would define the management protocol. We chose not to take this step because we wanted to retain the largest possible list for the experts to evaluate.

6.4.5 Analysis of the Elicited Data Using the SDM Method

To select the focal species using SDM, we used five habitat characteristics scored by the experts that could be estimated from landscape-level data: proximity to coast, water type, water depth, forest type, and canopy (Table 6.2; Fig. 6.3b). For example,

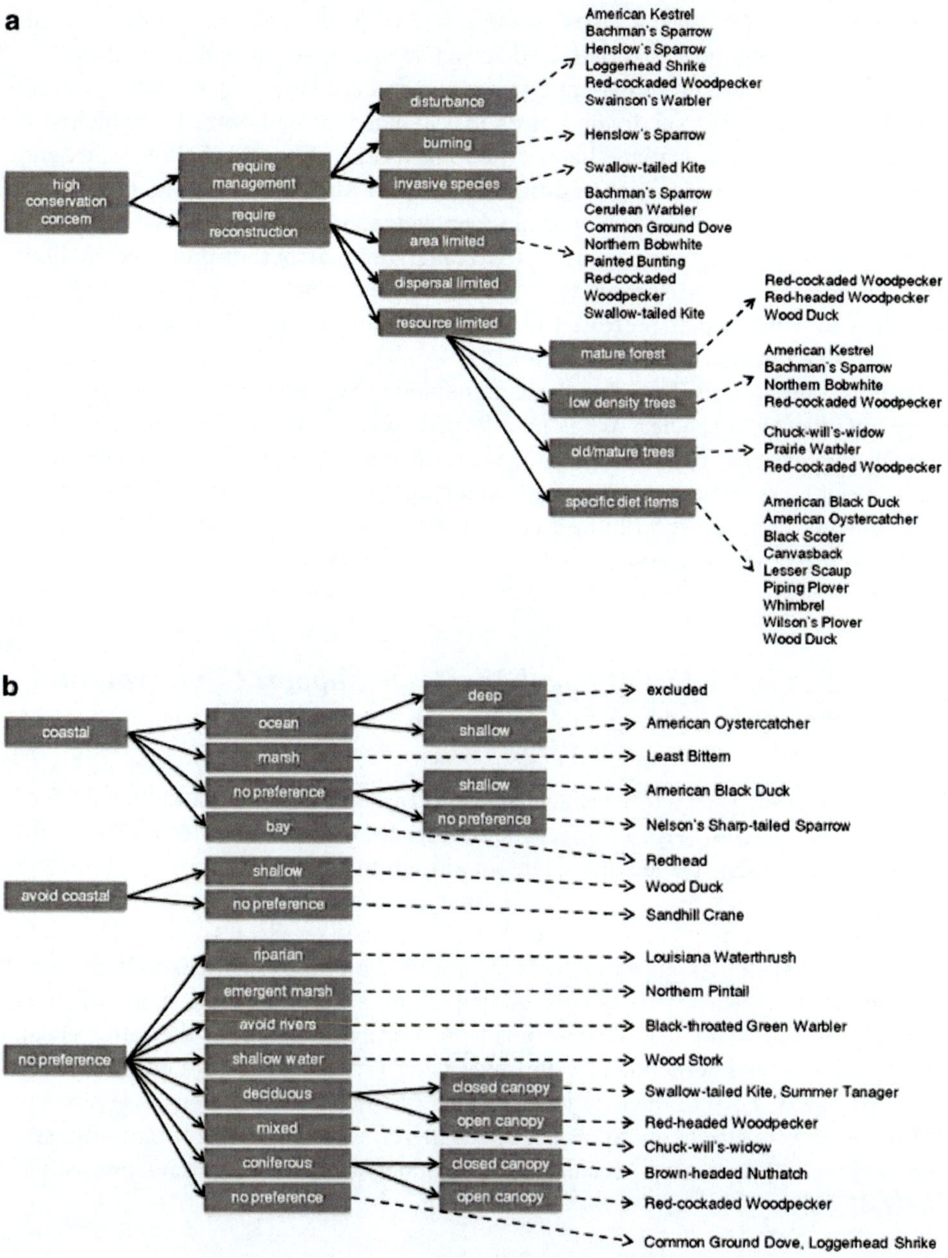

Fig. 6.3 We used two different processes to select focal species using the results of our elicitation of expert opinion. (**a**) A process based on that of Lambeck (1997). (**b**) A process based on SDM. Species common names are those designated in American Ornithologists' Union (1998)

some species prefer coastal areas, some avoid coastal areas, and some have no preference. Among those that preferred coastal areas, we further subdivided species according to their preferred habitat types such as intertidal beach, coastal marshes, and shallow areas. We subdivided birds that avoided coastal areas but which still required open water or wetland habitats, into birds that preferred shallow water and those that had no preference regarding water depth. Birds exhibiting no preference in their use of coastal and non-coastal areas were subdivided into those that used riparian areas, avoided riparian areas, used emergent marshes, or preferred shallow water. Forest-associated species were divided by the type of forest they preferred and then into those that preferred closed and open canopy. Finally, some birds preferred open habitats.

After grouping birds based on these associations, we selected one species as the representative focal species. We generally picked species with the most complete habitat associations. For example, we selected the American Oystercatcher as a focal species associated with shallow water along beaches. Species with similar requirements included the Piping Plover, the Red Knot, the Whimbrel, the Least Tern, and the Black Skimmer.

6.4.6 Results: A List of Focal Species to Support Conservation Planning in the SAMBI Region

The focal species selected using the SDM and Lambeck methods included 35 of the initial 65 species, with 11 species common to both lists (Fig. 6.3). The SDM method selected ten species that were not chosen using the Lambeck method, and the Lambeck method selected 14 species that were not chosen using SDM.

To create a list of species that would be validated (see Sect. 6.4.7) and used to develop the decision-support tool, we used species common to both lists, all selected land birds that appeared in both lists, and waterbirds, waterfowl, and shorebirds that appeared in the SDM list. We excluded the Lambeck list from the latter group because the experts agreed that waterfowl, waterbird, and shorebird habitat tended to overlap at the level of the data we used, and the SDM method let us assess where habitat overlaps were likely to occur; it was therefore a better list for our purposes. We retained all land birds because we had no reason to prefer either selection method.

6.4.7 Validation Through Online Surveys

To validate the list of focal species described in Sect. 6.4.6, we created a follow-up survey using online survey software. We began our online surveys by informing the experts that our list required revision, and engaging the experts in this way let them criticize more freely. The online survey included supporting documentation and was introduced to respondents during a conference call. We asked the experts to review and rank the selected focal species and to add or remove species as necessary. We provided criteria for evaluating whether a species was a suitable focal species. The focal species could meet more than one of the following criteria: representative of other species, well-known biology, easily sampled or observed, sensitive to disturbance, umbrella species, flagship species, habitat specialist, dietary specialist, or keystone species (Caro and O'Doherty 1999). Our questionnaire listed species associated with each habitat type in the SAMBI Plan (Watson and Malloy 2006), with focal species highlighted, although we did not state the selection method used to select them. The participants scored the suitability of each species as a focal species using ranks ranging from 1 (very poorly) to 5 (very well); they could also respond that they had insufficient personal knowledge to rank the species.

The scoring process let us create a "focal species value" and a measure of uncertainty that we used to assign species weights in the landscape model (Table 6.3). The mean score provided a measure of the relative value of each focal species and the variation in scores provided us with a measure of uncertainty. For example, if all participants assigned a score of 4 to a species, we were confident that the species was a good focal species. In contrast, we had less confidence if participants assigned an equal number of 3s, 4s, and 5s. For example, experts differed in their opinions of the Black-Throated Green Warbler as a focal species for alluvial forested wetlands: 4 of 12 experts thought it was a good or very good focal species. In contrast, 12 of 13 experts scored the Prothonotary Warbler, which was not included in our focal species list, as a good or very good focal species; the other expert declared insufficient personal knowledge. We did not remove any species from the focal species list based on the online validation, but we did add 11 species (Table 6.2) to our landscape model based on the expert scores.

6.5 Discussion

Neither selection method produced a list that we considered entirely suitable for conservation planning. Each method selected at least one species per habitat included in the SAMBI Plan, but the online validation survey included several species that were not included by either method and several that were not suitable focal species. For example, experts gave Bachman's Sparrow, the Cerulean Warbler, the Redhead, the Canvasback, and the Sandhill Crane an average focal species value less than 2 (poor). However, the Redhead and the Canvasback were added to the initial focal species list because they were resource-limited species according to Lambeck's (1997) definition. When re-evaluating the species, experts may have reduced the value of these species because we did not indicate that resource limitation was a criterion. Species values may also have decreased if they were uncommon in the study region, such as the Cerulean Warbler (Hamel 2000) and the Sandhill Crane (Tacha et al. 1992), or if they only overwintered in the region, such as the Redhead (Woodin and Michot 2002) and the Canvasback (Mowbray 2002).

Table 6.3 Example of the tabular summaries generated from our online survey to validate the selection of focal species. Species groups for 11 additional habitat types were assessed and similarly summarized

		How well did the species work as a focal species within the habitat?							
		Very poorly	Poorly	Neutral	Well	Very well	Insufficient		Weighted
Habitat	Species	1	2	3	4	5	knowledge	Total score	score
Alluvial forested wet-lands	Prothonotary Warbler	0	0	0	1	10	1	53.64	3.3
	Swainson's Warbler	**0**	**1**	**1**	**2**	**7**	**1**	**47.76**	**3.0**
	Yellow-throated Warbler	0	1	0	5	5	1	47.28	2.9
	Wood Duck	**0**	**1**	**2**	**1**	**7**	**1**	**46.68**	**2.9**
	Swallow-tailed Kite	**0**	**1**	**2**	**1**	**5**	**3**	**37.08**	**2.6**
	Cerulean Warbler	**2**	**1**	**3**	**4**	**1**	**1**	**33.60**	**2.1**
	Black-throated Green Warbler	**2**	**1**	**4**	**3**	**1**	**1**	**32.64**	**2.0**
	Mallard	2	1	3	3	0	3	24.96	1.8

Numbers represent the number of experts who chose a given score. Totals represent the sum of each score multiplied by the number of respondents who chose that score. The weighted score equals the total score divided by the number of experts who felt they had sufficient knowledge to assign a score. Bold species were focal species selected by either the Lambeck (1997) or the SDM method. Common names conform to those in the American Ornithologists' Union (1998) standard

Bachman's Sparrow had a low value in only one habitat (early successional and shrub-scrub) of the three in which it occurs; it had a high value in the other two habitats [longleaf pine–slash pine (*Pinus elliotti*) flatwoods and mature open pine].

In our list of potential focal species, we only used species in the SAMBI plan (Watson and Malloy 2006) that were associated with particular habitats, although experts could add species during the meetings. This gave us 65 species, out of a total of 172 species rated as being of highest, high, and moderate concern (see Table 1 in Watson and Malloy 2006). It was important that the habitats of our focal species represent the full suite of habitats used by all species identified in the SAMBI Plan, and we believe we accomplished this because the focal species we chose cover all habitats in the SAMBI Plan.

There may be concerns about the repeatability of our selection process because we asked experts to score bird–habitat associations without referring to published materials. We made this choice rather than using references to complete the tables ourselves because elicitation of knowledge not found in the published literature was a key goal of the process (Pierce et al. 2005; Pressey and Bottrill 2009). A different set of experts may provide different knowledge, thereby limiting the repeatability of the results. However, using a large group of experts and limiting answers to a discrete qualitative scale improved the reliability of the process. Using a simple scoring process likely also improved the ability to reach consensus. For example, asking experts to quantify canopy heterogeneity would produce a wide array of values, but similar focal species would be selected as long as there was general agreement on the direction and strength of the relationship between habitat quality and factors such as canopy heterogeneity. Insisting on consensus can eliminate potentially important differences of opinion among experts, but it was appropriate for our project. Grouping the experts (e.g., land versus water birds) probably increased the repeatability of our results by eliminating outlier answers that would arise when experts speculated about species–habitat combinations they were not truly familiar with.

Although neither the Lambeck method (Lambeck 1997) nor the SDM method was ideal for selecting a suite of focal species, combining expert opinion with these processes had benefits for selecting focal species. Both methods provided an initial list of species we could subsequently ask the experts to validate. Many expert knowledge studies have not included detailed information about their process (e.g., Coppolillo et al. 2004). We hope that our experience will help others who are considering a focal species approach based on expert elicitation. To improve such a process, we suggest the following.

6.5.1 Quantitative Versus Qualitative Data

Qualitative data is easier to explain to experts and does not require extensive analytical knowledge (Low-Choy et al. 2009; Chap. 2). Requesting qualitative rather than quantitative data probably increased our response rate because more experts would have felt sufficiently confident to participate, and this also decreased the time it took us to collect and review the data. Eliciting quantitative data would have provided more detailed data, but our project did not require such detailed information. However, care should be taken to ensure that questions are well defined. Prevalidation of the survey in a practice session with qualified people who will not be part of the final expert group is recommended.

6.5.2 Visualizing the Data

Flow diagrams (Lambeck 1997) and influence diagrams (McCann et al. 2006) are commonly used to visualize data. However, it would have been difficult and time-consuming to identify focal species by developing such tools during the meetings. Asking experts to complete tables of species–habitat associations provided information about a large number of species (65) and habitats in a short period of time (2 days). Without this approach, gathering the expert data would have taken much longer. We do not believe that influence diagrams would have been useful, since they are typically used to characterize beliefs based on the relationships among system states and objectives, and we lacked sufficient information to characterize all those relationships. By focusing experts on entering data in tables, we reduced variability and increased consensus. Although we wanted consensus answers, that may not be appropriate for projects with different goals.

Although we could have used influence diagrams or flowcharts developed prior to the meetings, we wanted the experts to guide the process rather than reacting to tools that we presented. Using unfamiliar visualization tools would have required the experts to understand our process for diagramming the important relationships.

6.5.3 Online Surveys

When time or money is limited, online surveys can rapidly and inexpensively collect data from experts. However, if reaching a consensus among experts is an objective, as it was for us, this would be difficult to accomplish using an online survey. The individual, anonymous nature of online surveys facilitates gathering of independent ideas and avoids "groupthink," which results from inappropriate group cohesion (Janis 1972), but eliminates the dialogue required to seek consensus. Online surveys facilitate quantifying values and related uncertainty even with qualitative scoring systems, but require relatively large numbers of participants.

We found the online survey program SurveyMonkey economical, easy to use and sufficiently flexible to structure our questions effectively, but it seemed designed for simpler surveys and smaller groups of respondents. If online surveys will be used to gather data from experts, their design should be modified so they will be more suitable for this type of research.

6.5.4 *Implementing the Results*

Using experts in our planning process filled data gaps in the published literature, ensured that we had appropriately defined the problems and objectives (e.g., population goals versus specific management actions), and will increase user confidence in our final products (Cowling and Pressey 2003; Younge and Fowkes 2003). The list of focal species that we developed will be used to prioritize areas for bird conservation in the SAMBI area. However, the Southeastern United States is home to many other at-risk species, including amphibians, reptiles, and mammals (Van Lear et al. 2005), that were not included in our planning process. The selection process described in this chapter can be extended to include these species, and expert opinion may be even more valuable because so little published information is available about some of these species.

References

American Ornithologists' Union (1998) Check-list of North American birds, 7th ed. American Ornithologists' Union, Washington

Amici V, Geri F, Battisti C (2010) An integrated method to create habitat suitability models for fragmented landscapes. J Nat Conserv 18:215–223

Andelman SJ, Fagan WF (2000) Umbrellas and flagships: efficient conservation surrogates or expensive mistakes? Proc Natl Acad Sci USA 97:5954–5959

Caro TM (2001) Species richness and abundance of small mammals inside and outside an African national park. Biol Conserv 98:251–257

Caro TM, O'Doherty G (1999) On the use of surrogate species in conservation biology. Conserv Biol 13:805–814

Cipollini KA, Maruyama AL, Zimmerman CL (2005) Planning for restoration: a decision analysis approach to prioritization. Restor Ecol 13:460–470

Coppolillo P, Gomez H, Maisels F, Wallace R (2004) Selection criteria for suites of landscape species as a basis for site-based conservation. Biol Conserv 115:419–430

Cowling RM, Pressey RL (2003) Introduction to systematic conservation planning in the Cape Floristic Region. Biol Conserv 112:1–13

Dame R, Alber M, Allen D et al (2000) Estuaries of the South Atlantic Coast of North America: their geographical signatures. Estuaries 23:793–819

Didier KA, Wilkie D, Douglas-Hamilton I et al (2009) Conservation planning on a budget: a "resource light" method for mapping priorities at a landscape scale? Biodivers Conserv 18:1979–2000

Favreau JM, Drew CA, Hess GR et al (2006) Recommendations for assessing the effectiveness of surrogate species approaches. Biodivers Conserv 15:3949–3969

Fleishman E, Blair RB, Murphy DD (2001) Empirical validation of a method for umbrella species selection. Ecol Appl 11:1489–1501

Gregory RS, Keeney RL (2002) Making smarter environmental management decisions. J Am Water Resour Assoc 38:1601–1612

Griffith JA, Stehman SV, Loveland TR (2003) Landscape trends in mid-Atlantic and southeastern United States ecoregions. Environ Manage 32:572–588

Guyer C, Bailey MA (1993) Amphibians and reptiles of longleaf pine communities. In: Hermann SM (ed) The longleaf pine ecosystem: ecology, restoration and management. Proc. 18th Tall Timbers Fire Ecology Conf. Tall Timbers Research, Inc., Tallahassee, pp 139–158

Hamel PB (1992) Land manager's guide to the birds of the South. The Nature Conservancy, Chapel Hill
Hamel PB (2000) Cerulean Warbler (*Dendroica cerulea*). In: Poole A (ed) The birds of North America, vol 511. Cornell Lab of Ornithology, Ithaca
Hess GR, King TJ (2002) Planning open spaces for wildlife. I. Selecting focal species using a Delphi survey approach. Landsc Urban Plan 58:25–40
Hodges JD (1997) Development and ecology of bottomland hardwood sites. For Ecol Manage 90:117–125
IPCC (2007) Climate change 2007: synthesis report. Contribution of Working Groups to the Fourth Assessment Report of the Intergovernmental Panel on Climate Change. Cambridge University Press, Cambridge
Janis IL (1972) Victims of groupthink. Houghton Mifflin Company, Boston
Kuhnert PM, Martin TG, Griffiths SP (2010) A guide to eliciting and using expert knowledge in Bayesian ecological models. Ecol Lett 13:900–914
Lambeck RJ (1997) Focal species: a multi-species umbrella for nature conservation. Conserv Biol 11:849–856
Landres PB, Verner J, Thomas JW (1988) Ecological uses of vertebrate indicator species: a critique. Conserv Biol 2:316–327
Low-Choy S, O'Leary R, Mengersen K (2009) Elicitation by design in ecology: using expert opinion to inform priors for Bayesian statistical models. Ecology 90:265–277
MacNally R, Fleishman E, Thomson JR, Dobkin DS (2008) Use of guilds for modelling avian responses to vegetation in the Intermountain West (USA). Glob Ecol Biogeogr 17:758–769
Margules CR, Pressey RL (2000) Systematic conservation planning. Nature 405:243–253
McCann RK, Marcot BG, Ellis R (2006) Bayesian belief networks: applications in ecology and natural resource management. Can J For Res 36:3053–3062
Mowbray TB (2002) Canvasback (*Aythya valisineria*). In: Poole A (ed) The birds of North America, vol 659. Cornell Lab of Ornithology, Ithaca
North American Bird Conservation Initiative, U.S. Committee (2009) The state of the birds, United States of America, 2009. U.S. Department of the Interior, Washington
Noss RF (1990) Indicators for monitoring biodiversity: a hierarchical approach. Conserv Biol 4:355–364
Noss RF, Carroll C, Vance-Borland K, Wuerthner G (2002) A multicriteria assessment of the irreplaceability and vulnerability of sites in the Greater Yellowstone ecosystem. Conserv Biol 16:895–908
Olson DM, Dinerstein E (1998) The global 200: a representation approach to conserving the Earth's most biologically valuable ecoregions. Conserv Biol 12:502–515
Outcalt KW, Sheffield RM (1996) The longleaf pine forest: trends and current conditions. USDA Forest Service, Southern Research Station, Asheville, Resource Bulletin SRS-9
Pierce SM, Cowling RM, Knight AT et al (2005) Systematic conservation planning products for land-use planning: interpretations for implementation. Biol Conserv 125:441–458
Power ME, Tilman D, Estes JA et al (1996) Challenges in the quest for keystones. Bioscience 46:609–620
Pressey RL, Bottrill MC (2009) Approaches to landscape- and seascape-scale conservation planning: convergence, contrasts and challenges. Oryx 43:464–475
Pressey RL, Ferrier S, Hager TC et al (1996) How well protected are the forests of north-eastern New South Wales? – Analyses of forest environments in relation to formal protection measures, land tenure, and vulnerability to clearing. For Ecol Manage 85:311–333.
Richardson CJ (2003) Pocosins: hydrologically isolated or integrated wetlands on the landscape? Wetlands 23:563–576
Roberge JM, Angelstam P (2004) Usefulness of the umbrella species concept as a conservation tool. Conserv Biol 18:76–85
Sanderson EW, Redford KH, Vedder A et al (2002) A conceptual model for conservation planning based on landscape species requirements. Landsc Urban Plan 58:41–56
Simberloff D (1998) Flagships, umbrellas, and keystones: is single species management passé in the landscape era? Biol Conserv 83:247–257

Tacha TC, Nesbitt SA, Vohs PA (1992) Sandhill Crane (*Grus canadensis*). In: Poole A (ed) The birds of North America, vol 31. Cornell Lab of Ornithology, Ithaca

Teck SJ, Halpern BS, Kappel CV et al (2010) Using expert judgment to estimate marine ecosystem vulnerability in the California Current. Ecol Appl 20:1402–1416

USGS (2010) Protected Areas Database of the United States (PAD-US) Version 1.1. U.S. Geological Survey, Gap Analysis Program, Moscow. Available from http://gapanalysis.nbii.gov/portal/server.pt/community/maps_and_data/1850/ (Accessed March 2011)

Valiela I, Martinetto P (2007) Changes in bird abundance in eastern North America: urban sprawl and global footprint? Bioscience 57:360–370

Van Lear DH, Carrol WD, Kapeluck PR, Johnson R (2005) History and restoration of the longleaf pine-grassland ecosystems: implications for species at risk. For Ecol Manage 211:150–165

Walker J (1998) Ground layer vegetation in longleaf pine landscapes: an overview for restoration management. In: Kush JS (ed) Proceedings of the longleaf pine ecosystem restoration symposium. Longleaf Alliance, Fort Lauderdale, Report no. 3, pp 2–13

Watson C, Malloy K (2006) The South Atlantic Migratory Bird Initiative implementation plan: an integrated approach to conservation of "all birds across all habitats". Atlantic Coast Joint Venture Report, Charleston

Wear DN (2002) Chapter 6: land use. In: Wear DN, Greis JG (eds) Southern forest resource assessment. USDA Forest Service, Southern Research Station, Asheville, Gen. Tech. Rep. SRS-53, pp 153–173

Wear D, Pye J, Ritters K (2004) Defining conservation priorities using fragmentation forecasts. Ecol Soc 9:4

Wiens JA, Hayward GD, Holthausen RS, Wisdom MJ (2008) Using surrogate species and groups for conservation planning and management. Bioscience 58:241–252

Woodin MC, Michot TC (2002) Redhead (*Aythya americana*). In: Poole A (ed) The birds of North America, vol 695. Cornell Lab of Ornithology, Ithaca

Younge A, Fowkes S (2003) The Cape action plan for the environment: overview of an ecoregional planning process. Biol Conserv 112:15–28

Chapter 7
An Expert-Based Modeling Approach to Inform Strategic and Operational Land Management Decisions for the Recovery of Woodland Caribou

R. Scott McNay

Contents

7.1 Introduction

Ungulates are a valuable natural resource due to their contribution to biodiversity (Ray 2005) and to their value as game animals for aboriginal peoples, guide outfitters, and hunters. For the past decade in British Columbia (BC), forest practices have been regulated to conserve the wildlife range that provides for the overwinter survival of ungulates. For the purposes of the regulations (http://www.env.gov.bc.ca/wld/frpa/uwr/), ungulates include moose (*Alces alces*), mule (or black-tailed) deer (*Odocoileus hemionus*), white-tailed deer (*Odocoileus virginiana*),

R.S. McNay (✉)
Wildlife Infometrics, Inc., Box 308, Mackenzie, BC, Canada V0J 2C0
e-mail: scott.mcnay@wildlifeinfometrics.com

A.H. Perera et al. (eds.), *Expert Knowledge and Its Application in Landscape Ecology*,
DOI 10.1007/978-1-4614-1034-8_7, © Springer Science+Business Media, LLC 2012

elk (*Cervus elaphus*), caribou (*Rangifer tarandus*), Stone sheep (*Ovis dalli stonei*), Dall sheep (*Ovis dalli dalli*), bighorn sheep (*Ovis canadensis*), and mountain goats (*Oreamnos americanus*). Wildlife range has also been regulated in BC to conserve areas used during other seasons by wildlife species considered by the BC government to be at risk of local extinction.

Woodland caribou (*Rangifer tarandus caribou*; hereafter, "caribou") throughout Canada have undergone a history of range reduction (de Vos and Peterson 1951; Spalding 2000; Thomas and Gray 2002), and populations in many herds are currently in decline (Rettie and Messier 1998; Schaefer et al. 1999; McLoughlin et al. 2003; Wittmer et al. 2005). The BC government considers caribou to be at risk (http://www.env.gov.bc.ca/atrisk/), therefore the BC conservation measures apply to caribou winter ranges as well as to their other seasonal ranges. Caribou have dynamic range requirements due to their broad distribution (the species is potentially found in as much as 30 million ha in BC alone), and wildlife managers lack the tools and specific understanding of how to manage for that set of dynamic requirements, let alone how to manage landscapes to assist the recovery of declining populations. Common responses to such uncertainty have included deferral of decisions, implementation of long-term research programs, development of strategic plans, and participation in a variety of management debates (Thomas 1985). These responses essentially delay or preclude effective management actions by consuming an enormous amount of time and resources.

In 2004, to eliminate this inefficient use of resources and to provide the information required to implement effective conservation regulations, the BC government produced legally binding, expert-based management guidelines for the amount, distribution, and attributes of the range required by each ungulate species in the province, including caribou. Managers were instructed to implement the interim guidelines until areas could be legally designated as Ungulate Winter Ranges (UWRs) or Wildlife Habitat Areas and until specific management actions for these designated areas could be provided. In 2005, the BC government also brought together a science team to provide expert technical advice specifically on how to promote the recovery of caribou in the southern portion of their range. Although such expert opinion may at times lack complete empirical scientific support, the implementation of guidelines based on expert advice may be justified because the potential consequence of inaction can be local extinction of a species (Hebblewhite et al. 2010). Hebblewhite et al. (2010) suggested that taking some action, even if it is based only on interim study results (e.g., expert-based information), could benefit the species and possibly lead to effective management.

In north-central BC, managers have used scenario modeling (Daum 2001), expert-based information, management simulations, and empirical testing to provide insights into the probability that woodland caribou, mountain goats, and mule deer will occupy a given range. This probabilistic approach was used to inform strategic decisions about recovery planning for woodland caribou and the formal operational identification of UWRs for all three species, and to provide a transparent framework for adaptation of current management regimes and tools for monitoring the effectiveness of the new management regimes. Using woodland caribou in north-central BC as a case study, my objectives in this chapter were to demonstrate the use of expert-based information at strategic and operational levels of management, and to reveal why the expert-based approach can help to resolve complex and important ecological problems.

7.2 Ecological and Management Context

My case study focused on threatened caribou herds that range over 3.7 million ha in north-central BC, and specifically the Chase (~550 animals), Wolverine (~375 animals), Takla (~125 animals), and Scott (~50 animals) herds (Giguère and McNay 2007; Wilson et al. 2004; Wildlife Infometrics, Inc., Mackenzie, BC, unpubl. data). Caribou in the area generally use lodgepole pine (*Pinus contorta*) forests at mid- to low-elevations (700–1,300 m asl) during the fall and early winter, and use alpine and subalpine areas (>1,300 m asl) during the late winter, spring, and summer (Terry and Wood 1999; Wood and Terry 1999; Johnson 2000; Poole et al. 2000). Except during the spring, their diet consists primarily of terrestrial forage lichens (*Cladina mitis*, *Cladina rangiferina*, *Cladina arbuscula* ssp. *beringiana*, *Cladonia uncialis*, and *Cladonia ecmocyna*), with an increased use of arboreal forage lichens (*Bryoria* spp.) during the late winter (Johnson et al. 2000). Because the early-winter range is located on relatively flat terrain at low elevations, it is at risk of significant anthropogenic disturbance; for example, extensive industrial development began in the study area after construction of the W.A.C. Bennett hydroelectric dam in 1961. Caribou also experience predation risk throughout their range, and predation is the most proximate factor in the general decline of caribou in BC (Seip 1992; Wittmer et al. 2005; Bergerud 2007). Landscape change as a result of anthropogenic disturbance is considered to be the ultimate cause of the decline in caribou populations through the resulting alteration of the relationships between predators and their prey (Golder Associates 2010).

The tendency for caribou to frequent high-elevation range, dispersed to create a low population density, is a common tactic for avoiding predators (Bergerud et al. 1984, 1992; Bergerud 1992), which in the study area are mostly wolves (*Canis lupus*). Aboriginal people have reported seasonal use of the area by wolves, but also described an increase in wolf abundance and a more persistent presence following the first appearance of moose in the early 1920s (McKay 1997). Other predators of caribou in this area include grizzly bear (*Ursus arctos horribilis*), black bear (*Ursus americanus*), and wolverine (*Gulo gulo*). The BC government considers the impact on caribou populations caused by hunting to be minor.

Although the Committee on the Status of Endangered Wildlife in Canada considered the herds in this case study to be at risk of a population decline (COSEWIC 2002), the BC Government did not consider the herds to be a priority for recovery planning. Strategic objectives to conserve caribou range were described in local land use plans (BCMSR 1999, 2000), but there was no legal authority provided to implement any management consistent with the strategic objectives. In 2003, an *ad hoc* caribou Recovery Implementation Group (RIG) initiated a "grass roots" agenda to provide the BC government with the information required to develop a recovery plan for caribou in the area (McNay et al. 2008a). Specific information about the RIG's function, including its meeting agendas and minutes, is available at the Recovery Initiatives Web site (http://www.centralbccaribou.ca).

7.3 Gathering and Formalizing Information

The RIG members chose a modeling approach to make spatially explicit predictions about the quality of seasonal ranges for caribou, using existing environmental conditions as well as those that would presumably occur under a variety of hypothetical simulated landscape disturbance scenarios. The scenarios were based on the disturbances expected to be caused by land management or by natural, unmanaged disturbances such as wildfire (McNay et al. 2006). The intent was to compare the results of the disturbance scenarios as a way to inform RIG members about the potential utility of alternative management regimes. However, no model of caribou seasonal ranges existed at that time, and although information was available from previous research, RIG members recognized the limitations of the information and the lengthy and costly research process that would be required to address those limitations. As an alternative to inaction while awaiting this research, the RIG members chose to develop an interim model and address its data limitations by eliciting information from knowledgeable professionals (hereafter, "experts").

7.3.1 Identifying Experts and Eliciting Their Knowledge

The RIG hosted professionally facilitated, 1- to 2-day workshops approximately every 2 months from January 2000 to January 2003 to gather and formalize information about caribou and their range requirements. Professional facilitation was deemed necessary by the RIG to effectively elicit information from the experts and to move the discussion as efficiently as possible through the initial steps of developing a model. Experts in relevant domains (e.g., ecosystem mapping, population dynamics, lichen ecology, climate, and land management) were chosen by RIG members based on their reputation and their ability to support the model development process. Some experts had primary roles in research projects that had been conducted on caribou herds in the study area or in adjacent areas. Other experts, although knowledgeable about their domains, knew relatively little about caribou or caribou habitat.

Once selected, experts became members of the RIG and attended each workshop, except when there was a need for unique or specific information that was not central to developing the model (e.g., provincial timber-supply modeling). Workshops were usually attended by 10–15 members, including 1 facilitator, 3 modelers, and 6–11 domain specialists or experts. There was no specific intent to balance specific affiliations or types of professional endeavor (e.g., academia versus government or consultants), but groups with a vested interest (i.e., government and industry) tended to have the strongest representation.

The elicitation of information followed a series of steps that began with the definition of seasonal range types. Within each seasonal range, the group then identified the most important life requisites that should be represented by the model and the ecological or biophysical factors (e.g., environmental conditions) most likely to be functionally related to these life requisites. The relationships among the life requisites were then depicted as "influence diagrams". The modeling team distinguished between environmental conditions that would or would not be changed by management in order to address the eventual need for simulating landscape disturbance. The conditions that would be changed were then termed "management levers." The elicited information was summarized and reviewed as each meeting progressed. When a difference in opinion arose among the experts, it was resolved by discussion leading to a consensus, guided by the facilitator; all final results were recorded in the meeting minutes for review by workshop participants subsequent to the meeting.

The resulting influence diagrams were then represented as Bayesian belief networks (BBNs; Cain 2001; Chap. 5), which were developed by a three-person modeling team and prepared for presentation to the RIG members at the next workshop. This approach was chosen to maximize the efficiency of the consultation time with experts during the RIG meetings. BBNs can be used to derive and visualize predicted responses (i.e., model outputs) based on information on the influence of environmental conditions (i.e., the model inputs). The nodes of BBNs are linked by conditional probability tables. Marcot et al. (2006) provided a detailed description of the use of BBNs in ecology. The specific probabilistic nature of each of the identified ecological relationships was elicited from the experts as another step in the model development process. Although it was possible for RIG members to misrepresent probabilistic relationships, and for the modeling team to misrepresent expert knowledge (Kuhnert and Hayes 2009), these potential errors were usually avoided by following specific guiding principles. These principles were developed by Bruce Marcot, and eventually become the basis for a journal paper (Marcot et al. 2006). Errors that were identified by the modeling team were corrected through subsequent consultation with the experts.

7.3.2 Ecological Relationships

McNay et al. (2002, 2006, 2008a) summarize the specific ecological relationships and associated conditional probabilities that resulted from this process. The BBNs covered the following seasonal-range combinations: high-elevation winter, pine-lichen winter, calving and summer, post-rut, and migration. Each range prediction was modified by accounting for a BBN based on predation risk. BBN outputs were expressed as the expected probability of occupancy of a site by caribou, which was subsequently classed for convenience into three states: low (0.00–0.33), moderate (0.34–0.66), or high (0.67–1.00). The modeling team felt that this summary would be easy for experts to understand and use to judge the fit of the model results to their expectations. The modeling team used equidistant division points among the classes because the experts were unable to provide a better alternative. Maps of the classified ranges were used by the modeling team to demonstrate the BBN results to the RIG members. Although it was possible to derive a measure of uncertainty in the model output, model developers did not provide this information, mostly due to perceived time constraints and real funding constraints.

7.3.3 Ecological Stressors

The RIG facilitator elicited information from experts and other RIG members regarding the stressors expected to alter environmental conditions and thereby change the probability of range use by caribou. Although the stressors were generally well documented in the scientific literature, their perceived importance and degree of interaction varied because their relative strength is still being debated. The work on stressors therefore tended to be a confirmation among the experts about their relative ranking of the known stressors as applied to the conditions of the study area. The debate and conclusions that resulted from this discussion were largely based on the published literature, but set within the context of the personal observations of the experts.

The RIG experts believed that where timber harvesting occurred, the resulting early-seral forests would support abundant moose interspersed through the adjacent older forest (Franzmann and Schwartz 1998) and, in turn, abundant wolves (Messier et al. 2004). Compounding the predation risk from increased wolf numbers was the development of roads associated with timber harvesting operations, which provide wolves with easier travel and potentially increase hunting efficiency (James and Stuart-Smith 2000). The experts assumed that caribou populations would generally experience greater mortality in areas where moose are interspersed throughout their range (Wittmer et al. 2005); this source of greater mortality therefore became a stressor, which was assessed using a BBN for predation risk.

A second stressor was the hydroelectric development in the area. Subsequent flooding of the Finlay, Peace, and Parsnip Rivers created BC's largest body of freshwater, which experts considered a barrier to caribou migration. This barrier has likely contributed to reductions of caribou populations, particularly for the Scott herd. The barrier effect of the reservoir was included as a variable in the migration BBN.

Timber harvesting and similar disturbances were considered to be a third group of stressors through their ecological effect on natural succession of vegetation communities, and hence on the abundance of terrestrial forage lichens. Forage lichens tend to dominate the understory of pine forests during distinct (but not all) stages of natural vegetation succession (Coxson and Marsh 2001). Winter ranges were therefore considered to require regular natural (i.e., wildfire) or managed (e.g., timber harvesting) disturbance to sustain the lichen supply; consequently, these disturbances had varying effects on the BBN results for the pine-lichen winter range.

7.3.4 Management Scenarios

Seasonal ranges for caribou were predicted and evaluated using five land-management scenarios defined by the RIG members:

1. The *potential range* was estimated by setting all input nodes to their most favorable condition for caribou.
2. The *current range* was estimated by setting the input nodes to use the existing environmental conditions.
3. The *managed range* was estimated based on forest management, such as timber harvesting and road construction, conducted under rules specifically intended to conserve caribou range.
4. The *natural unmanaged range without elevated predation* was estimated based on assumed natural patterns of wildfire without accounting for the moose–wolf predator–prey system.
5. The *natural unmanaged range with elevated predation* was estimated based on the same natural disturbance patterns as in scenario 4, but accounting for the moose–wolf predator–prey system.

The rules for conservation of caribou range were adopted from the local land-use plans (BCMSR 1999, 2000), which stated that 50% of the potential pine-lichen winter range should be in a condition usable by caribou at all times. The natural disturbance scenarios were based on historical patch sizes and return intervals for wildfire within the study area (Delong 2002). All scenarios were simulated over 290 years in 10-year time steps using the Spatially Explicit Landscape Event Simulator (Fall and Fall 2001), and the natural disturbance scenarios were repeated with random start positions to generate a range of results over those conditions. These scenarios are described in more detail by Fall (2003).

7.3.5 Validation and Verification of the Results

The modeling team considered *validation* to be an assessment of the model's implementation and *verification* to be an assessment of its accuracy. Validation assessments conducted by the modeling team included reviewing the mapped output for obvious errors (e.g., missing data, apparent background noise, unnatural boundaries between range classifications) and manually inspecting data and relationship calculations to confirm that the model inputs at specific, random locations led correctly to the specific output.

Preliminary verification of the model's performance was limited to a simple visual inspection of the mapped seasonal range predictions by the RIG caribou experts to verify that classified seasonal range results met with their expectations or knowledge of how caribou used their range. Peer reviewers were solicited to review the BBN structures and the associated conditional probability tables. The RIG members considered this limited verification to be sufficient for use in strategic planning (i.e., development of management actions to promote caribou recovery). In contrast, a more formal verification of the model's results was conducted before the results were used in operational planning (i.e., UWR identification). The original mapped results were first smoothed to facilitate their application to the landbase. An aerial reconnaissance was then conducted to verify the spatial locations of the predicted range. Relocations of radio-collared caribou were also used to help assess model validity using either a statistical test of range selection (Chesson 1983) or a simple measure of inclusion (the proportion of animals that used the range). The selection test was based on an analysis of winter (1 January to 30 April) relocations with the hypothesis that caribou would choose to use modeled ranges in direct proportion to their availability (i.e., selection was equivocal). Alternatively, we assumed that caribou selected the modeled range if they used the modeled range more than expected, and that caribou did not select the modeled range if use was less than expected. To assess correspondence to the hypothesis, the modeling team also used a confusion matrix (Provost and Kohavi 1998) of the selection observations to calculate standard performance criteria for the model. The proportion of inclusion was a simple and less formal measure of the relative proportion of relocations that could be enclosed by the modeled range while attempting to minimize the total amount of range predicted by the model.

7.3.6 Interpretation and Use of the Expert-Based Information

Following the workshops that were used for model building, the RIG hosted a second series of ten professionally facilitated workshops between December 2003 and February 2007. The purpose of these workshops was to develop a set of management actions intended to promote the recovery of caribou populations using the expert-based modeling results. At this stage, new members were added to the RIG who had a vested interest in how land management might unfold in the future (e.g., First Nations, recreational snowmobilers, guide-outfitters – "stakeholders" Chap. 1). As was the case for selecting the experts, the new members were chosen based on their reputation for being knowledgeable professionals and their perceived ability to support the planning process. The workshops proceeded using the following series of steps:

1. Confirm stakeholder dedication to the process and define the extent of the area in which recovery would be promoted.
2. Review the available knowledge for each herd, including the modeled range predictions.

3. Determine the general goals to set boundaries on the scope of the recovery planning.
4. Confirm the stressors identified by the previous series of workshops and identify potential mitigation measures.
5. Compose a set of specific management actions to promote recovery of the caribou populations.
6. Establish a basis to review the socioeconomic impacts of the anticipated management direction.

Each workshop was conducted following a standard protocol, which began with a meeting announcement and request for attendance. Agendas were then developed and final meeting arrangements were established based on the responses of the members. The RIG attempted to have all members attend, and this was usually achieved. Maps were used to help RIG members interpret the spatial results of the expert-based seasonal range models. Further, without specific information on seasonal range carrying capacity, the modeling team created a habitat index so RIG members could conveniently and consistently compare quantitative model results among seasonal ranges. The index, which was calculated by multiplying the amount of seasonal range by a seasonal range value weight (SRVW), effectively standardized original model results for each seasonal range based on a constant, hypothetical density of caribou that might be expected under conditions of sustainability (McNay et al. 2008a). The SRVW was calculated as:

$$\mathrm{SRVW} = -0.53 + 0.04\mathrm{RV} + 0.79\mathrm{RT} - 0.35\mathrm{RT}^2 + 0.04\mathrm{RT}^3,$$

where RT is the range type (i.e., pine-lichen winter, post-rut, high-elevation winter, or calving and summer) and RV is the range value (i.e., high, medium, or low) predicted by the BBN. Minutes were recorded by an RIG secretary and salient points (e.g., decision points and action items) were recorded by the facilitator. Minutes were prepared and sent to RIG members for review.

7.4 Results of the Expert-Based Modeling

Clear differences were revealed in the results for each herd area by applying the expert-based BBNs for the seasonal ranges. For example, whereas the potential for calving and summer range exceeded the potential for any other range in all areas, the potential for pine-lichen winter range was generally the lowest, though not in all herd areas (Fig. 7.1). Furthermore, the potential effect of predation risk varied across seasonal ranges and areas (Table 7.1), and the different scenarios also produced results that varied over the forecasted conditions for the simulation period (Fig. 7.2). In general, the results for seasonal ranges, herd areas, and management scenarios successfully provided the RIG members with opportunities to compare the existing availability of caribou range to the caribou range that would result from a variety of hypothetical forecasted future conditions.

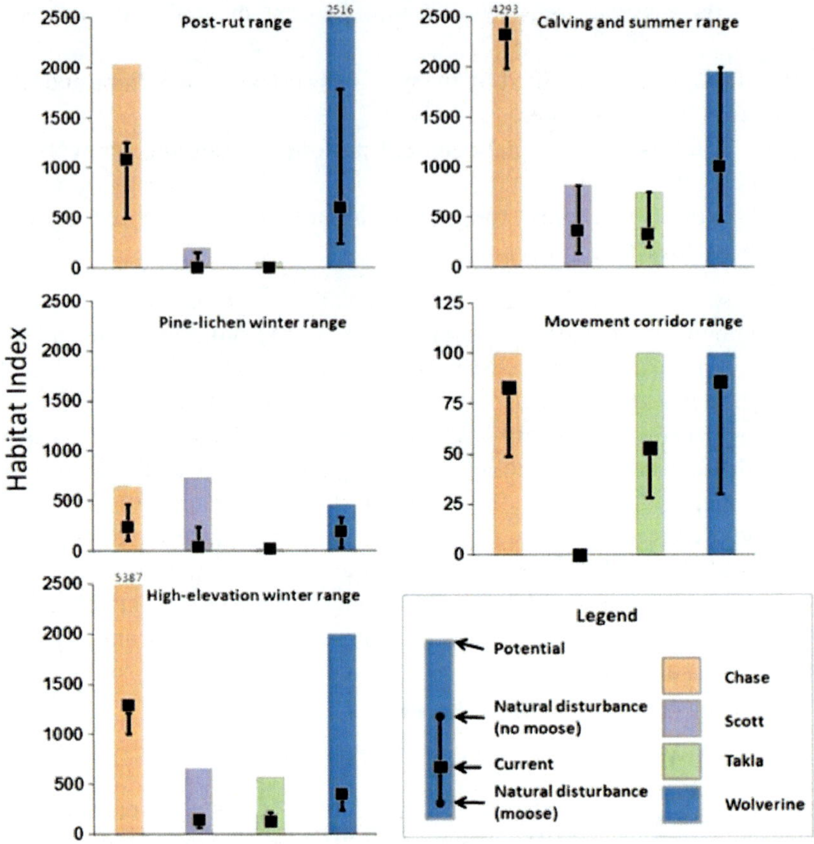

Fig. 7.1 The relative amount of seasonal range (i.e., the habitat index; see Sect. 7.3.6 for a description) modeled for conditions in four caribou herd areas (Chase, Scott, Takla, and Wolverine herds) of north-central British Columbia (from McNay et al. 2008a). Predictions were made for hypothetical simulated landscape scenarios representing the potential best conditions for caribou, current environmental conditions, and two natural disturbance scenarios (with and without accounting for elevated predation based on the abundance of moose as primary prey for wolves). A hypothetical management scenario was modeled but is not presented here because that scenario was dynamic through time and could not be characterized using a static estimate. The actual habitat index is placed above histograms whenever they exceed the limits of the Y-axis

7.5 Validation and Verification of the Modeled Results

The RIG arranged for an aerial reconnaissance that intersected 54 of 74 available patches of pine-lichen winter range in the Chase and Wolverine herd areas (Fig. 7.3). Terrestrial lichens were not abundant in nine of the 54 patches. In eight other cases along the flight line, false negatives were observed (i.e., there was abundant terrestrial lichen even though the BBN did not predict its occurrence). Winter relocations of 40 and 33 radio-collared caribou in the Wolverine herd area ($n=3{,}239$) and Chase herd area ($n=5{,}207$), respectively, were collected and used in the selection tests for the pine-lichen winter range. Selection or avoidance of range was stronger in the

Table 7.1 Area and percentage of total area (in parentheses) for different seasonal ranges in north-central British Columbia modeled as having a high or medium probability of occupancy by woodland caribou (from McNay et al. 2008a)

	Area of seasonal range (ha) and % of total area			
Herd area (Scenario)	Post-rut range	Pine-lichen winter range	High-elevation winter range	Calving and summer range
Chase herd (Scenario 1)	22,500 (1%)	17,184 (1%)	208,505 (12%)	1,094,879 (63%)
Percentage reduction 1–2	26	28	71	2
Scenario 2	16,679	12,407	59,462	1,069,999
Percentage reduction 2–4	56	63	21	47
Scenario 4	7,343	4,587	47,078	579,012
Percentage reduction 4–5	41	54	24	15
Scenario 5	4,324	2,100	35,997	492,419
Scott herd (Scenario 1)	2,319 (<1%)	21,883 (4%)	26,069 (4%)	204,831 (34%)
Percentage reduction 1–2	13	70	56	0
Scenario 2	2,009	6,525	11,419	204,060
Percentage reduction 2–4	100	86	53	56
Scenario 4	0	929	5,354	90,172
Percentage reduction 4–5	0	98	52	64
Scenario 5	0	21	2,556	32,312
Takla herd (Scenario 1)	492 (<1%)	835 (<1%)	22,420 (4%)	186,322 (38%)
Percentage reduction 1–2	3	3	53	0
Scenario 2	477	812	10,529	186,122
Percentage reduction 2–4	97	55	56	57

(continued)

Table 7.1 (continued)

Herd area (Scenario)	Area of seasonal range (ha) and % of total area			
	Post-rut range	Pine-lichen winter range	High-elevation winter range	Calving and summer range
Scenario 4	12	374	4,613	80,635
Percentage reduction 4–5	0	100	17	40
Scenario 5	12	0	3,827	48,741
Wolverine herd (Scenario 1)	26,703 (3%)	11,722 (1%)	78,785 (9%)	484,830 (57%)
Percentage reduction 1–2	30	6	68	1
Scenario 2	18,762	10,981	24,918	478,449
Percentage reduction 2–4	83	59	38	48
Scenario 4	3,101	4,545	15,430	249,703
Percentage reduction 4–5	35	87	41	55
Scenario 5	2,001	595	9,141	111,754

Predictions were made for hypothetical simulated landscape scenarios derived based on expert consultations: Scenario 1, potential best conditions for caribou; Scenario 2, current environmental conditions; Scenario 4, natural unmanaged conditions without accounting for elevated predation; and Scenario 5, natural unmanaged range accounting for elevated predation. The percentage reduction in range area compares the results in Scenarios 2 and 1 and the results in Scenarios 4 and 2. Scenario 3 was a dynamic forecast of managed range conditions over time and therefore could not be characterized by a static estimate

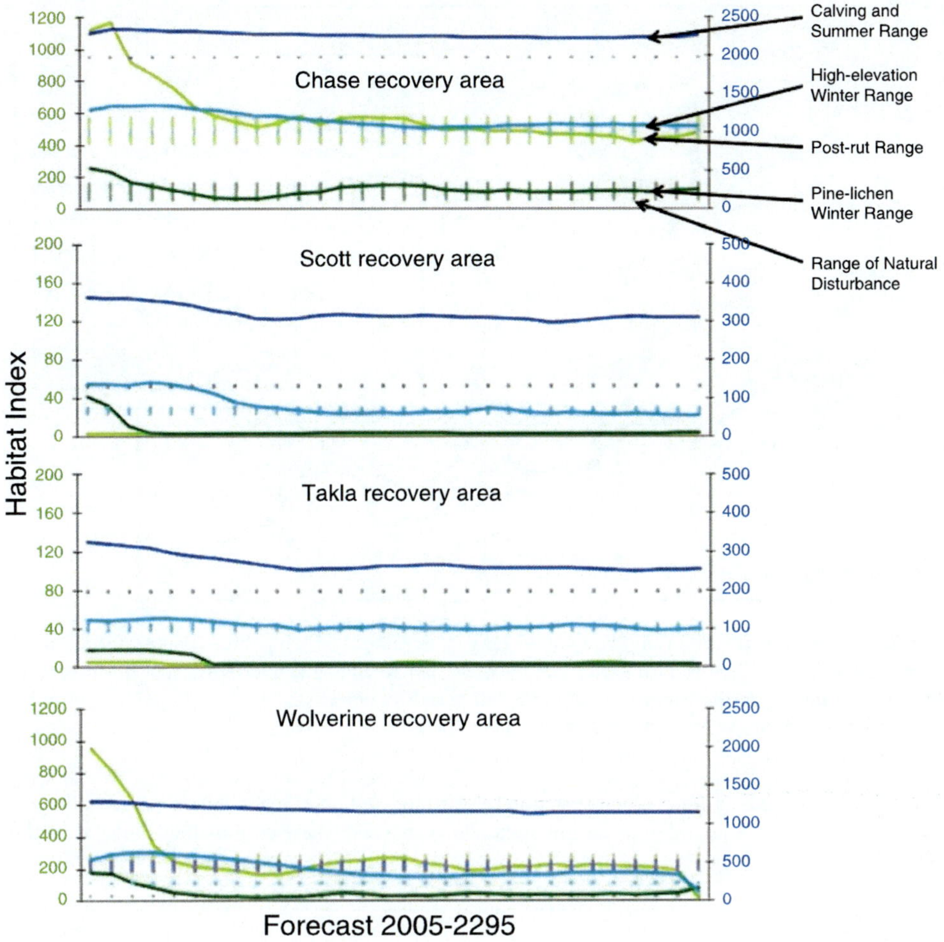

Fig. 7.2 The relative amount of seasonal range (i.e., the habitat index; see Sect. 7.3.6 for details) modeled to represent hypothetical simulated environmental conditions [a conservation scenario (*solid lines*) and a natural disturbance scenario (*vertical bars*)] in four caribou herd areas (Chase, Scott, Takla, Wolverine) of north-central British Columbia (from McNay et al. 2008a). See the text for a description of the modeling and landscape scenarios

Chase area than in the Wolverine area, although the rate was acceptably high (>70%) in both cases (Table 7.2). Overall accuracy was ≥75% in both areas (Table 7.3), but the prediction error in the Wolverine area was marginal (i.e., a false negative rate of nearly 30%).

In comparison with the relatively successful tests of pine-lichen winter range, the test of caribou selection for high-elevation winter range revealed a poor and inconsistent fit of the relocation data to the original modeled range. Reconnaissance surveys of the high-elevation winter range suggested that arboreal forage was not being predicted properly (Rankin and McNay 2007). This led to a more detailed study of the abundance of arboreal forage lichen in subalpine habitats within the study area,

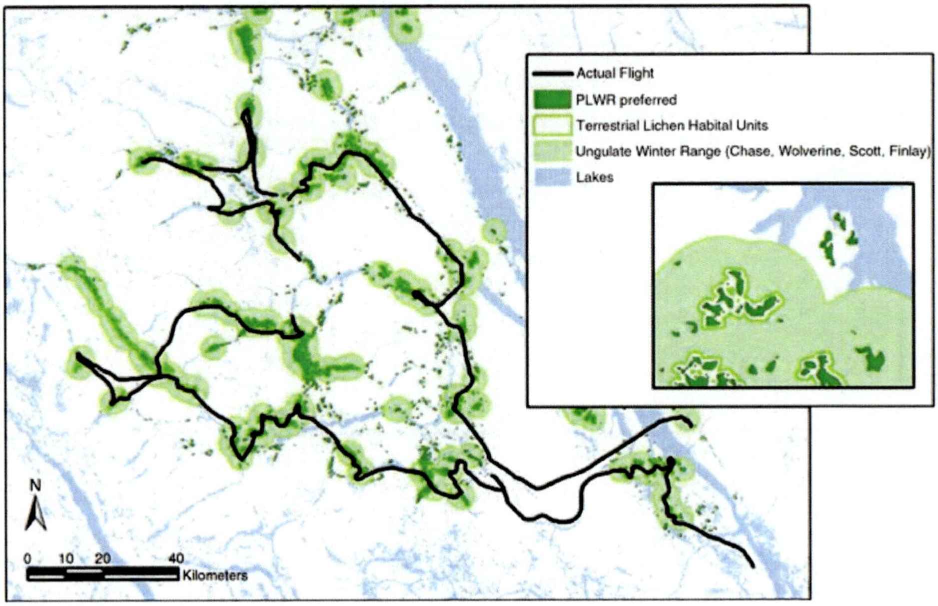

Fig. 7.3 Management units [ungulate winter ranges, terrestrial lichen habitat, and preferred pine-lichen winter range (PLWR)] for caribou in the Scott, Wolverine, and Chase herds of north-central British Columbia and the flight line depicting an aerial reconnaissance of the management units conducted in late 2003 (from McNay and Sulyma 2003)

Table 7.2 Observed caribou habitat selection for the modeled pine-lichen winter range estimated from relocations of radio-collared caribou in the Wolverine and Chase caribou herd areas of north-central British Columbia

Caribou herd area and modeled probability of range occupancy		Observed selection[a]				
		Avoided	Preferred	Total selected (avoided+preferred)	Equivocal	Total sample
Wolverine	Low	25	7	32	8	40
	High and medium	7	18	25	15	40
	Total	32	25	57	23	80
Chase	Low	27	4	31	2	33
	High and medium	4	22	26	7	33
	Total	31	26	57	9	66

[a]Estimates of selection were calculated for individual caribou based on the methods described by Chesson (1983)

a variety of updates to the BBN for the high-elevation winter range, and a reapplication of the model; the revised model included 63% of the observed winter relocations of caribou at higher elevations (McNay et al. 2009). Expert and peer review of the modeling and maps resulted in detailed documentation of the model (McNay et al.

Table 7.3 Key performance criteria summarized from a confusion matrix containing information about the observed selection for modeled pine-lichen winter range by radio-collared caribou in the Wolverine and Chase caribou herd areas of north-central BC

	Caribou herd area	
Performance criteria[a]	Wolverine	Chase
Selection rate (%)	71	86
Recall rate (%)	72	85
Accuracy (%)	75	86
Precision (%)	72	85
False-positive error rate (%)	22	13
False-negative error rate (%)	28	15

[a] Criteria definitions:

Selection rate = [(actual preferred observations + actual avoided observations)/total observations] × 100%

Recall rate = (number of predicted preferred choices that were actually observed as preferred/total actual preferences) × 100%

Accuracy = [(number of predicted avoided choices that were actually observed as avoided + number of predicted preferred choices that were actually observed as preferred)/total selections] × 100%

Precision = (number of predicted preferred choices that were actually observed as preferred/total predicted preferences) × 100%

False-positive error rate = (number of predicted preferred choices that were actually observed as avoided/total actual avoided) × 100%

False-negative error rate = (number of predicted avoided choices that were actually observed as preferred/total actual preferred) × 100%

2002), a user manual for application of the model (Doucette et al. 2004), a published summary of the expert-based approach to modeling (McNay et al. 2006), and a series of herd-specific seasonal range maps that were used by the BC government and other managers to implement management actions that focused on promoting recovery of the caribou (McNay et al. 2008a).

7.6 Practical Applications

7.6.1 Recovery Planning

The use of BBNs allowed a systematic and transparent use of expert-based information to support planning of management actions that would promote the recovery of caribou herds in the study area. The transparency of the method and its reliance on multiple experts helped to establish agreement about this complex problem among land managers and other RIG members, and encouraged a collective approach to the identification of specific management priorities. For example, one important agreement among RIG members fundamental to the recovery plan was the primary assumption that the distribution of moose (and therefore of wolves) in the plan area had likely expanded due to historical processes that encouraged natural colonization of parts of the area by moose, but that a further increase in moose numbers resulted from an abundance of moose forage following recent forest harvesting. Based on this agreed-upon assumption, management actions to promote recovery of the caribou populations were therefore given the following specific priorities:

1. Restore critical range by reducing the extent and value of the range for moose, but not below the level that would be likely to occur normally as a result of natural disturbance regimes.
2. Implement priority 1 and provide interim controls or limits on the abundance of the wolves' primary prey (e.g., implement an aggressive hunting season to control moose populations).
3. Implement priority 1 and provide interim controls or limits on the abundance of wolves.

If range restoration was ultimately found to be unsuccessful within a herd area, and ongoing management of other proximal factors (i.e., moose and wolves) were required to maintain a herd, then the caribou herd was considered to be not self-sustainable and recovery of caribou to self-sustaining levels would not be ecologically feasible (McNay et al. 2008a).

Another complex ecological problem considered by the RIG experts focused on the fact that moose were an important, albeit recent, resource in the area. The change in moose abundance meant that even natural, unmanaged conditions are unlikely to support caribou today as well as they did historically. This is because as moose populations increase (hence, as more wolves enter the study area), the incidental predation on caribou would increase. This logic was supported by the BBN results (Fig. 7.1), so the RIG experts decided that it would not be efficient or perhaps even feasible to artificially create range conditions for caribou that would equate to historical conditions. Therefore, the population levels to which caribou herds may recover would likely be lower than historical levels. Similarly, the irreparable barrier to migration created for the Scott herd by the reservoir meant that high-elevation range was permanently separated from low-elevation range in that herd's area; therefore, RIG experts assumed that this barrier would limit the likelihood and feasibility of recovery by the Scott herd.

Consistent application of the expert-based BBNs across the herd areas revealed two results that otherwise might have been considered counterintuitive (Table 7.1): a general lack of potential for low-elevation range in the Takla herd's area and a high risk of predation in low-elevation range in the Scott herd's area, even under natural disturbance conditions. The RIG members therefore de-emphasized restoration of caribou range in these areas. In contrast, the BBN results indicated that the best recovery opportunities existed for caribou in the Chase and Wolverine herd areas (Fig. 7.1). Without considering metrics other than the seasonal range habitat index (i.e., other metrics might include patch size and connectivity), the conservation policy that was modeled for these areas appeared able to provide for a sustainable supply of caribou range consistent with the conditions expected under an assumed natural disturbance regime (Fig. 7.2). That said, the pine-lichen winter range and post-rut range were likely to decrease from their current levels and undergo some decline over the next 2–3 decades. If this occurs, the caribou populations will decline as well, but RIG experts were uncertain whether the decline could be mitigated by the caribou by increasing their use of high-elevation range. Another uncertainty that became obvious to the RIG experts pertained to the feasibility of implementing the recommended policy given the pending outbreak of mountain pine beetle (*Dendroctonus ponderosae*). This expected episodic rather than chronic forest disturbance had no apparent precedent in BC's natural systems, and it was unclear whether forest licensees could manage their forests in the manner intended by the caribou conservation policy in the context of the insect outbreak. Since the original study was conducted, the episodic nature of mountain pine beetle disturbance has killed most of the overstory tree layer in many of the UWRs.

7.6.2 Designations of Ungulate Winter Range

The first submission to the BC government for conservation of UWR was made by the RIG for pine-lichen winter range in 2005 using the expert-based BBN results. The submission was preceded by a collaborative workshop to develop management actions for the UWR (which totaled 360,029 ha). Because workshop participants were mostly RIG members, the participants were familiar with the background to the submission and the meeting progressed with little preamble. A subsequent submission to the BC government for conservation of UWR is still being prepared for high-elevation winter range totaling 877,087 ha. This second submission was developed under contract rather than during a collaborative workshop. The difference in approach was largely due to the BC government's perception that the contract would be more efficient. Although it is too early to say for sure, it seems as though the anticipated efficiency may not be realized because the consultation phase of the process has yet to begin.

7.7 Discussion

In general, planning for the recovery of endangered wildlife is a difficult problem with no easy solutions, particularly where the availability and future supply of natural resources are fundamental to the species as well as to the economic or recreational development undertaken by licensed stakeholders, aboriginal peoples, and the general public. Competing demands on natural resources may mean that insufficient management options exist to allow for full implementation of desirable actions to promote recovery of animal populations. Also, incomplete understanding of key ecological relationships may add to management uncertainty and to the perceived risk of failure. However, in the example described in this chapter, several intended mechanisms and a variety of unintended coincidental activities led to almost universal acceptance of using expert-based modeling as the foundation for planning and implementing management actions to promote the recovery of local caribou herds.

Expert-based information was depicted using influence diagrams and later quantified using BBNs through a series of workshops that were inclusive rather than exclusive of participants. In competitive social or economic settings, exclusion is common, but the "grass-roots" nature of the RIG initiative led to a more inclusive team environment. Professional facilitation insured that debate among experts was welcome and expected, but that consensus was eventually achieved. The graphical nature of the influence diagrams assisted this process by allowing the participants, regardless of their education or experience, to grasp at least the conceptual nature of the ecological relationships. The inclusiveness led to a sense of shared ownership of the results by stakeholders and scientists alike. Ownership was important because all stakeholders could claim pride in the product while also being in a position to defend its implementation. Open discussion and peer review enabled consensus on the final form of the BBNs for the seasonal ranges and acceptance of the mapped results. The systematic approach provided by regular workshops and formal modeling, combined with the transparent use of the expert information by using the influence diagrams, instilled confidence in and understanding of a complex ecosystem, leading to a more rational and focused discussion than what might otherwise have occurred. The BBN approach to depicting expert information provided the ability to discuss and model a comprehensive description (i.e., not limited by incomplete empirical data) of how caribou relate to their environment and of how stressors may affect their populations and their use of seasonal ranges. Although we recognized that the results could likely be sensitive to the inherent properties of the BBN (Kuhnert and Hayes 2009), the modeling team did not have time to fully evaluate the potential implications. Rather, the RIG members relied on recommended BBN construction standards (Marcot et al. 2006). Nonetheless, this combination of a formal approach with transparency led to acceptance of the expert knowledge and subsequently allowed workshop participants to identify, discuss, and make decisions about the potential implications of certain management actions (or lack thereof).

There are many alternative approaches to the implementation of management actions for conservation of seasonal range for ungulates. For example, government biologists could have simply taken the results from recent studies, determined a habitat-use model that best fit the observed relocation data (e.g., resource-selection functions; Johnson et al. 2006), and used those results to designate conservation areas (e.g., UWRs and Wildlife Habitat Areas). Such an inductive modeling approach may provide more precise identification of seasonal ranges, but the accuracy is restricted to the environmental conditions under which the animal relocations were observed. Such models are not particularly well suited for scenario planning in which the environmental conditions change, because the interactions among the descriptor variables in the model are not robust across all environmental conditions. Also, in the specialized case of declining populations, it's unlikely that inductive model results are a desirable representation of habitat-use patterns; moreover, the resulting algorithms rarely offer transparency about the actual causal ecological relationships, making it difficult for some stakeholders to understand (and therefore accept) the model. Lastly, the application described by the inductive model excludes other resource-use interests and is therefore unlikely to address the multiple-use objectives of a broader government agenda. For all these reasons, expert judgment and deductive or abductive thinking may be more suited to addressing complex environmental problems (Douglas 2006; Martin 2007).

Although the use of expert-based information may be expedient and well suited to resolving complex problems, mistakes can be made (e.g., the high-elevation winter range model was initially inadequate). Protocols for the use of expert-based information should therefore include a dedication to testing (validating and verifying) the models prior to use, at least at operational management levels (e.g., Chap. 5 and Chap. 14). Future applications of the expert-based approach used in north-central BC would benefit from a prior understanding of the potential influence of the inherent structure of the BBNs and by making measures of uncertainty more explicit in the information provided to decision-makers. Although it is sometimes impossible to envision future catastrophic changes, the RIG process would have benefited from a more serious consideration of the potential effects of the mountain pine beetle outbreak (McNay et al. 2008b).

References

BCMSR (1999) Ft. St. James land and resource management plan. British Columbia Ministry of Sustainable Resource Management, Land Use Coordination Office, Prince George, Internal Rep

BCMSR (2000) Mackenzie land and resource management plan. British Columbia Ministry of Sustainable Resource Management, Land Use Coordination Office, Prince George, Internal Rep

Bergerud AT (1992). Rareness as an antipredator strategy to reduce predation risk for moose and caribou. In: McCullough DR, Barrett RH (eds) Proceedings of Wildlife 2001: Populations. Elsevier Applied Sciences, London, pp 1008–1021

Bergerud AT (2007) The need for the management of wolves—an open letter. Rangifer Spec Issue 17:39–50

Bergerud AT, Butler HE, Miller DR (1984) Antipredator tactics of calving caribou: dispersion in mountains. Can J Zool 62:1566–1575

Cain J (2001) Planning improvements in natural resources management: guidelines for using Bayesian networks to support the planning and management of development programmes in the water sector and beyond. Centre for Ecology and Hydrology, Crowmarsh Gifford, Wallingford

Chesson J (1983) The estimation and analysis of preference and its relationship to foraging models. Ecology 64:1297–1304

Coxson DS, Marsh J (2001) Lichen chronosequence (post-fire and post-harvest) in lodgepole pine (*Pinus contorta*) forests of northern-interior British Columbia. Can J Bot 79:1449–1464

COSEWIC (2002) Canadian species at risk, November 2000. Committee on the Status of Endangered Wildlife in Canada, Environment Canada, Ottawa

Daum J (2001) How scenario planning can significantly reduce strategic risks and boost value in the innovation chain. http://www.juergendaum.com/news/09_08_2001.htm (accessed February 2011)

Delong C (2002) Natural disturbance units of the Prince George Forest Region: Guidance for sustainable forest management. British Columbia Ministry of Forests, Prince George, Internal Rep

De Vos A, Peterson RL (1951) A review of the status of caribou (*Rangifer caribou*) in Ontario. J Mammal 32:329–337

Doucette AM, McCann RK, Barrett T, Caldwell J, Fall A (2004) Caribou habitat assessment and supply estimator (CHASE): User's Guide Version 3. Wildlife Infometrics Inc., Mackenzie, Wildlife Infometrics Rep 61

Douglas G (2006) Achieving sustainable development: the Integrative Improvement Institutes™ project. http://www.jpb.com/creative/ACE_Douglas20070206.pdf (accessed February 2011)

Fall A (2003) Omineca northern caribou project harvest schedule and disturbance models for the Wolverine caribou herd area. Wildlife Infometrics Inc., Mackenzie, Wildlife Infometrics Rep 73

Fall J, Fall A (2001) A domain-specific language for models of landscape dynamics. Ecol Model 141:1–18

Franzmann AW, Schwartz CC (eds) (1998) Ecology and management of the North American moose. Smithsonian Institution Press, Washington

Giguère L, McNay RS (2007) Abundance and distribution of woodland caribou in the Chase, Wolverine, and Scott recovery plan areas. Wildlife Infometrics Inc., Mackenzie, Wildlife Infometrics Rep 225

Golder Associates (2010) Caribou "State of the Science" backgrounder: Caribou—"state of the science" 2010 update (biology, impact pathways and next steps; Alberta and British Columbia). Canadian Association of Petroleum Producers, Calgary

Hebblewhite M, White C, Musiani M (2010) Revisiting extinction in National Parks: Mountain caribou in Banff. Conserv Biol 24:341–344

James ARC, Stuart-Smith AK (2000) Distribution of caribou and wolves in relation to linear corridors. J Wildl Manage 64:154–159

Johnson CJ (2000) A multi-scale behavioural approach to understanding the movements of woodland caribou. PhD Thesis, Univ. of Northern British Columbia, Prince George

Johnson CJ, Nielsen SE, Merrill EH et al (2006) Resource selection functions based on use-availability data: Theoretical motivation and evaluation methods. J Wildl Manage 70:347–357

Johnson CJ, Parker KL, Heard DC (2000) Feeding site selection by woodland caribou in north-central British Columbia. Rangifer Spec Issue 12:159–172

Kuhnert PM, Hayes, KR (2009) How believable is your BBN? In: Anderssen RS, Braddock RD, Newham LTH (eds) Proceedings of the 18th World IMACS/MODSIM Congress, Cairns, Australia, 13–17 July 2009 pp 4319–4325 http://mssanz.org.au/modsim09

Marcot BG, Steventon JD, Sutherland GD, McCann RK (2006) Guidelines for developing and updating Bayesian belief networks applied to ecological modelling and conservation. Can J For Res 36:3063–3074

Martin R (2007) The opposable mind. Harvard Business School Press, Boston

McKay B (1997) Valleau Creek caribou study. British Columbia Ministry of Water, Land, and Air Protection, Prince George, Internal Rep

McLoughlin PD, Dzus E, Wynes B, Boutin S (2003) Declines in populations of caribou. J Wildl Manage 67:755–761

McNay RS, Brumovsky V, Sulyma R, Giguère L (2009) Delineating high-elevation ungulate winter range for woodland caribou in north-central British Columbia. Wildlife Infometrics Inc., Mackenzie, Wildlife Infometrics Rep 299

McNay RS, Heard D, Sulyma R, Ellis R (2008a) A recovery action plan for northern caribou herds in north-central British Columbia. Forest Research Extension Partnership, Kamloops, FORREX Series 22

McNay RS, Marcot BG, Brumovsky V, Ellis R (2006) A Bayesian approach to evaluating habitat suitability for woodland caribou in north-central British Columbia. Can J For Res 36:3117–3133

McNay RS, Sulyma R (2003) Aerial reconnaissance of modeled terrestrial lichen habitat units in the Scott, Wolverine, and Chase caribou herds of north-central British Columbia. Wildlife Infometrics Inc., Mackenzie, Wildlife Infometrics Rep 94

McNay RS, Sulyma R, Voller J, Brumovsky V (2008b) Potential implication of beetle related timber salvage on the integrity of caribou winter range. BC J Ecosyst Manage 9:121–126

McNay RS, Zimmerman K, Ellis R (2002) Caribou Habitat Assessment and Supply Estimator (CHASE): Using modelling and adaptive management to assist implementation of the Mackenzie LRMP in strategic and operational forestry planning. Wildlife Infometrics Inc., Mackenzie, Wildlife Infometrics Rep 55

Messier F, Boutin S, Heard D (2004) Revelstoke mountain caribou recovery: An independent review of predator-prey-habitat interactions. Revelstoke Caribou Recovery Committee, Revelstoke

Poole K, Heard D, Mowat G (2000) Habitat use by woodland caribou near Takla Lake in central British Columbia. Can J Zool 78:1552–1561

Provost F, Kohavi R (1998) Guest editors' introduction: On applied research in machine language. Machine Learning 30:127–132

Rankin ML, McNay RS (2007) An assessment of modeled high-elevation winter range in woodland caribou herd areas of north-central British Columbia. Wildlife Infometrics Inc., Mackenzie, Wildlife Infometrics Rep 224

Ray J (2005) Large carnivores and the conservation of biodiversity. Island Press, Washington

Rettie WJ, Messier F (1998) Dynamics of caribou populations at the southern limit of their range in Saskatchewan. Can J Zool 76:251–259

Schaefer JA, Veitch AM, Harrington FH et al (1999) Demography of decline of the Red Wine Mountains caribou herd. J Wildl Manage 63:580–587

Seip DR (1992) Factors limiting caribou populations and their interrelationships with wolves and moose in southeastern British Columbia. Can J Zool 70:1494–1503

Spalding DJ (2000) The early history of woodland caribou (*Rangifer tarandus caribou*) in British Columbia. British Columbia Ministry of Environment, Lands, and Parks, Victoria, Wildl Bull B-100

Terry E, Wood M (1999) Seasonal movements and habitat selection by woodland caribou in the Wolverine Herd, North-central BC Phase 2: 1994–1997. Peace/Williston Fish and Wildlife Compensation Program, Prince George, Rep 204

Thomas DC, Gray DR (2002) COSEWIC assessment and update status report on the woodland caribou *Rangifer tarandus caribou* in Canada. In: COSEWIC (ed.) Committee on the Status of Endangered Wildlife in Canada, Ottawa, pp 1–98

Thomas JW (1985) Toward the managed forest—going places that we've never been. Wildl Soc Bull 13:197–201

Wilson L, Schmidt K, McNay RS (2004) Aerial-based census results for the Takla caribou herd February 2004. Wildlife Infometrics Inc., Mackenzie, Wildlife Infometrics Rep 105

Wittmer H, McLellan B, Seip D et al (2005) Population dynamics of the endangered mountain ecotype of woodland caribou (*Rangifer tarandus caribou*) in British Columbia, Canada. Can J Zool 83:407–418

Wood M, Terry E (1999) Seasonal movements and habitat selection by woodland caribou in the Omineca Mountains, north-central British Columbia. Phase 1: The Chase and Wolverine herds (1991–1994). Peace/Williston Fish and Wildlife Compensation Program, Prince George, Rep 201

Chapter 8
Using Expert Knowledge Effectively: Lessons from Species Distribution Models for Wildlife Conservation and Management

Chris J. Johnson, Michael Hurley, Eric Rapaport, and Michael Pullinger

Contents

C.J. Johnson (✉)
Ecosystem Science and Management Program, University of Northern British Columbia, 3333 University Way, Prince George, BC, Canada V2N 4Z9
e-mail: johnsoch@unbc.ca

M. Hurley
The Natural Resources Institute, University of Manitoba, 303-70 Dysart Road, Winnipeg, MB, Canada R3T 2N2

E. Rapaport
Faculty of Architecture and Planning, School of Planning, Dalhousie University, PO Box 1000, Halifax, NS, Canada B3J 2X4

M. Pullinger
School of Geography and Environmental Studies, University of Tasmania, Private Bag 78, Hobart, TAS 7001, Australia

A.H. Perera et al. (eds.), *Expert Knowledge and Its Application in Landscape Ecology*, DOI 10.1007/978-1-4614-1034-8_8, © Springer Science+Business Media, LLC 2012

8.1 Introduction

The spatial and temporal relationships between organisms and their environments are fundamental to both theoretical and applied ecology. The heterogeneous distribution of organisms in space and time will influence most ecological relationships, including predation, competition, and resource use, and, ultimately, population dynamics and evolution (Turchin 1996). Recognizing that the science and practice of ecology involves a consideration of spatial processes, much recent research has focused on formally representing and quantifying the spatial and temporal relationships between organisms and their environments (Morales et al. 2010). One prominent area of investigation for landscape ecologists has been the development of statistical models and associated analyses that empirically represent those relationships (Elith and Leathwick 2009). This set of methods has become known as "species distribution models" (SDMs). Guisan and Thuiller (2005) define SDMs as "… empirical models relating field observations to environmental predictor variables, based on statistically or theoretically derived response surfaces."

The first development and application of SDMs can be traced back to the conceptually and mathematically simpler habitat suitability index (HSI; U.S. Fish and Wildlife Service 1981). From the late 1970s through the early 1990s, HSI models, often developed using expert knowledge or opinion, were the most common technique for describing wildlife–habitat relationships, including the expected responses to anthropogenic factors and the implications for species distributions. Beginning in the mid-1990s, spatial databases of species occurrence began to become more common. In combination with the mainstream application of geographical information system (GIS) data and associated software, and the increase in the accessibility of multivariable statistical methods, SDMs rapidly evolved in complexity, utility, and application and have been increasingly used by researchers (Fig. 8.1). Currently, practitioners can choose among a wide variety of techniques to understand and map the distribution of a species (Guisan and Zimmermann 2000; Johnson and Gillingham 2005; Johnson et al. 2006; Franklin 2010). Where empirical data are lacking or difficult to collect in a timely or cost-effective way, expert knowledge or previously published studies can be used to quantify species–environment relationships.

Consistent with the rapid growth in the application of SDMs to problems in applied and theoretical ecology has been a growing body of studies that investigated the methodological elements of the approach. Such work has focused on uncertainty and sensitivity analyses of model inputs and predictions and on the strengths and weaknesses of competing modeling frameworks (Karl et al. 2000; Burgman et al. 2001; Brotons et al. 2004; Gu and Swihart 2004; Johnson and Gillingham 2004, 2005). Although expert-based SDMs have been widely applied across taxonomic groups, are technically simple, and are sufficiently intuitive for use by management and conservation agencies, the practice and methods for developing and evaluating

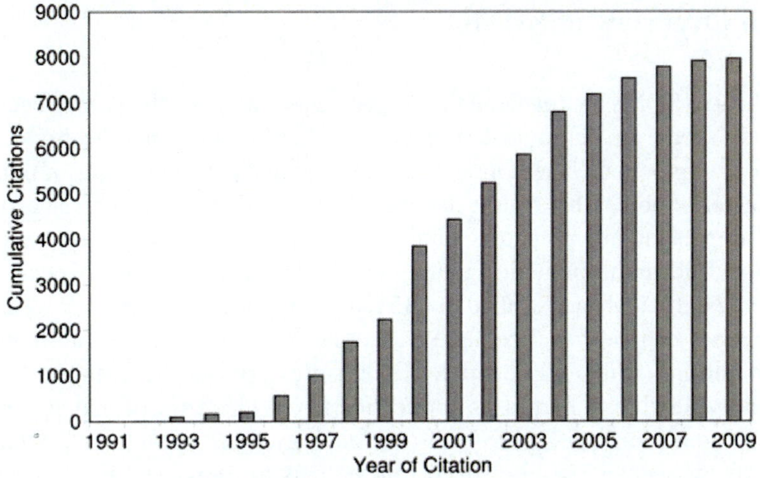

Fig. 8.1 Cumulative citations of peer-reviewed papers focused on the application or development of species distribution models. Cited papers were identified in the Web of Science database (http://www.isiknowledge.com) using the search terms "species," "distribution," "model," and "GIS," and likely only represent a portion of the total number of papers on this topic

these models have received considerable criticism (Bender et al. 1996; Roloff and Kernohan 1999; Burgman et al. 2001; Johnson and Gillingham 2004). Of particular relevance to this book, others have compared the predictive performance of expert-based versus empirical SDMs (Cowling et al. 2003; Pullinger and Johnson 2010), assessed the efficacy of combining both expert and empirical knowledge systems in a single modeling framework (Pearce et al. 2001; Mouton et al. 2009; Chap. 5), and evaluated methods for eliciting and presenting expert knowledge (Al-Awadhi and Garthwaite 2006; Czembor and Vesk 2009; Murray et al. 2009; O'Leary et al. 2009). Despite the large number of published applications of SDMs and the growing number of studies designed to evaluate and contrast methods and approaches for collecting and applying distribution data, the best practices remain uncertain.

In this chapter, we draw on our experience with SDMs to provide further guidance on developing and evaluating defensible and transparent SDMs based on both expert knowledge and empirical data. Drawing on the insights gained from three projects, we illustrate appropriate methods and pitfalls and contrast the strengths and weaknesses of the two data sources. We conclude with recommendations on how best to apply expert knowledge to maximize a study's rigor and ensure scientifically credible results that support reliable management and conservation practices. In the context of this chapter, we equate "rigor" with study methods that are systematic and repeatable and scientifically defensible, which means that the results will be logical and capable of being validated using empirical data.

8.2 An Overview of SDMs

The products of SDMs are mathematical coefficients that relate the known occurrence of the study species to ecological attractants such as a nutritional resource or some other habitat component. These models can also quantify the responses of organisms to some factor or object that results in their avoidance of some place in an ecological space. Examples include predators, habitats modified by human activities or natural disturbance, and anthropogenic features of the landscape (Johnson et al. 2005). When these coefficients are applied to GIS data, the spatial and temporal distribution of an organism can be visualized and extrapolated to new areas or future environmental conditions (Elith and Leathwick 2009). When presented as maps, the output is sufficiently flexible to permit its application to a wide range of management and conservation questions specific to landscape ecology. Such models have been developed for many species, ranging from rare lichens to extinct reptiles (Raxworthy et al. 2003; Radies et al. 2009).

8.2.1 Models Dependent on Empirical Data

SDMs can be formulated using a large variety of statistical modeling frameworks; however, all of the empirical approaches have two fundamental data requirements: species distribution data and data on environmental variables that are correlated with species occurrence (Boyce 2010). Modes of collection of distribution data include GPS collars, ground- or aircraft-based surveys of direct or indirect signs of presence, natural history databases, and georeferenced specimens from museum collections. These point locations are then related to environmental variables that are hypothesized to influence the observed variation in the organism's distribution (Bio et al. 2002; Raxworthy et al. 2003; Newbold et al. 2009; Radies et al. 2009; Hebblewhite and Haydon 2010). A statistical technique or model is then chosen to quantify the correlation between the two datasets. Depending on the sampling protocol and the research question, models can range from simple descriptions of the average niche space of the organism (Rose and Burton 2009) to more complex procedures based on maximum-likelihood estimation (Chetkiewicz and Boyce 2009).

8.2.2 Models Dependent on Expert Knowledge

When empirical data are absent, species–environment relationships can be derived from expert knowledge or through an analysis of published habitat relationships. HSIs are a good example of how expert knowledge has been applied to questions of species distribution when empirical data were unavailable. Using this technique, a number of experts who understand the ecology of the focal species are asked to provide a quantitative score representing the importance of each attribute of the environment for the organism's persistence, reproduction, productivity, or simple occupancy of a patch of habitat. Similarly, scores can be calculated from published values. Scores typically range from 0 to 1, with larger scores representing more favorable habitat conditions. When combined additively or geometrically, the scores provide a relative weighting for the combination of habitat attributes within each patch or GIS polygon found throughout the study area. Although this is an intuitive approach with hundreds of applications to a wide range of species, the methods for eliciting or deriving those scores have often been *ad hoc* and nonrepeatable. Furthermore, the final index was often not validated, or validation occurred some time after the collection of empirical data (Brooks 1997; Roloff and Kernohan 1999; Mitchell et al. 2002; Bowman and Robitaille 2005; but see Tirpak et al. 2009).

Numerous methods are available for eliciting the expert knowledge necessary to construct HSIs. These range from structured written surveys and semi-structured interviews to Delphi approaches, in which a combination of individual and group-based instruments are used to obtain each expert's perspective and develop a dialogue that might lead to a modified understanding of species–habitat relationships (MacMillan and Marshall 2006; Grech and Marsh 2008; Hurley et al. 2009). Development of effective elicitation methods is an active area of research, with the objective of testing techniques that allow experts to explore and accurately document their knowledge (Chap. 3). Ultimately, research teams should strive to increase the rigor of their study design and implementation, since this would provide a controlled and repeatable set of methods for collecting, analyzing, and presenting expert knowledge (Sutherland 2006).

8.2.3 Expert Knowledge and Multi-criteria Evaluation

Multi-criteria evaluation is a broad set of approaches for ensuring that expert-based SDMs conform to the principles of scientific rigor. These approaches are concerned with the standardization and combination of several values or criteria that can be recorded using different measures or at different scales, but that in combination influence a decision. Multi-criteria evaluation formalizes and structures the implicit decision-making process used by experts so that evaluation of the criteria is consistent across experts and is made explicit (Gal et al. 1999).

One form of multi-criteria evaluation that we have explored is the analytical hierarchy process (AHP). Using this method, researchers identify experts and ask them to consider and rank a set of hypothesized categorical criteria or variables that explain some process or pattern. In the context of SDMs, experts would be asked to relate the set of variables to the habitat or to the expected presence or absence of an organism in a particular place. Before ranking begins, the participants are asked to draw on their knowledge of the process being studied and, if necessary, amend the set of proposed variables. After finalizing the set of variables, each expert provides a pairwise ranking of each combination of explanatory variables. An ordinal scale (i.e., the number of each variable in the sequence) is used to assess the strength of one variable relative to the other variable in each paired comparison. Weights are generated from those scores and applied to GIS data to create a measure of the value of habitat or likely occurrence of a species throughout a study area. In this approach, a measure of precision can be associated with the average or median score for each variable across the experts.

Although the AHP can improve the rigor of the elicitation process, it is difficult to validate the predictions from the final model without an independent dataset. Essentially, one must attempt to determine whether the expert knowledge adequately describes the observed patterns or processes. In such situations, one approach for exploring the reliability of the final expert-based SDM is through uncertainty and sensitivity analysis. Uncertainty analysis differs from error assessment or model validation. Whereas the two latter approaches relate model predictions to observed data (i.e., the "truth"), uncertainty analysis explores the precision or range of predictions produced by the SDM. Sensitivity analysis allows the modeler to identify the parameters or inputs that are most influential in reducing the precision of model predictions. Thus, uncertainty analysis can better represent the precision of our predictions, whereas sensitivity analysis suggests areas of improvement for data inputs or the model's structure (Crosetto and Tarantola 2001).

In this chapter, we present three case studies that illustrate both poor practices and rigorous methods for developing expert-based SDMs. The first case study explores a series of HSIs from British Columbia, Canada, and reveals the pitfalls of poor study design and implementation. We discuss how uncertainty and sensitivity analyses can assist with an evaluation of the utility of model predictions and the diagnosis of flawed methods. Next, we present two studies from British Columbia that apply the AHP: first, to better understand the locations where collisions between moose (*Alces alces*) and motor vehicles occur and the factors that resulted in these collisions, and second, to predict the location of corridors used by migrating woodland caribou (*Rangifer tarandus caribou*). Sets of parallel empirical data let us compare the performance of expert-based and empirical SDMs. Based on these case studies, we discuss the strengths and weaknesses of the AHP, and compare the advantages and drawbacks of using expert-based or empirical data to construct SDMs.

8.3 Expert Knowledge and SDMs: A Plea for Rigor

Across much of British Columbia, SDMs are used to guide industrial development that accounts for ecological values (Chap. 7). The location and maintenance of habitats for regionally important or protected species is an important consideration during the exploration for and extraction of oil and gas reserves, the harvesting of forest products, and the development of mineral resources. In an effort to address the habitat needs of wildlife, the provincial government has developed a two-pronged inventory approach that maps vegetation communities and rates those communities in terms of their habitat potential for select wildlife species (RIC 1999). The vegetation mapping is based on existing ecosystem and topographical data, in combination with field investigations and the interpretation of large-scale aerial photos or other remotely sensed land cover data. Rating the habitat value for those polygons is a complex process. Given a lack of empirical data for the majority of wildlife species, the provincial environmental agency has developed a standardized wildlife habitat ranking system based mostly on existing information and, when knowledge of a species is insufficient, contributions from experts.

For each regionally important species, seasonal habitat attributes are identified through a review of existing studies, available data, and expert knowledge or opinion. At this point in the process, a species "account" is generated to document the ecology and habitat requirements of the focal species. Following the production of the species account, rating tables are generated that give each habitat attribute a score that expresses its suitability based on the criteria in the account. Scores are combined to provide an overall rating for a GIS polygon. The habitat rating is meant to represent the potential of part of an ecosystem to support a particular species during a specified season and for a specified activity (e.g., reproduction) compared to the best habitat for this purpose in the province (RIC 1999). In British Columbia, this process is known as "Wildlife Habitat Ratings," but the logic and approach is similar to the HSI approach.

Much effort was expended by the provincial government to develop a computerized template for generating the rating tables. Presented as a spreadsheet, the tables let biologists and land managers easily review the scores for each habitat attribute and determine the overall influence of each attribute on the overall rating for a polygon. Unfortunately, the process for generating the ratings is not nearly as well developed or transparent. As demonstrated below, there is no repeatable method for eliciting expert knowledge or analyzing existing information. The practitioner therefore has no way of evaluating the final SDMs or the degree of uncertainty in their predictions.

In northern British Columbia, exploitation of significant deposits of natural gas has resulted in rapid modification of landscapes. Activities associated with seismic exploration and the extraction and transport of gas can potentially displace sensitive wildlife species and reduce or adversely modify their habitats (Sorensen et al. 2008). That broader zone of industrial activity encompasses a collection of parks and special management zones known as the Muskwa-Kechika Management Area (MKMA). At 6.4 million ha, the MKMA is one of the largest intact parcels of wilderness south of the 60th parallel. Although this area is subject to greater levels of conservation planning and protection, there is considerable interest in the rich gas deposits found within and adjacent to the MKMA.

Generation of species accounts and rating tables for 1.2 million ha of the MKMA was performed using three sources of information. A small team of non-local biologists reviewed the existing information, which was largely composed of peer-reviewed published studies, and developed species accounts to define the basic life-history strategies and ecology of ten regionally important species. Preliminary rating tables were generated from this analysis of the existing information. The team of analysts then used the data collected at 116 field plots and during a 1-day workshop with local experts to refine the rating tables (EBA Engineering 2002a).

Despite the apparent thoroughness of the rating tables and the final predictive maps, the process for populating the ratings tables cannot be considered rigorous (EBA Engineering 2002a, b). First, there was no systematic method for developing the ratings. The species accounts were based on the available literature, but there was no process for translating the knowledge or data within that literature into quantitative scores. Likewise, there was no method for summarizing the information elicited during the expert workshop or the field data, and no method for applying that information to calculation of the ratings. This lack of consistent methods prevented a systematic review of the approach and ultimately meant that the habitat ratings were accepted based on the best judgment of the authors of the report. Neither industry, government, and conservation groups, nor concerned citizens could investigate the inherent precision of any of the values in the rating tables or, more fundamentally, their origin.

One could also criticize the intensity of the sampling applied during the knowledge and data collection. Given the size of the study area, the range of ecosystems it contains, and the number of target species, it is difficult to accept that 116 field plots could document the ecology of ten species as diverse as the mountain goat (*Oreamnos americanus*), American marten (*Martes americana*), and bay-breasted warbler (*Dendroica castanea*). Such efforts might provide highly specific data that resolve a few uncertainties, but are unlikely to be generally applicable to the range of attributes captured in the rating tables for thousands of ecosystem associations. Similarly, the 1-day workshop designed to consult experts was unstructured and lacked any of the attributes of a rigorous method. No definition of "expert" and no qualifications for participation were defined before selecting the experts. Discussion of the habitat requirements of each focal species was limited to approximately 40 min, and was insufficient in depth, length, and structure to allow for an adequate evaluation of the ratings for each habitat attribute used to construct the rating tables (EBA Engineering 2002a). Based on the agenda and transcriptions of the conversations during the workshop, it is impossible to link the expert knowledge to the final ratings. The failure to adopt a systematic method of data collection and analysis is disturbing from scientific and public-policy perspectives: the MKMA has significant wilderness value, the proposed resource development is known to negatively affect wildlife (Sorensen et al. 2008), and considerable financial resources were required to complete the project.

The method used to develop the habitat ratings for the MKMA appears to have been based on a subjective interpretation of the existing information as well as on informal expert knowledge and/or opinion. Despite the best intentions of the authors and supporting agencies, and considerable innovation shown in some aspects of the work, this is likely to be a worst-case example in terms of the methods used for the elicitation and analysis of expert knowledge. Fortunately, this is not an inherent characterization of all such studies. Existing data and structured elicitation of expert knowledge can have much utility in landscape ecology (Clevenger et al. 2002; Chap. 5 and Chap. 6). Furthermore, when researchers develop and implement appropriate methods, these information sources can meet the test of rigor as applied to empirical work. In the absence of a thoughtful study design, however, one can question the validity and usefulness of findings based solely or primarily on expert knowledge. This data source can be biased so that it under- or overrepresents certain geographic areas or time periods, and many elicitation methods allow a single dominant personality to influence the knowledge provided by other study participants during group discussions (MacDougall and Baum 1997). Thus, it is important to correctly identify experts with a sufficient breadth or depth of experience to address the research question, to use methods that effectively capture their knowledge, and to identify the uncertainty inherent in the elicitation process and in the resulting data obtained from the experts (Czembor and Vesk 2009; Chap. 2).

As is the case with the elicitation of expert knowledge, there are limitations and risks when collecting and using existing secondary information. Definitions and key terminology must be understood to confirm whether the data are directly comparable between studies. The precision and accuracy of the reported data and, ultimately, the conclusions that are based on it, will vary across studies. Probably of greatest concern for the development of SDMs is the geographic and temporal scope of inference in published research. Wildlife populations of the same species vary both spatially and temporally in their ecology; thus, the results of even the most rigorous scientific study may not apply to a different geographic population or time period.

Recognizing the potential conservation implications of incorrect habitat ratings and the impacts of the resulting resource development activities, we used uncertainty and sensitivity analyses to explore the range of plausible predictions from the rating tables developed for the MKMA (Johnson and Gillingham 2004). We assumed that as uncertainty in the predictions from the models increased, the specificity of the predictions for regulating the timing or location of development activities would decrease. Uncertainty and sensitivity analyses are often conducted using approaches such as Monte Carlo simulations. For this method, many predicted scores are generated via repeated sampling from a range of plausible parameter values defined by some distributional parameters (e.g., the mean and standard deviation). The distribution of the predicted scores then provides insights into the most influential variables and the uncertainty inherent in the predictions. The rating tables for the MKMA included only point estimates. Poor study design prevented the authors who created the tables from reporting a variance for their estimates of species responses. As such, we were forced to assume a range of values for each parameter and to test a number of possible distributions for those values.

Following our uncertainty and sensitivity analyses, we found that the mean ratings generated through our simulation diverged considerably from the reported ratings (Johnson and Gillingham 2004). When uncertainty was introduced, reported ratings that were near 1 (i.e., excellent habitat) or 0 (i.e., unsuitable habitat) were consistently biased towards values in the middle of the range. We also discovered that the sensitivity of habitat ratings to the model parameters varied across the study area. When applied to spatial data in the form of a species distribution map, we demonstrated that the inherent uncertainty could result in predictions for a patch of habitat that varied by up to two levels within a six-level rating system. This suggests, for example, that a polygon rated as valuable habitat could actually be marginal habitat, potentially leading to lost opportunities for oil and gas development

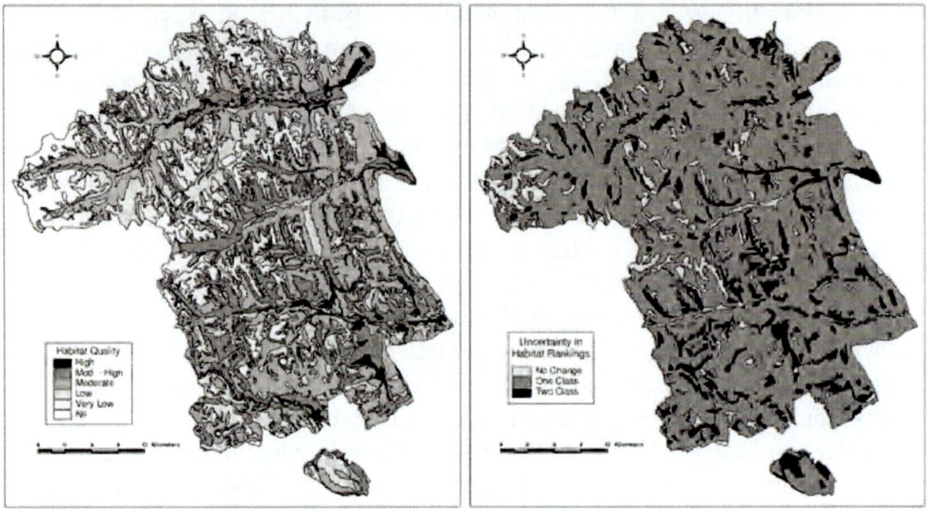

Fig. 8.2 Results of the spatial uncertainty analysis for expert-based ratings of woodland caribou habitat for one pre-tenure planning area for oil and gas exploration in the Muskwa-Kechika Management Area of northeastern British Columbia, Canada. *Left*: results without accounting for uncertainty. *Right*: magnitude of the change in the ratings when uncertainty was accounted for

(Fig. 8.2). More distressing from a conservation perspective, after considering the assumed uncertainty in the interpretation of the existing literature and expert opinion, areas ranked as good to excellent habitat could be incorrectly rated as being of sufficiently low quality to allow resource development, leading to serious negative implications for a species of conservation concern.

This case study illustrates the complexity of and difficulties in attempting to combine multiple sources of information when developing SDMs. Uncertainty and sensitivity analyses can reveal flaws in model development or uncertainty in expert opinion, but such approaches are not a substitute for proper ab initio study design. As a starting point, we recommend methods that force the expert or analyst to explicitly document the thought process used to score various habitat variables. The AHP is one such approach, but other, more sophisticated, methods are available (Chap. 3). The application of existing secondary information could involve similar structured processes, where experts are asked to use the existing literature directly to inform their estimates.

8.4 Expert Knowledge and Empirical Data: Systematic Methods and a Comparison

Although there have been great advancements in methods for collecting ecological data (Hebblewhite and Haydon 2010) there are still situations where an SDM is required, but suitable distribution data are not immediately available. Thus, a series of methodological and philosophical questions must be confronted before the SDM can be developed. Expert-based models may be less expensive and less time-consuming, but may suffer from inadequate validation (Doswald et al. 2007; Grech and Marsh 2008). Alternatively, there may be logistical constraints on collecting empirical data for species that occur at low densities or that are very difficult to capture or monitor. Expert-based and empirical data also differ in their capacity to guide population-level inferences. Expert knowledge may capture the behavior and habitat affinities of a species over a long time period for a relatively undefined area, whereas empirical data typically represents a much shorter period, but the spatial representation is explicit. Finally, as a product of their training and professional belief systems, researchers from the natural sciences may have a philosophical bias toward empirical data. Some see expert knowledge as useful for developing hypotheses or implementing complex management or conservation plans, but not as a source of repeatable quantitative data for testing hypotheses or parameterizing models (Hiddink et al. 2007). Bayesian methods have shown some promise for integrating expert knowledge with empirical data (Kuhnert et al. 2010; Chap. 5). However, time constraints and limited budgets, especially in the case of conservation, often prevent parallel data collection efforts.

Consistent with the efforts of others (Pearce et al. 2001; Bowman and Robitaille 2005), we developed two studies to better understand the relative merits and limitations of expert knowledge. First, we worked with wildlife managers and park wardens in southeastern British Columbia to understand the factors that influenced the likelihood of moose colliding with motor vehicles. We focused our efforts on the portion of the Trans-Canada Highway that traversed Mount Revelstoke and Glacier national parks. This was an excellent study location because we could engage a group of professionals who were legally mandated to monitor human and animal safety in the parks. In addition, the national parks maintain a relatively complete and accurate record of the location and timing of wildlife–vehicle collisions as well as the species involved.

As our second case study, we developed an SDM for woodland caribou found in central British Columbia. Here, we wanted to evaluate the efficacy of least-cost-path models to predict the location of corridors that sustain the movement of caribou (Pullinger and Johnson 2010). Least-cost-path models are a common tool for designing or identifying movement corridors that maintain or increase landscape connectivity for wide-ranging mammals (Chetkiewicz and Boyce 2009). The paths are constructed using a number of GIS algorithms that identify the lowest-cost routes for a species according to a landscape's permeability (i.e., openness to travel by the animal), which may be related to physical boundaries, topography or the distribution of predators and food. Landscape permeability is often the inverse of a species distribution map that is generated using empirical or expert-based data (Chetkiewicz and Boyce 2009). Woodland caribou travel long distances annually, and are a species of conservation concern. As such, the species is an excellent model for testing the accuracy of least-cost paths, and there are many professionals with considerable knowledge of the distribution and habitat requirements of caribou.

8.4.1 Developing SDMs Based on Expert Knowledge and Empirical Data

For both case studies, we developed and compared the predictive accuracy of paired empirical and expert-based SDMs. For the empirical models, we used resource selection functions to contrast the recorded locations of moose–vehicle collisions and large-scale movements of caribou with randomly selected locations (Hurley et al. 2007, 2009; Pullinger and Johnson 2010). For each modeling exercise, we tested a set of explanatory variables that we hypothesized could explain the distribution of monitored caribou and the location of moose collisions.

The first step in developing the expert-based models was to define and then identify experts who were willing to participate in our studies. For the study of moose–vehicle collisions, we defined an expert as an individual with career-based knowledge (i.e., *expert practitioners*; Chap. 1) of moose movements, habitat requirements, or the factors leading to moose–vehicle collisions. We used published literature and our knowledge of this issue to identify a core group of experts. Later, we called on that initial group to nominate additional experts for participation in the study (i.e., the snowball technique; Lewis-Beck et al. 2004). We hypothesized that local knowledge of a study area would increase an expert's understanding of the factors that influence moose–vehicle collisions, thus we divided the experts into local and non-local groups. In generating the list of experts for the study of least-cost paths for the caribou, we invited biologists or managers who had published a research report or paper on the ecology of northern woodland caribou or, through their professional duties, had spent >5 years contributing to the development of management guidelines for woodland caribou.

We used the AHP to formally elicit the knowledge of these experts relative to the factors that we believed would influence the location of moose–vehicle collisions and the distribution of caribou moving between their seasonal ranges. For both studies, we contacted the experts and briefed them on the objectives of the research. We then asked them to review the set of explanatory variables and complete a structured survey that resulted in a pairwise weighting for each combination of variables. The surveys were completed independently by each participant.

8.4.2 Comparing the Effectiveness of the SDMs

The majority of the experts accepted our invitation to participate in the two studies. For the moose–vehicle collision research, all five local and all five non-local experts participated. For the study evaluating the least-cost-path models, ten experts were invited, but six participated. We found that the experts were much better at predicting moose–vehicle collisions than they were at predicting the habitats used by caribou during large-scale movements (Hurley et al. 2009; Pullinger and Johnson 2010). Non-local experts were slightly more effective at predicting moose–vehicle collisions

Table 8.1 Predictive ability of expert-based and empirical species distribution models relating moose–vehicle collisions to driving and habitat conditions (Hurley et al. 2007, 2009)

Data type	Expert type	Model scenario	Area under curve
Expert	Local experts	Habitat – GIS data	0.829
Expert	Local experts	Driving conditions	0.767
Expert	Non-local experts	Habitat – GIS data	0.804
Expert	Non-local experts	Driving conditions	0.790
Empirical	NA	Habitat – GIS data	0.960
Empirical	NA	Habitat – field data	0.702
Empirical	NA	Driving conditions	0.630
Empirical	NA	Highway design	0.462

The predictive ability was measured using the area under the curve for the receiver operating characteristic (ROC). This statistic ranges from 0 to 1 and measures the ability of the predictive model to discriminate between locations where vehicle collisions were recorded and were not; a score >0.7 suggests a useful predictive model. *NA* not applicable

when they considered factors related to driving conditions, whereas local experts had slightly higher predictive scores when they considered the habitat conditions for moose (Hurley et al. 2009). The majority of the empirical models only slightly outperformed the expert models; there was little evidence to suggest that either source of data was superior (Table 8.1). However, the empirical SDM was much more effective at predicting the locations used by caribou for their large-scale movements. Using a set of independent validation data, we found no significant relationship between the observed locations of caribou movements and areas that experts had predicted caribou would use as corridors (Pullinger and Johnson 2010).

We were not surprised to find that the expert-based models performed inconsistently across projects. Where researchers have conducted similar paired studies, there was often conflicting evidence, with some support for and some refutation of the utility of expert-based models. Across a range of species, Yang et al. (2006), Rubin et al. (2009), and Ready et al. (2010) found that expert models performed well in both an absolute sense and relative to empirical models. In contrast, Pearce et al. (2001) and Mouton et al. (2009) found little improvement in the predictions of models that included expert knowledge. We noted similar results when predicting the distribution of woodland caribou during the winter (Johnson and Gillingham 2005).

8.4.3 Effective SDMs: Expert or Empirical Data?

Various factors may explain the difference in performance between the expert-based models that we tested. The number of experts was larger for the project on moose–vehicle collisions, and comprised a total of 185 years of experience. In comparison, the six experts who described the caribou movement routes had only 70 years of cumulative experience. Others have noted differences in the predictive capacity of local versus non-local experts. Doswald et al. (2007) found that depending on the study area, local experts were marginally better than authorities with a broader level of knowledge at predicting the distribution of lynx (*Lynx lynx*). Although we noted some differences in the predictive performance of the moose–vehicle collision models across expert groups, the differences were not striking. Given the regional variation in habitat and topographic conditions for individual caribou herds, we suspected that some knowledge of the study area was essential. Thus, we engaged only local experts for that project. The most parsimonious explanation for the variation in performance of experts across the two studies is likely that some ecological relationships are easier to address with expert knowledge. The socioeconomic effects of the moose–vehicle collisions, the long-term management emphasis, and perhaps the contribution of non-expert knowledge, such as the personal and non-professional observations of collision locations, may have led to better-informed experts.

Our results and those published by others do not provide compelling evidence to suggest that empirical models are superior to expert-based models for all research questions or applications. There are many good examples of each model type that fit species distribution data well or poorly. Thus, validation or sensitivity and uncertainty analysis are an important component of any SDM project. As a starting point, however, a transparent and repeatable method is essential when using expert knowledge. As we have demonstrated and others have noted, past efforts at applying expert knowledge to questions of conservation and management were burdened by unstructured approaches and a lack of rigor (Sutherland 2006). This is not to say that the AHP or any other method is beyond reproach. For a relatively small number of criteria, the AHP is quick and cost-effective to administer and the resulting weights are easy to implement as species distribution maps. Scores can be presented with a measure of central tendency and variance, and can be stratified between expert groups (e.g., local versus non-local). In most cases, this method is much more rapid than projects based on empirical species distribution data. There are a number of important considerations, however, when developing expert-based weights; most importantly, the weights must reflect the relative importance and scale of each criterion with respect to the other criteria under consideration (Edwards and Barron 1994). In addition, the choice and range of criteria can strongly influence the resulting weights and the final model predictions.

8.5 Experts and SDMs: Guidance for Best Practices

We have fit SDMs to a large number of datasets, using multiple methods for a range of species (e.g., Johnson et al. 2004, 2005; Hurley et al. 2007; Radies et al. 2009). In particular, we have taken considerable interest in exploring the methodological differences among and the advantages and drawbacks of individual techniques (Johnson and Gillingham 2005, 2008; Hurley et al. 2009; Pullinger and Johnson

2010). Although the ideal outcome from this work would be a general and broadly applicable recommendation of one technique, we found none that was optimal for all situations. Our conclusion is confirmed by the growing body of work that has demonstrated variable results depending on the context and choice of model (Elith and Graham 2009). There are numerous data sources, sampling strategies, and statistical techniques available to researchers. Each combination of data and technique may or may not result in a useful and defensible model. Where the criterion is the prediction of an organism's distribution or habitat, our work suggests that there is no inherent advantage to choosing models based on empirical data. Clearly, expert knowledge can result in effective SDMs. Furthermore, expert-based projects have a number of indirect secondary benefits, including engagement of a diverse set of experts, knowledge generation, and overall "buy in" from participants tasked with the job of applying results and recommendations (Chap. 6).

Based on our work, we can provide some general guidance for improving the practices used to elicit information from experts, thereby leading to more defensible SDMs. As a starting point, project teams should develop a set of methods and a study design that adhere to the principles of rigor: the elicitation and analysis should be well-documented, transparent, and repeatable, and each question used to elicit responses should be pre-tested (validated) to ensure that its meaning is unequivocal (Chap. 2). Experts are human subjects and deserve the same level of methodological insight as other plant and animal species. A very simple point that is often neglected is the precise definition of the nature of expertise. The inference from expert-based studies is often criticized as weak because there is no appreciation of the knowledge and, ultimately, of the implicit data held by the participants. Knowledge can be confused with opinion, further weakening the strength of findings from such studies.

Methods become more transparent and the study becomes more defensible if the researchers develop a rigorous definition of expertise and adopt a systematic method for identifying study participants. Experts should be selected to participate according to a set of criteria, provided those criteria do not reduce the diversity of the knowledge base and thereby bias the results. As we demonstrated, experts may nominate other experts. The latter non-random "snowball" approach can confirm the definition of expert and potentially avoid researcher bias during the selection process, but is not without criticism (Lewis-Beck et al. 2004).

After identifying experts and obtaining their participation, the research team must choose a method for eliciting and documenting their knowledge. We presented the AHP as one method that is applicable to a wide range of research questions and applications (Doswald et al. 2007). Others in this book have presented different, but equally rigorous approaches that have been applied to the development of SDMs (Chap. 5 and Chap. 7). Of primary importance is that the method be systematic and transparent. This includes a clear and unbiased process for quantifying the weights or coefficients for the various habitat attributes defined by the study participants. Our largest criticism of the SDMs developed for the MKMA was the failure to adopt a systematic method for populating the rating tables. Such *ad hoc* or subjective approaches do not permit outside review of the process or final results; the ability to permit such scrutiny is a hallmark of good science.

There is much opportunity and need for secondary investigations of expert-based data. We explored differences in knowledge reported by geographically distinct expert groups, but other sources of variance can be considered, such as the number of years of formal education or professional experience. Although such investigations have considerable benefits, a measure of inter- and intra-group variance is only possible if the research team employs confidential individual-based elicitation methods. Focus groups or Delphi approaches will likely smooth out or completely eliminate variation in the responses of experts; however, these differences of opinion often reveal fruitful areas for future research (e.g., to identify the factors that led to the differences of opinion). On the other hand, methods designed to seek consensus may assist with the implementation of expert knowledge for management or conservation because the dialogue process can help participants to clarify their reasoning and better understand the perspective of other practitioners (Chap. 6). An estimate of variance, however, is normally a requirement when reporting any measure of central tendency, such as the mean response. This simple summary statistic will reveal the degree of confidence we can place on the expert knowledge and can potentially reveal areas of improvement in the elicitation process.

The assessment of model performance is the final step in developing an SDM, especially where the goal is prediction rather than an explanation of a process. The methods for such approaches are extensive and have been reported elsewhere, but they all require empirical data (Fielding and Bell 1997; Allouche et al. 2006). We were fortunate to have such data when fitting the SDMs described in this chapter. Where data are lacking, an independent set of expert knowledge may provide some evidence that the elicited information is not idiosyncratic or that predictions generalize to other study areas (Chap. 13). Formal uncertainty and sensitivity analyses can provide approximate confidence intervals for model predictions and can reveal problematic model parameters.

As with any scientific endeavor, the choice and development of methods should be dictated by the research question or the application of the results, although it will also depend on the technical aptitudes of the modelers and of the practitioners who must apply the results. Where time and financial resources are not limiting and the research team has strong quantitative skills, empirical data and associated methods may be appropriate. Alternatively, where time and resources are limited and the research team has some experience working with human subjects, an approach based on expert knowledge may be preferred. However, as we have demonstrated in this chapter, and as others have noted, the need to use expert knowledge is not a reason to abandon basic scientific principles (Sutherland 2006). Elicitation and data extraction must be systematic, transparent, and repeatable, and must produce results that can be validated. Analysts must strive for rigor, as we would for any scientific process. Anything less reduces the certainty and reliability of the findings, creating the potential for incorrect inferences and damaging or inefficient management or conservation decisions.

References

Al-Awadhi SA, Garthwaite PH (2006) Quantifying expert opinion for modeling fauna habitat distributions. Computational Stat 21:121–140

Allouche O, Tsor A, Kadmon R (2006), Assessing the accuracy of species distribution models: prevalence, kappa and the true skill statistic (TSS). J Appl Ecol 43:1223–1232

Bender LC, Roloff GJ, Haufer JB (1996) Evaluating confidence intervals for habitat suitability models. Wildl Soc Bull 24:347–352

Bio AMF, De Becker P, De Bie E, et al (2002) Prediction of plant species distribution in lowland river valleys in Belgium: modeling species response to site conditions. Biodivers Conserv 11:2189–2216

Boyce MS (2010) Presence-only data, pseudo-absences, and other lies about habitat selection. Ideas Ecol Evol 3:26–27

Bowman J, Robitaille JF (2005) An assessment of expert-based marten habitat models used for forest management in Ontario. For Chron 81:801–807

Brooks RP (1997) Improving habitat suitability models. Wildl Soc Bull 25:163–167

Brotons L, Thuiller W, Araujo MB, Hirzel AH (2004) Presence-absence versus presence-only modelling methods for predicting bird habitat suitability. Ecography 27:437–448

Burgman MA, Breininger DR, Duncan BW, Ferson S (2001) Setting reliability bounds on habitat suitability indices. Ecol Appl 11:70–78

Chetkiewicz CLB, Boyce MS (2009) Use of resource selection functions to identify conservation corridors. J Appl Ecol 46:1036–1047

Clevenger AP, Wierzchowski J, Chruszcz B, Gunson K (2002) GIS-generated, expert-based models for identifying wildlife habitat linkages and planning for mitigation passages. Conserv Biol 16:503–514

Cowling RM, Pressey RL, Sims-Castley R, et al (2003) The expert or the algorithm? Comparison of priority conservation areas in the Cape Floristic Region identified by park managers and reserve selection software. Biol Conserv 112:147–167

Crosetto M, Tarantola S (2001) Uncertainty and sensitivity analysis: tools for GIS-based model implementation. Internat J Geogr Inf Sci 15:415–437

Czembor, CA, Vesk, PA (2009) Incorporating between-expert uncertainty into state-and-transition simulation models for forest restoration. For Ecol Manage 259:165–175

Doswald N, Zimmerman F, Breitenmoser U (2007) Testing expert groups for a habitat suitability model for the lynx *Lynx lynx* in the Swiss Alps. Wildl Biol 13:430–446

EBA Engineering (2002a) Ecosystem mapping with wildlife interpretations to support oil and gas pre-tenure planning in the Muskwa-Kechika Management Area—wildlife report. British Columbia Ministry of Energy & Mines, Fort St. John

EBA Engineering (2002b) Predictive ecosystem mapping (PEM) with wildlife habitat interpretations to support oil and gas pre-tenure planning in the Muskwa-Kechika Management Area. British Columbia Ministry of Energy & Mines, Fort St. John

Edwards W, Barron FH (1994) Smarts and smarter: improved simple methods for multi-attribute utility measurement. Organ Behav Hum Dec 60:306–325

Elith J, Graham C (2009) Do they? How do they? WHY do they differ? On finding reasons for differing performances of species distribution models. Ecography 32:66–77

Elith J, Leathwick JR (2009) Species distribution models: ecological explanations and prediction across space and time. Annu Rev Ecol Evol Syst 40:677–697

Fielding AH, Bell JF (1997) A review of methods for the measurement of prediction errors in conservation presence/absence models. Environ Conserv 24:38–49

Franklin J (2010) Mapping species distributions: spatial inference and prediction. Cambridge University Press, Cambridge

Gal T, Stewart T, Hanne, T (eds) (1999) Multicriteria decision making—advances in MCDM models, algorithms, theory, and applications. Kluwer Academic Publishers, Norwell

Grech A, Marsh H (2008) Rapid assessment of risks to a mobile marine mammal in an ecosystem-scale marine protected area. Conserv Biol 22:711–720

Gu W, Swihart RK (2004) Absent or undetected? Effects of non-detection of species occurrence on wildlife-habitat models. Biol Conserv 116:195–203
Guisan A, Thuiller, W (2005) Predicting species distribution: offering more than simple habitat models. Ecol Lett 8:993–1009
Guisan A, Zimmermann, NE (2000) Predictive habitat distribution models in ecology. Ecol Model 135:147–186
Hebblewhite M, Haydon DT (2010) Building the bridge between animal movement and population dynamics. Phil Trans Roy Soc B 365:2303–2312
Hiddink JG, Jennings S, Kaiser MJ (2007) Assessing and predicting the relative ecological impacts of disturbance on habitats with different sensitivities. J Appl Ecol 44:405–412
Hurley MV, Rapaport EK, Johnson CJ (2007) A spatial analysis of moose–vehicle collisions in Mount Revelstoke and Glacier National Parks, Canada. Alces 43:79–100
Hurley MV, Rapaport EK, Johnson CJ (2009) Utility of expert-based knowledge for predicting wildlife-vehicle collisions. J Wildl Manage 73:278–286
Johnson CJ, Boyce MS, Case RL, et al (2005) Quantifying the cumulative effects of human developments: a regional environmental assessment for sensitive Arctic wildlife. Wildl Monogr 160:1–36
Johnson CJ, Gillingham, MP (2004) Mapping uncertainty: sensitivity of wildlife habitat ratings to variation in expert opinion. J Appl Ecol 41:1032–1041
Johnson CJ, Gillingham, MP (2005) An evaluation of mapped species distribution models used for conservation planning. Environ Conserv 32:1–12
Johnson CJ, Gillingham, MP (2008) Sensitivity of species distribution models to error, bias, and model design: An application to resource selection functions for woodland caribou. Ecol Model 213:143–155
Johnson CJ, Nielsen SE, Merrill, EH, et al (2006) Resource selection functions based on use-availability data: theoretical motivation and evaluation methods. J Wildl Manage 70:347–357
Johnson CJ, Seip DR, Boyce MS (2004) A quantitative approach to conservation planning: Using resource selection functions to identify important habitats for mountain caribou. J Appl Ecol 41:238–251
Karl JW, Heglund PJ, Garton EO (2000) Sensitivity of species habitat-relationship model performance to factors of scale. Ecol Appl 10:1690–1705
Kuhnert PM, Martin TG, Griffiths SP (2010) A guide to eliciting and using expert knowledge in Bayesian ecological models. Ecol Lett 13:900–914
Lewis-Beck MS, Bryman A, Lia TF (eds) (2004) The Sage encyclopedia of social science research methods. Sage, Thousand Oaks
MacDougall C, Baum F (1997) The Devil's advocate: a strategy to avoid groupthink and stimulate discussion in focus groups. Qual Health Res 7:532–541
MacMillan DC, Marshall K (2006) The Delphi process—an expert-based approach to ecological modelling in data poor environments. Anim Conserv 9:11–19
Mitchell MS, Zimmerman JW, Powell RA (2002) Test of a habitat suitability index for black bears in the southern Appalachians. Wildl Soc Bull 30:794–808
Morales JM, Moorcraft PR, Matthiopoulos J (2010) Building the bridge between animal movement and population dynamics. Philos Trans Roy Soc B 365:2289–2301
Mouton AM, De Baets B, Goethals PLM (2009) Knowledge-based versus data-driven fuzzy habitat suitability models for river management. Environ Modell Softw 24:982–993
Murray JM, Goldizen AW, O'Leary, et al (2009) How useful is expert opinion for predicting the distribution of a species within and beyond the region of expertise? A case study using brush-tailed rock-wallabies *Petrogale penicillata*. J Appl Ecol 46:842–851
Newbold T, Reader T, Zalat S, El-Gabbas A, Gilbert F (2009) Effect of characteristics of butterfly species on the accuracy of distribution models in an arid environment. Biodivers Conserv 18:3629–3641
O'Leary RA, Low-Choy S, Murray JV, et al (2009) Comparison of three expert elicitation methods for logistic regression on predicting the presence of the threatened brush-tailed rock-wallaby *Petrogale penicillata*. Environmetrics 20:379–398

Pearce JL, Cherry K, Drielsma M (2001) Incorporating expert opinion and fine-scale vegetation mapping into statistical models of faunal distribution. J Appl Ecol 38:412–424
Pullinger MG, Johnson CJ (2010) Maintaining or restoring connectivity of modified landscapes: evaluating the least-cost path model with multiple sources of ecological information. Landsc Ecol 25:1547–1560
Radies D, Coxson D, Johnson CJ, Konwicki K (2009) Predicting canopy macrolichen diversity and abundance within old-growth inland temperate rainforests. For Ecol Manage 259:86–97
Raxworthy CJ, Martinez-Meyer E, Horning N, et al (2003) Predicting distributions of known and unknown reptile species in Madagascar. Nature 426:837–841
Ready J, Kaschner K, South AB, et al (2010) Predicting the distribution of marine organisms at the global scale. Ecol Model 221:467–478
RIC (1999) British Columbia wildlife habitat ratings standards. V.2.0. Resources Inventory Committee, Wildlife Interpretations Subcommittee, Ministry of Environment, Lands and Parks, Victoria.
Roloff GJ, Kernohan BJ (1999) Evaluating reliability of habitat suitability index models. Wildlife Soc Bull 27:973–985
Rose NA, Burton PJ (2009) Using bioclimatic envelopes to identify temporal corridors in support of conservation planning in a changing climate. For Ecol Manage 258:S64–S74
Rubin ES, Stermer CJ, Boyce WM, Torres SG (2009) Assessment of predictive habitat models for bighorn sheep in California's Peninsular Range. J Wildl Manage 73:859–869
Sorensen T, McLoughlin P, Hervieux D, et al (2008) Determining sustainable levels of cumulative effects for boreal caribou. J Wildl Manage 72:900–905
Sutherland WJ (2006) Predicting the ecological consequences of environmental change: a review of methods. J Appl Ecol 43:599–616
Tirpak JM, Jones-Farrand DT, Thompson FR III, et al (2009) Assessing ecoregional-scale habitat suitability index models for priority landbirds. J Wildl Manage 73:1307–1315
Turchin P (1996) Fractal analysis of movement: a critique. Ecology 77:2086–2090
U.S. Fish and Wildlife Service (1981) Standards for the development of habitat suitability index models. Department of the Interior, Washington, Ecological Services Manual 103
Yang XF, Skidmore AK, Melick DR, et al (2006) Mapping non-wood forest product (matsutake mushrooms) using logistic regression and a GIS expert system. Ecol Model 198:208–218

Chapter 9
Exploring Expert Knowledge of Forest Succession: An Assessment of Uncertainty and a Comparison with Empirical Data

Michael Drescher and Ajith H. Perera

Contents

9.1 Introduction

Landscape-scale forest succession models are often used to simulate forest dynamics, and the results of these simulations are used to forecast future forest states. Such forecasts are frequently the basis for strategic decisions about forest management policy and planning. However, large gaps in empirical data stemming from insufficient sampling of the landscapes or poorly understood processes often make it difficult to design and apply the models (Kangas and Leskinen 2005). In consequence, expert knowledge is often used to supplement empirical data during the design and application of the models (e.g., Forbis et al. 2006), though its use remains mostly implicit or inadequately explained by researchers.

M. Drescher (✉)
School of Planning, University of Waterloo, 200 University Avenue West, Waterloo, ON, Canada N2L 3G1
e-mail: mdresche@uwaterloo.ca

A.H. Perera
Ontario Forest Research Institute, 1235 Queen Street East, Sault Ste. Marie, ON, Canada P6A 2E5

A.H. Perera et al. (eds.), *Expert Knowledge and Its Application in Landscape Ecology*, DOI 10.1007/978-1-4614-1034-8_9, © Springer Science+Business Media, LLC 2012

Expert knowledge, like other sources of information, carries a level of uncertainty (e.g., O'Hagan et al. 2006; Chap. 2). Approaches are available to achieve optimal forest management decisions despite uncertainty (e.g., Kangas and Kangas 2004), but these approaches require an understanding of the origin, degree, and extent of the uncertainty. However, in the majority of applications of forest landscape models, no attempts were made to quantify the accuracy and uncertainty of the expert knowledge or to describe its other characteristics, such as variability.

Expert knowledge can be a valuable source of information for many applications, including landscape-level forest succession models (Kangas and Leskinen 2005). However, if the uncertainty in expert knowledge is not acknowledged and its characteristics are not assessed, then biases and overconfidence are likely (Kahneman et al. 1982). The full range of possible forest landscape forecasts may not be considered, and the reliability of management decisions based on the model's forecasts will be largely unknown. Moreover, not only are the forecasts and decisions made under such circumstances uncertain, but also their level of uncertainty is unknown. Since well-characterized uncertainties are a prerequisite for optimal decision-making under uncertainty (Morgan and Henrion 1990), this lack of knowledge will make it difficult to achieve optimal decisions. Given the widespread use of expert knowledge in forest landscape models and its importance for generating forecasts and supporting management decisions (Kangas and Leskinen 2005), a rigorous evaluation of expert knowledge must precede applications (Davies and Ruddle 2010) to ensure that the knowledge is used prudently and efficiently (Mackinson 2001).

We pursue two goals in this chapter. First, we explore the characteristics of expert knowledge by investigating its degree of uncertainty, analyzing the possible sources of variation in uncertainty, and assessing its veracity by comparing it with empirical data. Second, we integrate expert knowledge with empirical data and develop a comprehensive body of knowledge of forest succession. We pursue these goals using a case study from the boreal forest of Ontario (Canada).

9.2 Case Study

In boreal Ontario, expert knowledge of forest succession is applied in forest dynamics simulations to support the development of forest management policies and plans. We evaluated the expert knowledge of forest succession in the forest landscape simulation model BFOLDS that is used to develop context for forest management planning (Perera et al. 2008). Since extensive details of this study can be found in Drescher et al. (2006, 2008a, b), Drescher and Perera (2010a, b), and Drescher et al. (Chap. 4), here we provide only an overview of the methods.

We explored expert knowledge holistically, beyond the simple probability estimates provided by experts (e.g., O'Hagan et al. 2006), by investigating the characteristics of their knowledge and analyzing the integrated knowledge space generated in combination with empirical data. Central to our approach is the expression of expert knowledge and empirical data in equivalent forms (i.e., on an "equal footing"; Failing et al. 2007), thereby permitting a direct comparison and a meaningful evaluation. The key steps we followed are described in Fig. 9.1.

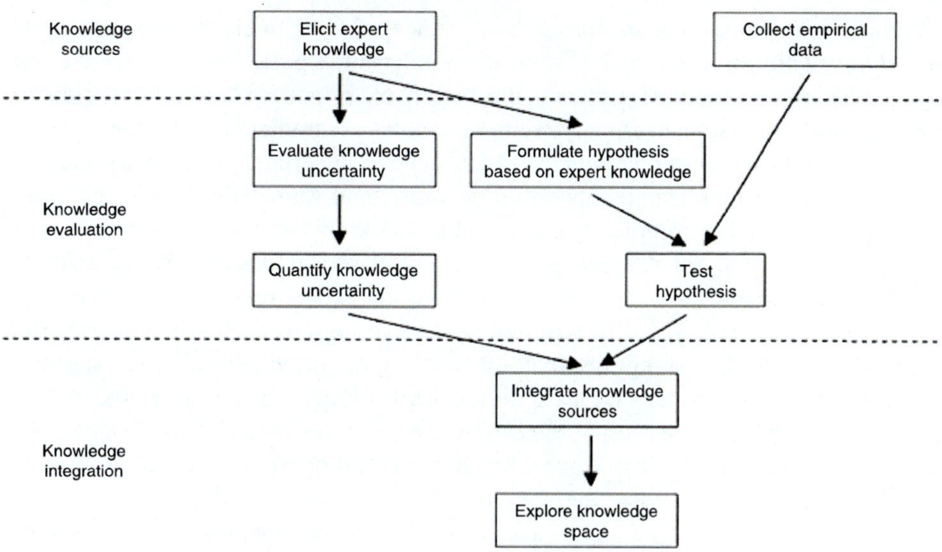

Fig. 9.1 The steps used to evaluate and test expert knowledge and integrate it with empirical observations

Forest management in boreal Ontario requires that knowledge of forests and forest succession be expressed as forest types and forest-type transitions. Transitions among forest types as a result of aging, canopy gaps, and regeneration (i.e., *natural succession*) and regeneration that occurs after stand-replacing wildfires (i.e., *post-fire succession*) are quantitatively expressed as *succession rules*, which indicate the proportions of forest stands that are transitioning among the forest types. These rules are used to parameterize state-and-transition models such as BFOLDS (Perera et al. 2008).

9.2.1 Assessment of Uncertainty

Knowledge uncertainty may stem from various sources, including insufficient knowledge about a process, inaccurate measurements, and the stochastic nature of the systems being managed (Kangas and Kangas 2004), each of which leads to different types of uncertainty. Here, we focused on uncertainty at the level of individual experts (*individual* uncertainty) and at the level of groups of experts (*collective* uncertainty). Individual uncertainty has two dimensions: *Epistemic* uncertainty is caused by a lack of knowledge about a process, and *aleatory* uncertainty arises from the inherent stochasticity and unpredictability of a process (Haenni 2009; Chap. 2). We addressed epistemic uncertainty through its antonym, *knowledge confidence*, which we defined as the degree of trust experts have in their own knowledge. We addressed aleatory uncertainty through *knowledge complexity*, which we defined as unexplained variation in successional processes caused by stochastic factors.

We chose a single dimension for collective uncertainty, namely the *among-expert variability*. This variability is based on the number of experts who disagreed about a given succession rule, and is similar to the concept of "source conflict" (after Smithson 1999), which occurs when equally credible sources contradict one another.

We derived two more components of knowledge uncertainty: individual knowledge uncertainty (a linear combination of individual knowledge confidence and individual knowledge complexity), and composite knowledge uncertainty (a linear combination of individual knowledge uncertainty and among-expert variability). We enlisted experts (see Chap. 4, for methods of expert recruitment and expert characteristics) who resided in two administrative regions of boreal Ontario: the Northeast and Northwest regions. In Northeast Region, the experts opted for a group consensus approach rather than an individual knowledge-elicitation approach (see Chap. 4, for further discussion of the difference). For this region, knowledge confidence and complexity were group-based, so we could not derive the among-expert variation and composite uncertainty.

To elicit expert knowledge, we asked the experts to generate succession rules (ca. 40 natural succession rules, ca. 40 postfire establishment rules) for various forest types under a variety of environmental conditions. First, experts reviewed examples of existing succession rules. Second, prompted by these examples, they predicted appropriate transitions between forest types and estimated the proportions of forests that transitioned to a different type. For each of these rules, the experts self-assessed their level of confidence in their knowledge and the complexity of the knowledge using three ranks (low, medium, and high). To investigate variation in the degree of uncertainty, we grouped knowledge confidence, complexity, and among-expert variability by succession process (i.e., postfire establishment versus natural succession) and examined the distributions for each group. We also examined the potential sources of uncertainty by investigating the associations between the uncertainty characteristics (knowledge confidence, complexity, and among-expert variability) and the forest characteristics and environmental conditions using Theil's U coefficient (Theil 1972). Theil's U is a measure of the nominal degree of association between two parameters, and ranges from 0 to 1, where 0 indicates independence and 1 indicates a perfect monotonic relationship. The forest characteristics we used in this comparison were the species composition and dominant leaf type (broadleaved, mixed, or coniferous), and the environmental conditions were moisture regime (dry, mesic, wet, or saturated) and nutrient regime (poor, mesotrophic, rich, or organic).

9.2.2 Distribution of Uncertainty and Association with Explanatory Factors

The patterns of association differed between transition types and between regions. In the Northwest Region (Fig. 9.2), the among-expert variability for postfire establishment rules was significantly associated with forest type and leaf type, but not with environmental conditions (Table 9.1). Among-expert variability was high for

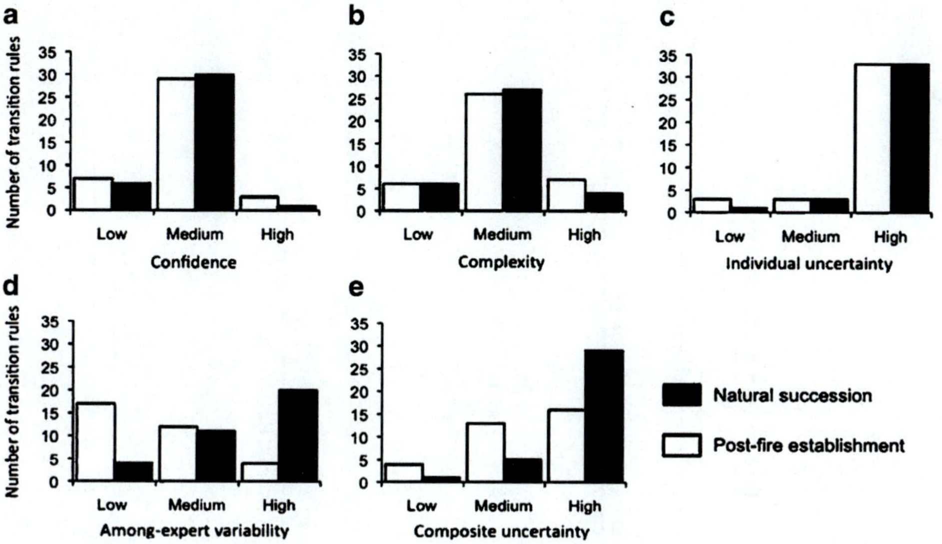

Fig. 9.2 Distribution of the components of expert knowledge uncertainty for postfire establishment and natural succession in the Northwest Region of Ontario: (**a**) individual knowledge confidence, (**b**) individual knowledge complexity, (**c**) individual uncertainty, (**d**) collective (among-expert) uncertainty, and (**e**) composite (individual plus collective) uncertainty

mixed forests, low for broadleaved forests, and very low for coniferous forests. For the natural succession rules, confidence, complexity, and among-expert variability were not significantly associated with forest characteristics or environmental conditions.

In the Northeast Region of Ontario, expert knowledge confidence for postfire establishment and natural succession was associated with forest type, but not with leaf type or environmental conditions (Fig. 9.3; Table 9.1). Confidence was lower for forests with mixed composition than for forests dominated by a single species. Knowledge confidence for postfire establishment was not significantly associated with forest characteristics or environmental conditions. However, knowledge complexity for natural succession was significantly associated with all forest characteristics and environmental conditions. Overall, knowledge complexity was considered to be lower for forests dominated by a single species and for coniferous forests, but was higher for mixed species and broadleaved forests. Knowledge complexity of natural succession was also lower under more extreme environmental conditions (e.g., oligotrophic) than under more moderate conditions (e.g., mesotrophic).

The overall distributions of confidence and complexity levels were similar in both regions: successional processes were mainly associated with medium confidence, though with a slight tendency toward higher confidence in the Northeast Region (for natural succession, $\chi^2=10.522$, df=2, $p=0.005$; for postfire establishment, $\chi^2=14.676$, df=2, $p=0.001$). At the same time, successional processes were mainly associated with medium complexity, though with a slight tendency toward lower complexity in the Northeast Region (for natural succession, $\chi^2=6.164$, df=2,

Table 9.1 Strength of the associations between expert knowledge uncertainty parameters and the forest characteristics and environmental conditions

	Strength of the association (Theil's *U*)					
	Confidence		Complexity		Among-expert variability	
	Postfire establishment	Natural succession	Postfire establishment	Natural succession	Postfire establishment	Natural succession
Northwest region						
Forest type	0.465^{ns}	0.221^{ns}	0.415^{ns}	0.295^{ns}	0.459*	0.341^{ns}
Leaf type	0.167^{ns}	0.097^{ns}	0.165^{ns}	0.044^{ns}	0.283**	0.056^{ns}
Moisture regime	0.049^{ns}	0.147^{ns}	0.024^{ns}	0.042^{ns}	0.044^{ns}	0.044^{ns}
Nutrient regime	0.022^{ns}	0.100^{ns}	0.064^{ns}	0.149^{ns}	0.021^{ns}	0.036^{ns}
Northeast region						
Forest type	0.748***	0.679**	0.546^{ns}	0.626**	–	–
Leaf type	0.123^{ns}	0.155^{ns}	0.065^{ns}	0.209*	–	–
Moisture regime	0.176^{ns}	0.062^{ns}	0.252^{ns}	0.275*	–	–
Nutrient regime	0.181^{ns}	0.100^{ns}	0.260^{ns}	0.483***	–	–

Association was quantified using Theil's *U* (Theil 1972), separated by region and succession process. Northeast Region experts only contributed as a group; it was not possible to estimate among-expert variability

ns, $P \geq 0.05$; *$P<0.05$; **$P<0.01$; ***$P<0.001$ (after sequential Bonferroni correction)

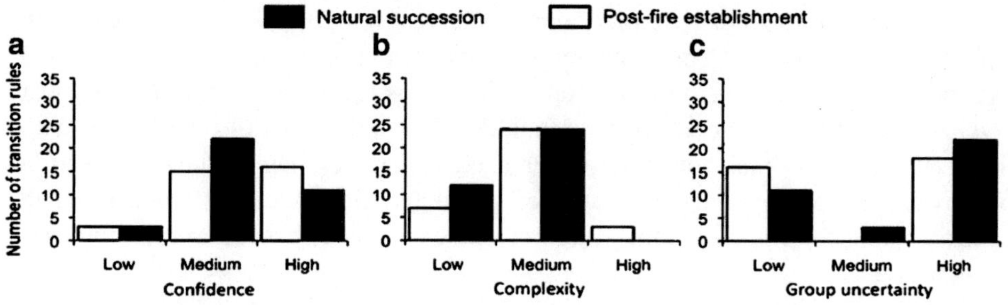

Fig. 9.3 Distribution of the components of expert knowledge uncertainty for postfire establishment and natural succession in the Northeast Region of Ontario: (**a**) group knowledge confidence, (**b**) group knowledge complexity, and (**c**) group uncertainty

$p=0.046$; for postfire establishment, $\chi^2=1.421$, df=2, $p=0.491$). However, the distribution of uncertainty differed strongly between regions (for natural succession, $\chi^2=16.039$, df=2, $p<0.001$; for postfire establishment, $\chi^2=10.522$, df=2, $p=0.005$). In the Northwest Region, individual uncertainty was high. In contrast, group uncertainty in the Northeast Region was bimodal, with a clear split between low and high levels of uncertainty. The reasons for this difference are not immediately evident. However, according to the experts, the two regions differ in their broad-scale spatial complexity. Although experts in the Northwest Region differentiate among dozens of ecologically different forest zones, those in the Northeast Region differentiate between just two zones. Of course, broad-scale complexity of forests is different from stand-scale complexity, which is the scale for which we elicited expert knowledge. However, the reported differences in broad-scale complexity between regions might nevertheless indicate general differences in the perception of expert knowledge at a variety of scales.

Among-expert variability differed between the two succession processes: it was mainly very low for postfire establishment and high for natural succession. This led to differences in composite uncertainty, which were medium to high for postfire establishment but high for natural succession ($\chi^2=9.060$, df=2, $p=0.011$). The combination of high individual uncertainty with low collective uncertainty for postfire establishment might stem from experts using a common but less reliable information source. In other words, individual experts might be less certain about their knowledge, but because they share the same knowledge source, they may all agree and appear more certain as a group.

A recurring pattern was the association between the components of uncertainty and forest characteristics. We hypothesize that this association might be the result of differences among forest types, since mixed forest types typically contain a larger number of species than coniferous and broadleaved forests. Therefore, mixed forests may have more diversity of possible succession trajectories than other forests. This may lead to lower confidence, higher complexity, and higher among-expert variability, which may increase uncertainty.

Significant associations of components of uncertainty with environmental conditions were less frequent. However, the level of knowledge complexity was associated with nutrient and moisture regimes, and was lowest for extremes in environmental conditions (e.g., wet, organic). We hypothesize that experts perceive successional dynamics under extreme environmental conditions as less diverse, resulting in lower levels of uncertainty. The stronger association of components of uncertainty with forest characteristics than with environmental characteristics may be due to perception bias by experts. We hypothesize that forest characteristics are easier to observe than environmental conditions and may therefore subconsciously lead to an overemphasis of their effects.

Our characterization of the expert knowledge space extends beyond *what* experts know to include *how well* they think they know it and *why* their knowledge may vary among experts. These findings can be used as the basis for general hypotheses about expert knowledge and to permit predictions about knowledge that has not yet been elicited. We hypothesize that uncertainty will be higher under conditions that lead to a greater diversity of successional trajectories, for example, when forests are species-rich or when environmental control over successional trajectories is weak.

9.2.3 Expert Knowledge as Hypotheses and Their Comparison with Empirical Data

We inspected the accuracy of expert knowledge of forest succession by comparing it with empirical observations; that is, we treated the elicited expert knowledge as hypotheses and tested them against empirical data. For such a test, both expert knowledge and empirical data must be expressed in an equivalent format, and the most appropriate format depends on the ecological unit of interest. Forest succession at the landscape level encompasses numerous interacting forest types and many successional trajectories that may be nonlinear and stochastic rather than linear and deterministic. Focusing on forest succession only as a bilateral process involving two forest types captures neither the complexity of this network of interactions nor the indirect successional effects when more than two forest types are sequentially connected. Capturing forest succession only at the lower level of individual forest types would therefore not adequately represent forest succession knowledge at the landscape level, which comprises all forest types and their interactions. Instead, it makes sense to express forest succession at a higher level, as a probabilistic network in which the successional transitions among all forest types are considered simultaneously.

Such a *forest succession network* represents succession at the entire landscape level. Because such a network is modular – consisting of simple building blocks of individual forest types and their bilateral interactions – it can be easily constructed from the existing, lower-level expert knowledge and empirical observations. Forest succession networks are ideally suited to integrate individual pieces of knowledge into a coherent whole and to represent it at the higher landscape level. Therefore, we chose forest succession networks as the format in which to express expert knowledge of succession as quantitative hypotheses.

We used concepts from graph theory (Harary 1969) to structurally describe and compare forest succession networks. The basic idea is to depict networks as graphs, consisting of *nodes* and *edges* that represent forest types and succession pathways, respectively. Because succession is directional, the edges between nodes have a direction, and because succession occurs with a certain probability, edges have weights, which represent the probability of succession.

An analysis of the degree distribution, which represents the number of edges entering and leaving each node, can be used to describe the network's structure. The number of edges entering a node is called its *in-degree*, and the number of edges leaving a node is called its *out-degree*. The sum of in- and out-degrees is the *total degree*, and indicates the overall level of connectedness of a forest type within the network. We compared expert depictions of forest succession networks with empirical descriptions of these networks by comparing their degree values (the in-, out-, and total degrees and their corresponding means and standard deviations). A positive difference (empirical minus expert) indicates a lower degree value for the expert network, whereas a negative difference indicates a higher degree value for the expert network.

The expert knowledge hypotheses were tested against empirical observations of forest succession based on data from forest inventory plots and forest inventory maps. Both the plots and the maps provided stand-scale information about the canopy composition and age. Using information about stand location and the year of observation, we identified repeated observations of individual stands and arranged them into several thousand time-series of canopy composition, stratified by forest zone. Based on these time series, we estimated the empirical probabilities of natural succession among forest types and connected them into forest succession networks using the same format that we used for expert knowledge. We tested expert knowledge hypotheses for seven forest zones. For each of these zones, we compared a zone-specific forest succession network based on expert knowledge with a zone-specific forest succession network based on the empirical observations.

9.2.4 *Similarity of Expert Knowledge of Forest Succession and Empirical Data*

Our analysis of the degree distribution produced many detailed results, so we have focused our discussion here on the results for a single representative forest zone surrounding Lake Nipigon, which is north of Lake Superior. For results from the remaining six forest zones, see Drescher et al. (2008a) and Drescher and Perera (2010b). The graphs of the expert and empirical forest succession networks (Fig. 9.4) appear visually similar: successional transitions in both networks move mainly from broadleaved forest types to mixed and coniferous forest types, with some successional pathways that occur exclusively among coniferous forest types.

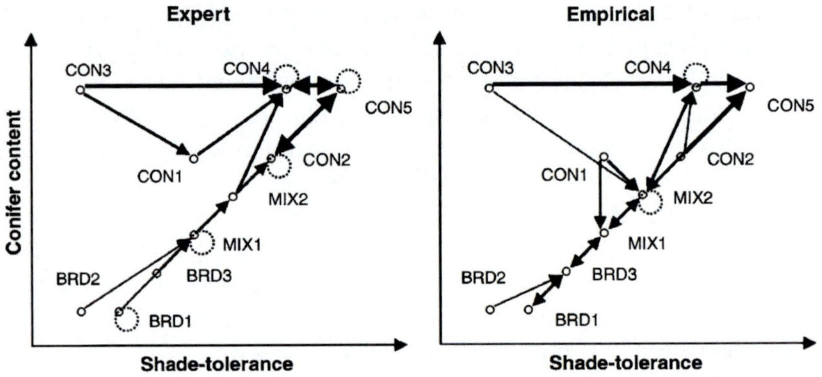

Fig. 9.4 Forest succession networks for the forests surrounding Lake Nipigon, Ontario, based on expert knowledge and empirical observations of forest succession. *Dots* indicate forest types, *arrows* indicate the direction of succession, and *broken circles* indicate self-replacement (i.e., no change in stand type). Forest types (broadleaved, BRD1–BRD3; mixed, MIX1 and MIX2; coniferous, CON1–CON5) were arranged based on their relative conifer content and shade-tolerance and their arrangement is only for illustrative purposes

However, our degree analysis indicated that the expert network was relatively simple, with low in-degrees (mean = 1.4) that showed only moderate variation (SD = 1.5) among forest types (Table 9.2). The out-degrees were also generally low (mean in- and out-degrees are always identical) and were similar among forest types (SD = 0.7). As was the case for the in-degrees, the total degrees showed moderate variation. They were generally low for broadleaved forest types, high for mixed forest types, and variable for coniferous forest types. The empirical network, on the other hand, was more complex (Table 9.2): the in- and out-degrees were higher than for the expert network (mean = 2.6), the in-degrees varied strongly among forest types (SD = 2.8), the out-degrees showed moderate variation among forest types (SD = 1.1), and the total degrees varied strongly among forest types (SD = 3.4) and were generally high for mixed forest types and either high or low for broadleaved and coniferous forest types.

The results of the degree analysis can be used to compare the two networks (Table 9.3). For most forest types, the degree differences between the expert and empirical networks indicated that the forest types in the expert network were less strongly connected than those in the empirical network. In-degree differences occasionally indicated higher connectedness for forest types in the expert networks, but out-degree differences did so only once. Degree differences were mostly positive for broadleaved and mixed forest types, suggesting lower connectedness for these forest types in the expert network. For coniferous forest types, degree differences often indicated higher connectedness in the expert network. Overall, the expert network was more uniform than the empirical network, as indicated by lower variation in all of the degree values.

Table 9.2 Results of the structural characterization of the expert and empirical networks of forest succession based on degree analysis

	Degree											
	BRD1	BRD2	BRD3	MIX1	MIX2	CON1	CON2	CON3	CON4	CON5	Mean	SD
Expert												
In-degree	0	0	0	3	1	1	2	0	4	3	1.4	1.5
Out-degree	1	1	1	1	3	1	1	2	1	2	1.4	0.7
Total degree	1	1	1	4	4	2	3	2	5	5	2.8	1.6
Empirical												
In-degree	2	0	4	5	8	0	0	0	2	5	2.6	2.8
Out-degree	4	1	4	4	2	2	2	2	2	3	2.6	1.1
Total degree	6	1	8	9	10	2	2	2	4	8	5.2	3.4

The different forest types are broadleaved (BRD1–BRD3), mixed (MIX1 and MIX2), and coniferous (CON1–CON5)

Table 9.3 Results of the structural comparison between the expert and empirical networks of forest succession based on degree analysis

	Degree difference (empirical minus expert)											
	BRD1	BRD2	BRD3	MIX1	MIX2	CON1	CON2	CON3	CON4	CON5	Mean	SD
In-degree	2	0	4	2	7	−1	−2	0	−2	2	1.2	2.8
Out-degree	3	0	3	3	−1	1	1	0	1	1	1.2	1.4
Total degree	5	0	7	5	6	0	−1	0	−1	3	2.4	3.1

The different forest types are broadleaved (BRD1–BRD3), mixed (MIX1 and MIX2), and coniferous (CON1–CON5)

9.2.5 Testing of Expert Knowledge of Forest Succession with Empirical Data

We tested the expert knowledge hypotheses by quantifying the level of similarity between the expert and empirical forest succession networks. We also wanted to investigate whether discrepancies between the two networks were mainly attributable to differences in their general structure or to differences in specific succession probabilities. Therefore, to test the similarity in the network structures and in the exact succession probabilities, we expressed the networks in binary form (i.e., with all edge weights set to one) and in probabilistic form (with each edge weight set to its exact probability), respectively. Using the empirical network as the reference point, the significance of the similarity between the two networks indicates the degree of support for or contradiction of the expert hypothesis.

We calculated the similarity between the expert and empirical networks using Pearson's product-moment correlation for networks (r). We tested the significance of this similarity by comparing it to a reference distribution of similarities between randomly created networks and the empirical networks using a Monte Carlo approach with 10,000 repetitions. The results suggested limited similarity between the expert and empirical forest succession networks, though the level of similarity depended on the network type: In binary form, the expert and empirical forest succession networks were not significantly similar ($r=0.10$; $p\geq 0.05$), but in probabilistic form, the two networks had low but significant similarity ($r=0.23$; $p<0.05$). This indicates that the expert and empirical networks differed in their general structure, and this difference was tempered by expression of forest succession in specific probabilities.

The main differences between the expert knowledge and empirical data were that (1) the empirical data showed that broadleaved and mixed forest types were linked to a larger number of other forest types than was suggested by expert knowledge, indicating a larger potential for variation in successional direction. (2) The empirical data indicated that forest types varied strongly in their centrality within the network, meaning that some forest types were much more important parts of successional pathways than other forest types. Expert knowledge, however, suggested low variation in the centrality of the forest types, meaning that most forest types had similar successional importance. (3) The empirical data suggested the presence of many successional pathways that occurred with low probability, whereas expert knowledge only indicated pathways with a similarly high probability.

Our empirical results did not strongly support the expert hypotheses. However, rather than discarding these hypotheses entirely, specific components of the hypotheses that were supported by the empirical data, such as particular successional transitions or groups of transitions, could be preserved whereas the components that received little or no support could be replaced by predictions based on empirical data. The resulting mosaic of expert knowledge and empirical data could then be tested using newly collected empirical data to support or refute the resulting predictions and lead researchers toward the best possible description of forest succession.

We can also interpret our results as indicating more general patterns of deviation between expert knowledge and empirical results. The empirical data indicated considerable diversity in forest succession at various levels, as evidenced by the large number of possible successional pathways, large variation in importance for landscape-level succession among forest types, and identification of many infrequent successional pathways. This diversity implies large variety in forest successional dynamics and limited predictability of future landscape forest states. Expert knowledge, on the other hand, points toward less diversity in forest succession, leading to lower diversity and higher predictability of future landscape forest states.

Though we did not explicitly investigate the reasons for the differences between expert knowledge and the empirical data, we suggest the following possible causes: Individual practical experience, which forms the backbone of expert knowledge, is limited by the spatial and temporal scales of each expert's personal observations (Fazey et al. 2006). Though experts may know a great deal about forest succession, their knowledge of the full range of successional transitions might be limited. Our empirical data, however, were collected over spatial and temporal scales that surpass the levels of personal experience. The empirical data might therefore capture a larger range of successional transitions, and this may explain the larger diversity of successional pathways found in the empirical data than in the expert knowledge. If this is true, then the reliable use of expert knowledge would depend on careful matching of the temporal and spatial scales of the expert knowledge with the scales of the application of this knowledge.

Another explanation may be that experts misjudged the frequency of certain successional transitions. Developing expertise involves continuous gathering of new experience and self-reflection about that experience (Fazey et al. 2006). This process, however, can be hindered by judgmental biases – what Sterman (1994) referred to as "anchoring." Given the dominance of some forest types and their successional transitions, experts may have inadvertently overemphasized these types and transitions and ignored infrequent ones. This may explain the difference in low-frequency successional transitions between the expert knowledge and empirical data.

Other approaches could be used to compare expert knowledge and empirical observations, with different levels of mathematical sophistication and consideration of different kinds of information (e.g., Kangas and Kangas 2004; Diaz-Balteiro and Romero 2008). For example, instead of rejecting components of expert knowledge that were not supported by empirical data, a Bayesian perspective could be adopted. Following this view, a hypothesis would not be falsified but instead, its degree of belief would be adapted and updated based on the empirical data (Chap. 5). Alternatively, information from both sources could be combined based on evidence theory (Kangas and Kangas 2004). This differs from Bayesian probability theory in that evidence theory does not consider the degree of support as a classical probability (e.g., probabilities of mutually exclusive events do not need to add up to 1.0). Nevertheless, based on belief and plausibility, evidence theory can define the probabilities of events. Either of these approaches differs from the mosaic of expert knowledge and empirical data that we suggested in this study, since the *mosaic* makes it possible to identify the components that stem from expert knowledge and those that stem from empirical data. In the *update* approaches based on Bayesian or evidence theory, expert knowledge and empirical data are combined, thereby losing the distinction between the two sources.

9.3 Conclusions

This assessment of expert knowledge uncertainty and accuracy provided useful insights:

1. Uncertainty is lower when abiotic control of successional dynamics is strong, such as under extreme environmental conditions.
2. Uncertainty is higher when the range of possible successional pathways increases, as in the case of mixed forest types.
3. Experts may have a simpler view of forest succession than what is empirically observed.
4. Experts may overlook rare events in forest succession.
5. In some aspects of forest succession, expert knowledge was similar to empirical data (e.g., succession of coniferous forest types), and in others, empirical data contradicted expert predictions (e.g., succession of broadleaved and mixed forest types).

The integration of expert knowledge with empirical data helped to quantify the patterns of forest succession and develop hypotheses at two hierarchical levels (low, at the level of forest types; high, at the forest landscape level), and under a range of soil nutrient and moisture regimes. Disagreements between expert knowledge and the empirical data point toward knowledge gaps and areas of limited knowledge certainty that can be points of departure for future research on forest succession. Because the body of knowledge investigated in this case study is now explicitly known, including its uncertainties, more judicious application of this knowledge is possible in forest landscape models, and this will lead to better-defined levels of confidence in forecasts of future forest landscapes.

Finally, we recommend that the method of integrating knowledge sources be quantitative, rigorous, and flexible enough to accommodate information that may differ in spatial or temporal scale, level of detail, and type of expression. Such an integrated body of knowledge, with known veracity, uncertainties, and gaps, will improve the reliability of applications, which otherwise rely on expert knowledge as *ad hoc* supplements or complements to a purely empirical body of information.

References

Davies A, Ruddle K (2010) Constructing confidence: rational skepticism and systematic enquiry in local ecological knowledge research. Ecol App 20:880–894

Diaz-Balteiro L, Romero C (2008) Making forestry decisions with multiple criteria: a review and an assessment. For Ecol Manage 255:3222–3241

Drescher MD, Perera AH (2010a) A network approach for evaluating and communicating forest change models. J Appl Ecol 47:57–66
Drescher M, Perera AH (2010b) Comparing two sets of forest cover change knowledge used in forest landscape management planning. J Environ Planning Manage 53:591–613
Drescher M, Perera AH, Buse LJ et al (2006) Identifying uncertainty in practitioner knowledge of boreal forest succession in Ontario through a workshop approach. Ontario Ministry of Natural Resources, Ontario Forest Research Institute, Sault Ste. Marie. Forest Research Report No 165
Drescher M, Perera AH, Buse LJ et al (2008a) Boreal forest succession in Ontario: An analysis of the knowledge space. Ontario Ministry of Natural Resources, Ontario Forest Research Institute, Sault Ste. Marie. Forest Research Report No. 171
Drescher M, Perera AH, Buse LJ et al (2008b) Uncertainty in expert knowledge of forest succession: A case study from boreal Ontario. For Chron 84:194–209
Failing L, Gregory R, Harstone M (2007) Integrating science and local knowledge in environmental risk management: a decision-focused approach. Ecol Econ 64:47–60
Fazey I, Fazey JA, Salisbury JG et al (2006) The nature and role of experiential knowledge for environmental conservation. Environ Conserv 33:1–10
Forbis TA, Provencher L, Frid L, Medlyn G (2006) Great Basin land management planning using ecological modeling. Environ Manage 38:62–83
Harary, F. (1969) Graph Theory. Addison-Wesley, Reading
Haenni, R. (2009) Non-additive degrees of belief. In: Huber F, Schmidt-Petri C (eds.) Degrees of belief. Springer Science+Business Media B.V., Dordrecht, pp 121–159
Kahneman D, Slovik P, Tversky A (eds) (1982) Judgment under uncertainty: heuristics and biases. Cambridge University Press, Cambridge
Kangas AS, Kangas J (2004) Probability, possibility and evidence: approaches to consider risk and uncertainty in forestry decision analysis. For Pol Econ 6:169–188
Kangas J, Leskinen P (2005) Modelling ecological expertise for forest planning calculations—rationale, examples, and pitfalls. J Environ Manage 76:125–133
Mackinson S (2001) Integrating local and scientific knowledge: an example in fisheries science. Environ Manage 27:533–545
Morgan MG, Henrion M (1990) Uncertainty: a guide to dealing with uncertainty in quantitative risk and policy analysis. Cambridge University Press, New York
O'Hagan A, Buck CE, Daneshkhah A et al (2006) Uncertain judgments: eliciting experts' probabilities. Wiley, Chichester
Perera AH, Ouellette MR, Cui W et al (2008) BFOLDS 1.0: A spatial simulation model for exploring large scale fire regimes and succession in boreal forest landscapes. Ontario Ministry of Natural Resources, Sault Ste. Marie. Forest Research Report No 152
Smithson M (1999) Conflict aversion: preference for ambiguity vs conflict in sources and evidence. Org Behav Human Decis Processes 79:179–198
Sterman JD (1994) Learning in and about complex systems. System Dynam Rev 10:291–330
Theil H (1972) Statistical decomposition analysis: with applications in the social and administrative sciences. North-Holland Publishing Company, Amsterdam

Chapter 10
Assessing Knowledge Ambiguity in the Creation of a Model Based on Expert Knowledge and Comparison with the Results of a Landscape Succession Model in Central Labrador

Frédérik Doyon, Brian R. Sturtevant, Michael J. Papaik, Andrew Fall, Brian Miranda, Daniel D. Kneeshaw, Christian Messier, Marie-Josée Fortin, and Patrick M.A. James

Contents

F. Doyon (✉)
Département de Sciences Sociales, Secteur Foresterie, Centre d'Étude de la Forêt (CEF), Université du Québec en Outaouais, C.P. 1250, Succursale Hull, Gatineau, QC J8X 3X7, Canada
e-mail: frederik.doyon@uqo.ca

B.R. Sturtevant • B. Miranda
Institute for Applied Ecosystem Studies, Northern Research Station, USDA Forest Service, 5985 Hwy K, Rhinelander, WI 54501, USA

M.J. Papaik
Department of Biology, Sonoma State University, 1801 E. Cotati Ave., Rohnert Park, CA 94928, USA

A. Fall
Gowlland Technologies Ltd, Tucker Bay Road, Lasqueti Island, BC V0R-2J0, Canada
and
Department of Resource and Environmental Management, Simon Fraser University, 8888 University Drive, Burnaby, BC V5A 1S6, Canada

D.D. Kneeshaw • C. Messier
Département des Sciences Biologiques, Centre d'Étude de la Forêt (CEF), Institut des Sciences Environnementales, Université du Québec à Montréal, C.P. 8888, Succursale Centre-Ville, Montréal, QC H3C 3P8, Canada

M.-J. Fortin
Department of Ecology and Evolutionary Biology, University of Toronto, 25 Harbord St., Toronto, ON M5S 3G5, Canada

P.M.A. James
Department of Biological Sciences, University of Alberta, 11455 Saskatchewan Drive, Edmonton, AB T6G 2E9, Canada

A.H. Perera et al. (eds.), *Expert Knowledge and Its Application in Landscape Ecology*,
DOI 10.1007/978-1-4614-1034-8_10,

10.1 Introduction

Sustainable forest management (SFM) recognizes that the spatial and temporal patterns generated at different scales by natural landscape and stand dynamics processes should serve as a guide for managing the forest within its range of natural variability (Landres et al. 1999; Gauthier et al. 2008). Landscape simulation modeling is a powerful tool that can help encompass such complexity and support SFM planning (Messier et al. 2003). Forecasting the complex behaviors of a forested landscape involving patterns and processes that interact at multiple temporal and spatial scales poses significant challenges (Gunderson and Holling 2002). Empirical evidence for the functioning of key elements, such as succession and disturbance regimes, is crucial for model parameterization (Mladenoff 2004). However, reliable empirical data about the forest vegetation dynamics that arise in response to forest management and other disturbances may be scarce, particularly in remote areas where harvesting activity has been historically limited.

Expert knowledge-based (EKB) modeling is receiving more attention as a companion approach to empirical modeling, and attempts are now being made to formalize the process of eliciting and including expert knowledge during the development of decision-support systems (Johnson and Gillingham 2005; Murray et al. 2009; Chap. 3; and Chap. 4). Forestry experts with local knowledge collectively have considerable knowledge about forest succession and disturbance. Such collective knowledge can contribute greatly to our understanding of the vegetation transitions within a landscape that are so critical for informed SFM planning (Drescher et al. 2008).

Eliciting scientifically precise information from the collective knowledge of a group of experts remains a significant challenge. However, rigorous expression of latent knowledge that can be incorporated into an expert model can be achieved using a structured information-mining procedure. By examining convergent and divergent expert opinions about specific forest dynamics questions, researchers can obtain insights into uncertainties, knowledge gaps, and where complexity lies (Drescher et al. 2008). Comparisons between EKB models and other knowledge sources can offer a more comprehensive examination of the potential bias underlying each technique, and can reveal uncertainty and knowledge ambiguity that will suggest logical avenues for additional research and monitoring to support SFM planning.

Recognizing knowledge ambiguity is particularly important in natural resource planning because it lets planners assess the degree of uncertainty in the outcomes of various management options (Drescher and Perera 2010a).

In this chapter, we compare and contrast postdisturbance (fire and clearcut harvesting) vegetation transition probabilities (including the regeneration delay) based on knowledge derived from local experts in central Labrador with analogous information derived from a process-based landscape-dynamics model (LANDIS-II) that has been parameterized for the same area by scientists with expertise in boreal forests outside of Labrador. Expert self-assessment of their degree of uncertainty, combined with our analysis of similarities and differences among expert opinions and the relative agreement between the EKB model and LANDIS-II, can reveal the magnitude of the knowledge ambiguity.

10.2 Methods

10.2.1 Study Area

The study area is a 1.9 million ha forest management district (District 19a) in south-central Labrador (52°18′–54°0′ N, 62°05′–59°11′ W; Fig. 10.1) located at the transition between the closed-canopy boreal forest and the open-canopy taiga (Bajzak 1973; Bajzak and Roberts 1984). The central valley in District 19a contains the majority of Labrador's boreal forests, which are dominated by black spruce (*Picea mariana*) and balsam fir (*Abies balsamea*) (Foster 1984; Forsyth et al. 2003). Spruce–fir stands are embedded within a diverse mosaic of open sphagnum forest, lichen woodlands, mixed hardwoods (*Betula* spp., *Populus* spp.), black spruce bogs (with *Larix laricina*), lakes, and open wetlands. The topography is characterized by a moderate relief underlain by the bedrock geology of the Grenville formation, which is covered by glacial moraines and drumlins that support mostly podzols and gleysols (Batterson and Liverman 1995; Roberts et al. 2006). The climate is primarily continental, though it is moderated by the presence of Lake Melville, with long harsh winters, heavy snow accumulation, and annual precipitation ranging between 900 and 1,100 mm (Roberts et al. 2006). Fire is the dominant natural disturbance (Foster 1983) in Labrador.

The forestry potential of the region is impeded by slow growth of the forest (mean increment <1.0 m^3/ha/year), a long regeneration delay (sometimes lasting many decades), and conversion of productive forest into nonproductive forest after disturbance (Mallik 2003; Simon and Schwab 2005a, b). Some empirical studies have examined the response to different disturbances – for wildfire, Foster (1985), Foster and King (1986), and Simon and Schwab (2005a, b); for clearcutting, Simon and Schwab (2005a) and Elson et al. (2007). However, no studies have examined the mid- and long-term dynamics following disturbances (Roberts et al. 2006).

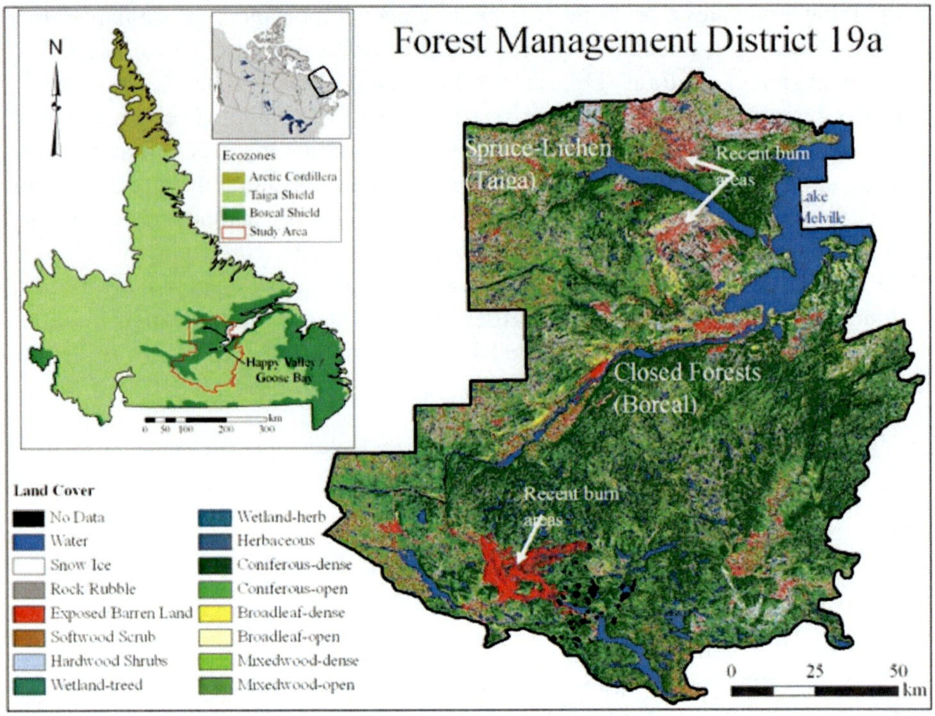

Fig. 10.1 The location of Forest Management District 19a in Central Labrador (Canada) straddles a major ecotone between closed-canopy boreal forest and open-canopy taiga systems

Although commercial harvesting has been limited in the district, a new forest management plan designed to stimulate economic growth and balance cultural and ecological values was recently approved (Forsyth et al. 2003). As clearcutting has been the only silvicultural system used in Labrador up to now, and will be according to the FM plan, expert knowledge was limited to this treatment.

10.2.2 Analytical Approach

We conducted two parallel analytical procedures for building succession models: the first (the EKB model) was based on expert knowledge, and the second used LANDIS-II, a well-established process-based succession model. Both approaches let us compare the predictions for postdisturbance succession, including transition probabilities, regeneration delays, and conversion of productive forest into nonproductive forest, and let us assess the uncertainty and level of agreement or disagreement among the experts and between the experts and LANDIS-II (Fig. 10.2).

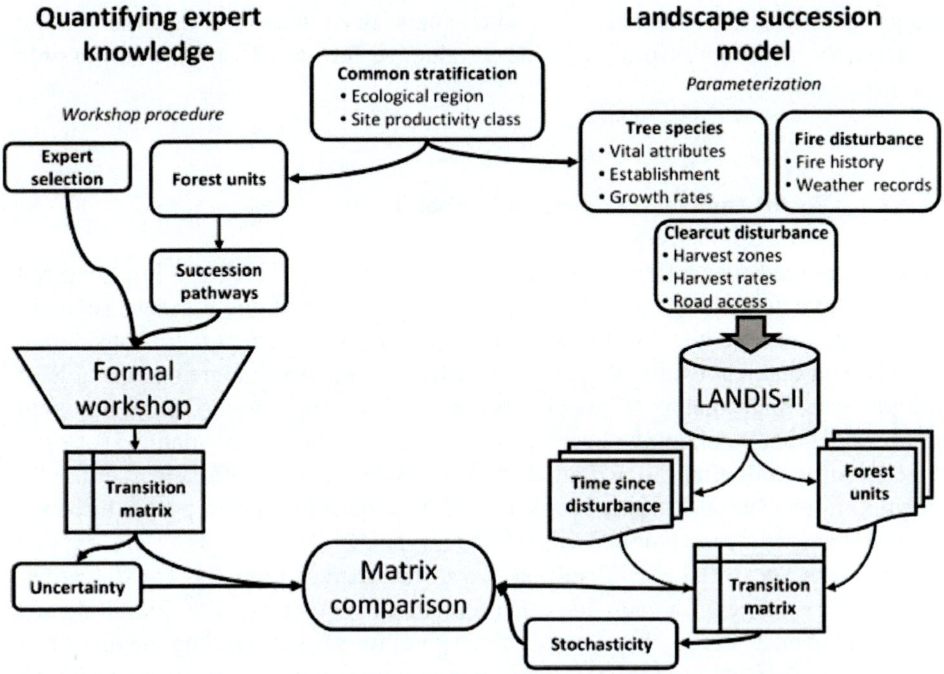

Fig. 10.2 Flowchart illustrating the process underlying the parallel development of vegetation transition matrices based on expert opinion (*left*) and the LANDIS-II model (*right*)

10.2.3 Estimating Vegetation Transition Probabilities Using Expert Opinion

10.2.3.1 Forest Units

The forest units were based on the Newfoundland and Labrador Department of Forestry stand-level inventory, in which forest units are defined by the combination of ecological region, site quality class, and forest type. Two broad ecological regions were defined (the High Boreal ecozone and the Low and Mid-Subarctic ecozone; Wiken 1986), each of which covered about half of District 19a. Of the two, the High Boreal ecozone is warmer and more productive. At finer (i.e., stand) spatial scales, forest productivity was classified into three site quality levels (good, medium, and poor). The forest types were pure balsam fir (Bf), balsam fir–black spruce (BfBs, with balsam fir dominant), black spruce–balsam fir (BsBf, with black spruce dominant), pure black spruce (Bs), softwood-dominated mixedwood (SwHw), hardwood-dominated mixedwood (HwSw), and pure hardwood (Hw). We restricted our modeling exercise to the 16 most prominent forest units (ten in the High Boreal ecozone and six in the Subarctic ecozone), which covered 95.4% of the forested landscape. In Labrador, a nonregenerated state can persist for decades after disturbance

(Simon and Schwab 2005a). To allow the experts to express such dynamics in the District 19a landscape, we added a nonproductive forest (NF) type as a possible vegetation state.

10.2.3.2 Postdisturbance Transition Probabilities

Experts assessed two different postfire situations: one in which the fire occurred during a sexually immature (nonseed-producing) stage and one during a sexually mature (seed-producing) stage. Only stand-replacing disturbances were considered. We assumed that harvesting occurred only when a stand was mature and capable of producing seed. Postharvest replanting is almost absent in Labrador, so we assumed in the model that no planting occurred. Vegetation transitions were defined for each selected forest unit after the two fire disturbance situations and after clearcutting by assigning a probability that each unit would develop into a given postdisturbance forest type (16 forest units × 3 disturbance types = 48 transitions). The experts (described in Sect. 10.2.3.4) identified two different transitions (potential postdisturbance forest types) for each disturbed forest unit. We provided no formal details of the disturbance size or shape, or of the residual forests surrounding the disturbed forest units; experts therefore had to assume that each unit had average conditions for the District 19a landscape.

10.2.3.3 Regeneration Delay

The postdisturbance regeneration delay and conversion of productive forest into nonproductive forest, which are believed to be important phenomena in this region (Bajzak 1973; Bajzak and Roberts 1984; Simon and Schwab 2005a), were also estimated by the experts. Regeneration delay was defined as the time following a disturbance that was required before the stand had sufficient regeneration to develop into a future merchantable stand. Experts were instructed to define the regeneration delay associated with a vegetation transition in 5-year classes. A regeneration delay of 60 years or more was defined as effectively "permanent" in terms of future SFM planning purposes, and was then considered to represent a conversion from productive forest into nonproductive forest (the NF type).

10.2.3.4 Workshop Procedures

Making expert knowledge explicit requires the use of an elicitation method that can help experts communicate their tacit knowledge in explicit terms (Ford and Sterman 1998). For this exercise, we used simultaneous interviews of several experts to elicit their knowledge of the forest vegetation dynamics of the study area during a 2-day workshop (similar to Drescher et al. 2006). The experts who we invited to the workshop were defined as individuals with a minimum of 10 years of local expertise in forestry or natural resource management. Seven experts participated, all of whom had a college degree in natural resources or in environmental education; they represented a combined total of 121 years of forest-related experience. Two-thirds of this experience was acquired in Labrador, primarily as part of their professional work and secondarily through other outdoor activities. On average, the experts spent 34.5 days in the field each year. All were familiar with the concept of forest succession and with the processes underlying forest dynamics, as well as with the autecology of Labrador's forest species. One aboriginal expert, in addition to contributing traditional Innu knowledge, also had conventional Western training. We used cross-validation of peer-recognized expertise among the individuals to ensure that we had successfully selected true experts.

The workshop was organized in three phases. In Phase I, the experts reviewed the workshop procedures, the definitions of terms, and local scientific studies on forest ecology and dynamics. In Phase II, they assigned vegetation transition probabilities to the different vegetation types in response to disturbance, as outlined by a workbook provided by the workshop coordinator (Doyon). In Phase III, the experts described their expertise in forestry, forest ecology, and succession via a questionnaire that determined the kinds of activities they had engaged in (professional, educational, academic, or nonprofessional) and the number of years of experience in each area.

10.2.3.5 Self-Assessment of Uncertainty

Uncertainty about an expert's opinion of any given transition probability arises from two components: the expert's degree of confidence in their knowledge and the perceived uncertainty (variability) in the system (Drescher et al. 2008). Confidence reflects the expert's knowledge, experience, and background, specifically with respect to the succession transition being assessed. System variability reflects the natural stochasticity of conditions that influence the processes involved in any given transition. Experts had to jointly evaluate these two components to "qualify" the level (low = 1, moderate = 2, and high = 3) of uncertainty in their opinion. This self-assessment was accomplished for all postdisturbance vegetation transitions, and included the step in which they estimated the regeneration delay.

10.2.3.6 The EKB Succession Model

Vegetation transition probabilities were calculated by averaging the estimates provided by all of the experts. Probabilities were then assembled into a vegetation transition matrix organized by disturbance type versus forest type, and further stratified by ecological region and site quality.

10.2.4 Deriving Vegetation Transition Probabilities from LANDIS-II

10.2.4.1 Overview and Parameterization

LANDIS-II is a process-based, spatially dynamic model of forest succession and disturbance in which the landscape is represented as a grid of interacting cells (Scheller et al. 2007; http://www.landis-ii.org). Cells have homogeneous light environments and are aggregated into "land types" with similar environmental conditions. In this study, the land types were defined based on the same ecological regions and forest site quality classes that we used to stratify forest units within the EKB succession model. Forest composition at the cell level was represented as age cohorts of individual tree species that interact via a suite of vital attributes (i.e., shade tolerance, fire tolerance, seed dispersal, ability to sprout vegetatively, and longevity) to produce nondeterministic successional pathways that are sensitive to disturbance type and severity.

We applied version 2 of the Biomass Succession extension (Scheller and Mladenoff 2004), which calculates competition among cohorts and their respective aboveground dynamics. We modified this extension to explicitly simulate the light environment that would affect species establishment and to better capture the light gradient from open- to closed-canopy forests in central Labrador. Tree species cohorts become established on new sites in the model based on a spatially explicit algorithm for seed dispersal (Ward et al. 2005) and based on establishment probabilities specific to each land type. The latter probabilities were estimated based on two soil properties (texture and available nitrogen) and two climate parameters (temperature and precipitation). Initial conditions were defined by assigning inventory sample plots to cells stratified by forest type, age class, and site quality class using a combination of classified satellite imagery, stand inventory data, and records of disturbance history. Tree species biomass information was translated into the standard fuel types (Forestry Canada Fire Danger Group 1992) used by version 1.0 of the Dynamic Fire extension to estimate fire spread rates, burn patterns, and resulting tree cohort mortality (Sturtevant et al. 2009). Timber harvesting was simulated using version 1.0 of the Harvest extension (Gustafson et al. 2000). We assumed that young (10-year-old) cohorts of balsam fir survived the clearcutting disturbance as advance regeneration. Each process was simulated using a 10-year time step and a 1-ha cell size.

10.2.4.2 Converting LANDIS-II Output into Vegetation Transition Matrices

Three 250-year simulations were run with LANDIS-II and the outputs were summarized by decade. Species cohort information from the simulations was converted into forest types and 20-year age classes using the rules for defining stand types in the Temporary Sample Plot Program (Newfoundland Forest 1995), but with biomass substituted for basal area. Sites with total biomass values (ignoring shrubs) that were less than the stocking threshold of 39.34 Mg/ha (i.e., the minimum value recorded within the District 19a stand inventory) were assigned to the NF type. We used the time since the last disturbance to assign the stand age class, and for each disturbance type, we recorded the forest types before the disturbance and for each decade of a 60-year period following the disturbance. Postfire observations were restricted to those cells that experienced a stand-replacing fire, which would result in a biomass less than the minimum stocking threshold. We applied a threshold age of 30 years for all forest types to distinguish fire disturbances that occurred within seed-producing (>30 years) versus nonseed producing (<30 years) situations. Postdisturbance transition probabilities 60 years after the disturbance were used for comparison with the transition matrix produced by the expert panel. The postdisturbance regeneration delay was estimated by recording the time after disturbance required for at least 75% of the cells to switch from an NF type to a given forest type based on the minimum stocking threshold. Cells that were still classified as NF after 60 years were considered to indicate a conversion from productive forest into nonproductive forest.

10.2.5 Data Analysis

We quantified the extent of the agreement both among the experts and between the EKB and LANDIS-II models using Pearson's correlation coefficient of the probability value for all pairwise transitions between forest types before and after disturbance. Correlations were computed for all transitions together, then by ecological region, by site quality class, by disturbance type, and by forest types before disturbance. Nonsignificant correlations were considered to represent disagreement between the sources of knowledge. An expert involved in many (more than half) nonsignificant correlations was considered an outlier of the group, and was excluded for the comparison analysis between the EKB and LANDIS-II.

We used the transition probabilities obtained from averaging of the estimates provided by all of the experts to compute a Shannon–Weaver diversity index (Shannon 1948) for each postdisturbance transition ($n=48$):

$$\text{Diversity} = -\sum_{i=1}^{8} p_i \log(p_i) \quad (10.1)$$

where p_i is the averaged probability among all experts of transiting to forest type i after a disturbance for a specific transition.

This measure of concentration of information is used to express the relative agreement among experts for a given transition; low diversity would indicate that experts have chosen to assign similar probabilities to the same forest types, showing a common understanding of the forest dynamics for a particular successional transition, whereas a high diversity would indicate disagreement among the experts. Uncertainty perceived by the experts within the transitions was summarized by averaging the rank order of their individual uncertainty assessments (low uncertainty=1, moderate=2, high=3). We used ANOVA (PASW v.18.0.0; SPSS Inc. 2009) to assess the effects of the ecological region, individual expert, site quality class, disturbance type, forest type before transition, and forest type after transition on uncertainty and on the probability of having an expert assigning a regeneration delay to a given transition. We evaluated mean differences using the *post hoc* least-significant-difference test where significant differences were indicated by the ANOVA.

10.3 Results

10.3.1 Postdisturbance Transitions

10.3.1.1 Expert-Based Transition Probabilities

Postdisturbance transition probabilities estimated by the experts varied considerably both by disturbance type and site quality, but varied less by ecological region (Table 10.1). The assigned transition probabilities suggested that clearcutting favored a transition to stands with a higher balsam fir content, whereas fire in mature stands favored black spruce. The experts agreed that postdisturbance succession on sites with good quality tended toward increased balsam fir, whereas postdisturbance succession on poor sites favored black spruce, irrespective of the ecological region. The EKB succession model clearly identified more conversion of productive forest into the NF forest type on poor-quality sites and after fires in immature stands. Indeed, in the Subarctic ecological region, the experts expected some conversion to a nonforested state for two-thirds of the 18 transitions.

10.3.1.2 Variability and Uncertainty in Expert Opinion

The diversity index of transition paths did not differ between ecological regions ($P=0.582$), among site classes ($P=0.196$), or among forest types before disturbance ($P=0.309$), but did differ significantly among disturbance types ($P=0.006$). The diversity in transitions given by the experts was greater after clearcutting and fire in immature stands than after fires in mature stands (Fig. 10.3 and Table 10.1), suggesting a better agreement among the experts for succession after fires in mature stands. Variability in opinion among experts was lower for transitions on medium-quality sites than on sites with good quality and on poor sites and for those involving forest types before disturbance dominated by black spruce (data not shown).

We found that expert opinions were significantly ($P<0.05$) correlated, with Pearson's r ranging from 0.17 to 0.70 and a mean of 0.47 (Table 10.2). Although the average correlation coefficient among the experts was higher for the Subarctic

Table 10.1 Comparison of the mean postdisturbance transition probabilities between the EKB and LANDIS-II succession models for all transitions, by disturbance type and by site quality class

			Transition probability (%)									
Disturbance	Site quality class	Model	Bf	BfBs	BsBf	Bs	SwHw	HwSw	Hw	NF	SWI[a]	Uncertainty[b]
All transitions		EKB	2	12	11	43	16	4	4	7	0.415	1.92
		LANDIS	3	2	0	43	1	1	6	43	0.564	
Clearcut	Good	EKB	13	37	20	5	23	1	0	1	0.500	1.68
		LANDIS	14	4	1	46	3	3	7	21	0.577	
	Medium	EKB	1	50	20	7	19	1	0	1	0.398	1.92
		LANDIS	9	4	1	50	2	2	4	27	0.585	
	Poor	EKB	1	8	42	34	6	0	0	9	0.486	2.10
		LANDIS	6	7	1	68	0	1	1	14	0.518	
Fire in immature stands	Good	EKB	0	3	7	44	24	10	11	1	0.508	1.99
		LANDIS	0	0	0	37	2	1	12	49	0.472	
	Medium	EKB	0	2	5	53	17	6	10	7	0.486	2.01
		LANDIS	0	0	0	36	1	1	8	54	0.448	
	Poor	EKB	0	0	0	58	0	0	6	35	0.346	1.94
		LANDIS	0	0	0	49	0	1	2	47	0.387	
Fire in mature stands	Good	EKB	0	0	5	62	22	6	5	0	0.375	1.74
		LANDIS	0	0	0	36	2	2	11	49	0.408	
	Medium	EKB	0	2	0	68	19	7	2	1	0.298	1.97
		LANDIS	0	0	0	37	1	1	5	55	0.586	
	Poor	EKB	0	0	0	71	4	0	0	25	0.301	1.98
		LANDIS	2	2	0	50	0	1	2	43	1.249	

Bf pure balsam fir, *BfBs* balsam fir–black spruce, *BsBf* black spruce–balsam fir, *Bs* pure black spruce, *SwHw* softwood-dominated mixedwood, *HwSw* hardwood-dominated mixedwood, *Hw* hardwood

[a]Shannon–Weaver index

[b]Uncertainty = mean expert uncertainty score based on individual self-assessments. Uncertainty values ranged from low (1) to high (3)

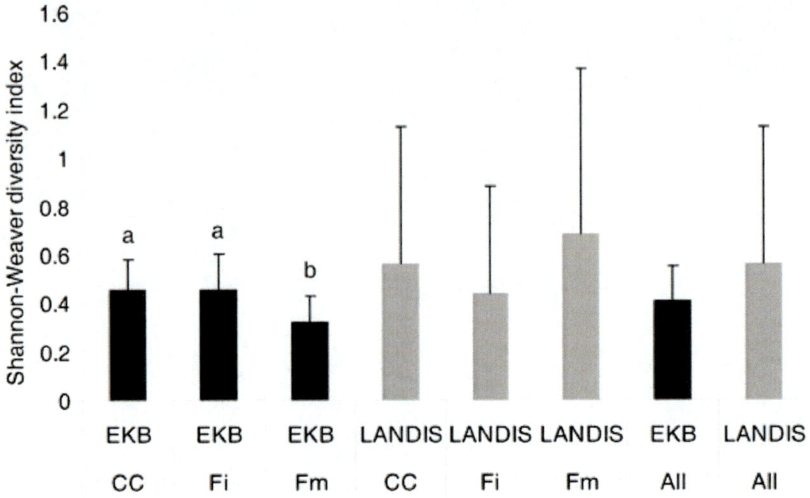

Fig. 10.3 Shannon–Weaver diversity index for transitions after three types of disturbance based on expert opinion (*EKB*) and the *LANDIS-II* model for Labrador's District 19a. Values (mean ± SD) labeled with *different letters* differ significantly ($P<0.05$). *CC* clearcutting, *Fi* stand-replacing fire in a sexually immature stand, *Fm* stand-replacing fire in a sexually mature stand

ecological region than for the Boreal region, the range of coefficients was also wider and included five nonsignificant correlations. Expert G was least experienced and was also rated as least knowledgeable by the other experts. This may explain why Expert G's answers were consistently different from the others, to the extent that removing Expert G increased the overall average correlation from 0.47 to 0.58. We therefore considered Expert G to be an outlier, and removed his assessments from the database for all subsequent analyses. It is possible, however, that this may have eliminated a perspective that was important to the understanding of this system.

The uncertainty perceived by the experts differed significantly among the participants ($F5{,}566=5.35$, $P<0.001$), with some experts significantly less certain than others. Uncertainty was significantly lower ($F2{,}566=7.10$, $P<0.001$) for sites with good quality than for those with medium or poor quality (Table 10.1 and Fig. 10.4). Experts had significantly less confidence (perceived more inherent variability) when they were assigning transition probabilities to disturbed SwHw stands or when a stand transitioned to mixedwood forest types ($F7{,}566=7.07$, $P=0.045$). Surprisingly, uncertainty was not correlated with the Shannon–Weaver diversity index of transition paths (Pearson's $r=0.19$, $P=0.19$).

10.3.1.3 Comparison Between the Experts and LANDIS-II

Postdisturbance vegetation transitions derived from expert opinion and LANDIS-II were significantly ($P<0.001$) correlated, but the Pearson's r was low (0.33; Table 10.2). Agreement between the two transition matrices was higher for the Subarctic ecological region than the Boreal region, for poor-quality sites, and for

Table 10.2 Correlation coefficients among experts for the postdisturbance vegetation transition probabilities, and correlations between the average expert response and the LANDIS-II prediction

		Among experts				Experts vs. LANDIS-II	
		Pearson's *r* [Mean (range)]	NS[a] corr.	Outlier	*n*	Pearson's *r*	*n*
Overall		0.47 (0.17 to 0.70)	0	–	192	0.33*	384
Ecological regions	High Boreal	0.45 (0.15 to 0.69)	2	G	125	0.24*	240
	Subarctic	0.52 (0.04 to 0.82)	5	G	67	0.47*	144
Site quality class	Good	0.39 (0.08 to 0.69)	6	F and G	80	0.17*	144
	Medium	0.56 (0.28 to 0.79)	0		73	0.24*	144
	Poor	0.46 (–0.16 to 0.88)	6	G	39	0.62*	96
Disturbance	Clearcut	0.37 (0.06 to 0.66)	4	G	68	0.04	128
	Fire in immature forest	0.39 (–0.14 to 0.76)	6	G	67	0.44*	128
	Fire in mature forest	0.65 (0.34 to 0.88)	0	–	57	0.45*	128
Forest type before disturbance	SwHw	0.36 (–0.02 to 0.84)	8	A, B and D[b]	26	0.16	48
	BfBs	0.25 (–0.25 to 0.64)	12	B, D and G	26	0.46*	144
	Bs	0.55 (0.12 to 0.85)	4	G	68	0.34*	144
	BsBf	0.54 (0.22 to 0.77)	1	G	72	–0.14	48

BfBs balsam fir–black spruce, *BsBf* black spruce–balsam fir, *Bs* pure black spruce, *SwHw* softwood-dominated mixedwood

*Significant at $P<0.05$

[a]Number of nonsignificant correlations

[b]Expert G did not assess this forest type

Fig. 10.4 Effect of site quality class on the uncertainty score associated with transitions among forest types after disturbance proposed by the experts for Labrador's District 19a. Values (mean ± SD) labeled with *different letters* differ significantly ($P < 0.05$)

succession after a fire (Table 10.2). The successional states produced by the experts and by LANDIS-II following clearcutting were not significantly correlated ($P = 0.69$).

Despite quantitative differences between the transition outcomes, we observed many qualitative similarities (Table 10.2). Both the experts and LANDIS-II indicated higher postdisturbance forest type diversity following clearcutting and fire in immature stands, and this diversity increased from poor-quality sites to sites with good quality (Table 10.1). The probability of postdisturbance transitions to the Bs forest type increased on poor sites in both the expert opinions and LANDIS-II. However, LANDIS-II indicated a much higher likelihood of the Bs forest type following clearcutting than was predicted by the experts, but the experts predicted a higher likelihood of the Bs forest type after fires. This difference resulted mainly from (1) lower transition probabilities to mixed forest types after clearcutting in LANDIS-II, (2) the absence of any Bf forest types after a fire in LANDIS-II, and (3) a much higher importance (frequency and probability) of conversion of productive forest into nonproductive forest after fires within LANDIS-II. The latter is probably the most important difference between the two methods. In LANDIS-II, 60 years after disturbance, all of the 48 transitions had some probability of conversion into NF, whereas the experts predicted this for only 40% of the transitions. Moreover, the average probability of transitioning into the NF type was much higher in LANDIS-II (43%) than was predicted by the experts (7%) (Table 10.1). This may explain why diversity in succession pathways was much higher and more variable among the transitions in LANDIS-II than in the EKB succession model (Fig. 10.3).

10.3.2 Regeneration Delay

Experts assigned a regeneration delay to 55% of the transitions. The probability was significantly ($P < 0.001$) higher for poorer site classes (Fig. 10.5) and significantly higher ($P = 0.005$) after a fire (in both immature and mature stands) than after a clearcut (Fig. 10.6). In addition, the higher the proportion of black spruce after the disturbance, the higher the likelihood of having an expert assign a regeneration delay to that transition ($P < 0.001$).

Fig. 10.5 Effect of site quality class on the likelihood (between 0 and 1) that experts would associate a regeneration delay with the transition probability after a disturbance in Labrador's District 19a. Values (mean ± SD) labeled with *different letters* differ significantly ($P < 0.05$)

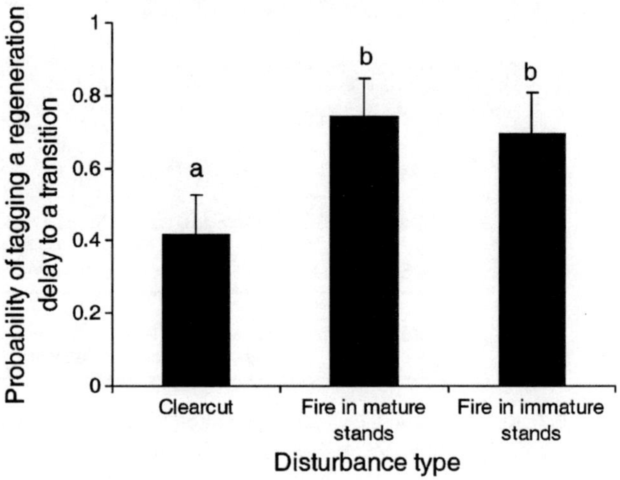

Fig. 10.6 Effect of disturbance type on the likelihood (between 0 and 1) that experts predicted a regeneration delay when assigning succession transition probabilities by disturbance type in Labrador's District 19a. Values (mean ± SD) labeled with *different letters* differ significantly ($P < 0.05$)

10.3.2.1 Variability Among Experts

The likelihood of assigning a regeneration delay to a transition after disturbance differed significantly among the experts ($P < 0.001$). Most of the experts assigned a regeneration delay in about 50% of the transitions, but one expert almost always assigned a regeneration delay to the postdisturbance succession. However, where experts agreed that a regeneration delay would occur after disturbance, the estimated duration did not differ significantly (an average of 16 years).

10.3.2.2 Comparison Between Experts and LANDIS-II

LANDIS-II predicted a regeneration delay for 96% of the postdisturbance transitions. If we consider only the transitions for which experts assigned a transition probability, the regeneration delay after the disturbance was 26 years shorter than that predicted by LANDIS-II. Hence, there was poor agreement between LANDIS-II

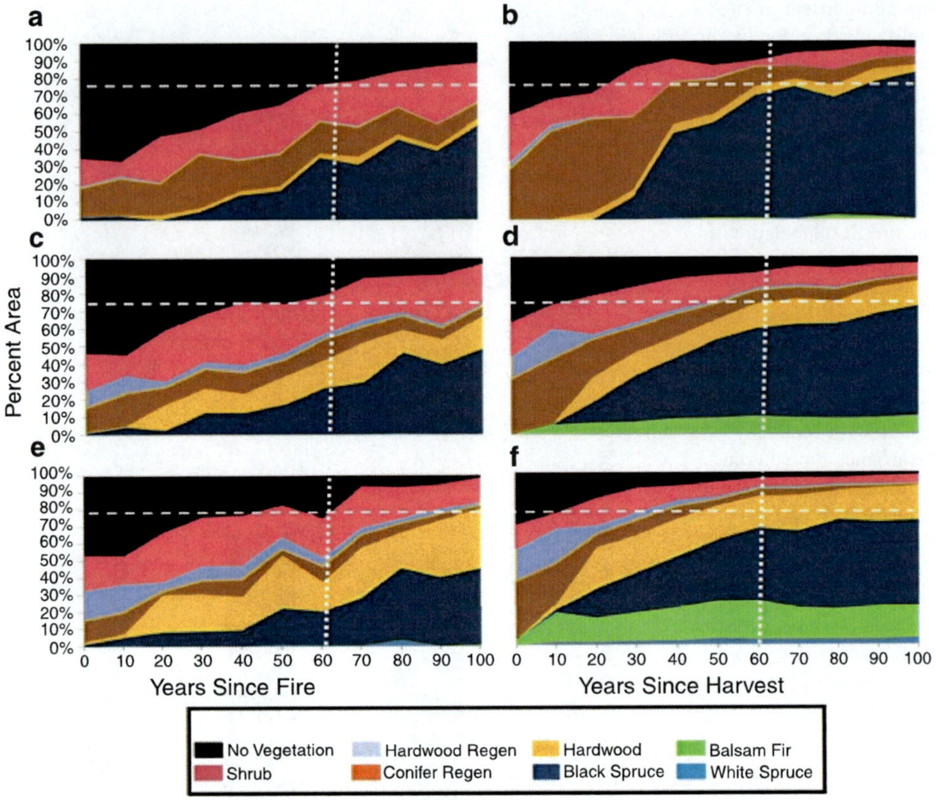

Fig. 10.7 Regeneration delay and composition trends following a fire or clearcutting on sites with (**a**, **b**) poor, (**c**, **d**) medium, and (**e**, **f**) good quality based on the LANDIS-II output. The *dashed horizontal line* indicates the 75% threshold used to assign a regeneration delay (NF) to a forest transition 60 years (*vertical line*) after the disturbance. *Regen* regenerating forest vegetation that remains below the minimum stocking level

and the experts on the regeneration delay; the overall correlation was significant but low ($r=0.24$, $P=0.002$). Nonetheless we observed similar qualitative trends from both methods: the regeneration delay was predicted to be frequent and of longer duration on poorer sites and after fires (Fig. 10.7).

10.4 Discussion

10.4.1 Insights from the Expert Workshop

In general, expert knowledge was strongly convergent; that is, the experts generally agreed about the postdisturbance transitions. Their level of agreement was only moderate for the postdisturbance regeneration delay. Informal discussions by the experts after the workshop suggested that such agreement was not the result of groupthink, which can lead to "bandwagon" reasoning, as they were often referring to different examples they had experienced during their career. It was also apparent during the workshop that experts paid strong attention to the consistency of their own mental models by looking back at their previous answers while answering the workshop questionnaire; they often asked to change an answer after they had had time to think about subsequent questions.

In general, group opinion on succession was more variable for the mixed forest types (BfBs, BsBf, and SwHw) than for the Bs forest type. Experts recognized that variability in the successional response increased with increasing diversity of the forest cover. Such variability translated into both a higher diversity of expert opinion and greater expert uncertainty for the transitions involving mixed forest types. Agreement among experts varied depending on ecological factors and the diversity of the forest community. Drescher et al. (2008) found that succession of monospecific stands under relatively extreme environmental conditions (dry sands and wet bogs) was more predictable than succession of mixed stands on sites with a good site quality. We observed similar results within the resource-limited landscape of central Labrador, where more species can become established on richer and warmer sites, leading to more variability in the successional response.

Although the diversity of expert opinion on postdisturbance succession was higher for richer sites, we were surprised that the expert uncertainty was lowest for sites with good site quality. We suspect that this inversion was due to the important roles that the regeneration delay and the conversion from productive forest into nonproductive forest play on poor sites and the variability of these phenomena perceived by the experts. Such insights demonstrate the importance of identifying whether the variation among the experts arises from perceived variability in the system's response to disturbance or from a lack of expert knowledge or experience with specific ecological patterns, as Drescher et al. (2008) pointed out.

Agreement was also lower after clearcutting and fire in immature stands than after fire in mature stands. Commercial harvesting in this district has been limited to a few thousand hectares thus far. Experts therefore had little experience with succession after clearcutting, but considerable experience with catastrophic wildfires in mature stands. Moreover, harvesting has mostly occurred at the most productive sites, which are closest to Goose Bay and Happy Valley. Since better site quality is expected to lead to more variability in the successional response, this disturbance history may have introduced additional uncertainty in the EKB model.

Distinguishing between postfire succession in mature seed-producing stands and that in immature nonseed producing stands proved to be important for the experts. First, the transition probabilities for conversion into the NF type and the regeneration delay differed somewhat between the two types of fire. Perhaps more importantly, agreement among experts was lowest and uncertainty was highest for fires in immature stands. The conversion process from productive forest into nonproductive forest occurs over long time scales that likely fall outside the experience of the experts. Nonetheless, the experts recognized that such conversions occur, perhaps frequently, and that they have important implications for forest sustainability.

10.4.2 Similarities and Differences Between Expert Opinion and LANDIS-II

We found more agreement among the experts than between the mean expert response and the predictions of LANDIS-II. Quantitative disagreements between the two models were attributed primarily to the length of time required for forest regeneration, resulting in greater conversion of productive forest into nonproductive forest, a higher probability of assigning a regeneration delay to a transition, and a longer average regeneration delay period in LANDIS-II. The longer regeneration delay in LANDIS-II can be partly explained by the model's more conservative definition of a "regenerated stand," which uses a minimum biomass threshold that we suspect would require more time to reach than the experts' mental picture of a regeneration state sufficient to produce a future merchantable stand. Including a clarification of fuzzy terms such as "regenerated stand" or "stand-replacing disturbance" in Phase I of the knowledge elicitation procedure would probably have helped to narrow the semantic differences, leading to less ambiguous thresholds in the EKB model used for comparison with the LANDIS-II model. A higher probability of regeneration delay might also have resulted from differences in the scale of the assessment; the LANDIS-II results were based on 1-ha cells, whereas the experts were generally thinking in terms of forest stands (tens of hectares). Such discrepancies highlight the difficulty of adequately comparing models that have been developed under frameworks that use different scales and concept definitions. Nonetheless, the fact that the qualitative trends in the regeneration delay were consistent between the EKB model and LANDIS-II (i.e., longer after a fire than after a harvest; more prevalent delays at sites with lower quality) suggests that these trends are both robust and important to forest dynamics in Labrador.

Both LANDIS-II and the experts predicted higher proportions of the balsam fir and hardwood species as components of the landscape after clearcutting, leading to higher forest type diversity after this disturbance type (Table 10.1 and Fig. 10.7). Similarly, both methods predicted that postdisturbance tree species diversity should increase from poor to good sites. Such a convergence of results gives us confidence in the general trends represented by these relationships. Nonetheless, there were key differences in the response to different disturbances that will have consequences for the future composition of District 19a's landscape. The experts anticipated that a relatively high proportion of clearcut sites would contain some balsam fir component, whereas LANDIS-II predicted that the majority of these sites would be dominated by black spruce (Table 10.1). Central Labrador is located at the northern range limit for balsam fir, which explains its low probability of establishment within LANDIS-II (Sturtevant et al. 2007). Although local experience suggested the range limits for balsam fir in LANDIS-II may have been too conservative, sites dominated by balsam fir nonetheless produced the highest disagreement among the experts (Tables 10.1 and 10.2). The experts also predicted a much higher frequency of mixed stand conditions than LANDIS-II predicted. Part of this discrepancy may be a consequence of differing resolution (i.e., small adjacent cells with different vegetation types within LANDIS-II may be aggregated into larger "mixed" stands by the experts). However, it may also be explained by the current dominance of black spruce in the landscape, and the explicit simulation of regeneration limited by seed sources within the LANDIS-II software. Such spatial interactions tend to reinforce the inertia of the current landscape composition within the process-based model, whereas experts, by excluding any spatial context for their transition probabilities, may have missed an important process (i.e., seed source limitation) that would affect the likelihood of transition to a mixed forest condition.

10.5 Conclusions

This study is among the first to formally compare EKB and process-based ecological succession models. Moreover, it is the first time that an EKB model was developed to address regeneration delays and the conversion of productive forest into non-productive forest during forest succession. Comparing models has proven to be an important heuristic process in science to develop a broadly accepted body of knowledge that can support decision-making (Robertson et al. 2003; Drescher and Perera 2010b). In this case study, understanding the convergences and divergences between the two methods helped to identify the limitations, uncertainties, and needed improvements in both models, as well as the gaps in our knowledge of Labrador's forest dynamics.

The knowledge ambiguity we identified concerning the relative importance of balsam fir after clearcutting, of the conversion of productive forest into nonproductive forest, of the effects of spatial heterogeneity in seed sources on future forest composition, and of the regeneration delay after disturbance all have important consequences for our ability to sustainably manage these forests. Indeed, we believe the choice of which transition matrix (expert-based or LANDIS-II) to use in developing SFM strategies would lead to very different sustainable timber yields and would have important impacts on all the other decisions that follow, as was observed by Drescher and Perera (2010a). This is particularly true if one considers that most of the coming changes that will result from application of the SFM plan to District 19a will lead to increased use of clearcutting; important knowledge ambiguities have not yet been resolved about the succession that will occur after this disturbance. Part of the stakeholder debate over the SFM plan stems from the unknown impact of clearcutting on both timber and nontimber values (Berninger et al. 2009). Because of the importance of the differences of opinion and the high uncertainties expressed by the experts on these processes, Newfoundland and Labrador should prioritize acquiring scientific knowledge on the conversion to the NF type, on regeneration delays, and on succession after clearcutting.

Expert modeling provides a complementary approach that can support data-driven development of scientific models. It offers an opportunity to quickly identify, during a first step, the critical elements of uncertainty that must subsequently be scrutinized by empirical research and other modeling approaches. In this study, we demonstrated that our comparison of the two models provided new insights that could not be achieved by either knowledge source alone. EKB models are often easier and less costly to develop than empirically validated models; with a surprisingly limited amount of resources and effort, we were able to derive the critical inputs necessary to drive a complete succession model using only expert knowledge, and the results showed strong consistency among the experts. The value of the insights we gained amply justified the investment in conducting parallel EKB modeling. We believe that there are many other situations in natural resource planning that could benefit from this approach. However, given the limited number of studies of such a combined approach, it appears that such benefits have not yet been fully recognized.

Developing an EKB model also brings indirect benefits. Involving stakeholders and planners in the process of developing a model enhances the likelihood of its use, since experts who participated in the model development are often involved in its subsequent use (Gustafson et al. 2006), and increases the likelihood that the model will be used properly, based on an improved understanding of its scope and limitations (as pointed out by Drescher and Perera 2010a). This approach also helps to structure and formalize the exchange of knowledge among participants. Hence, after such collective heuristic exercises, the experts can better express their mental models of the processes involved and better understand the mental models of other experts. Such a shared understanding facilitates further development of forest management planning. Finally, EKB models encourage formal retention of the expertise of all participants in a way that makes this knowledge easier to transfer to younger workers with less experience. In remote areas such as Labrador, where there is often a rapid turnover of personnel and where retaining expertise is a real challenge, this collection and sharing of knowledge becomes an important asset.

References

Bajzak D (1973) Biophysical land classification of the Lake Melville area, Labrador. Environment Canada, Canadian Forest Service, St. John's, Information Report No. NX-88

Bajzak D, Roberts BA (1984) Mapping land types for forest evaluation in Lake Melville area, Labrador, Canada. Paper presented at the Joint Meeting of IUFRO Working Parties No. 1.02-06 and No. 1.02-10 on Qualitative and Quantitative Assessment of Forest Sites with Special Reference to Soil, 10–15 September 1984, Birmensdorf, Switzerland. IUFRO, Vienna

Batterson M, Liverman D (1995) Landscapes of Newfoundland and Labrador. Department of Natural Resources, Government of Newfoundland and Labrador, St. John's, Geological Survey Report 95–3

Berninger K, Kneeshaw D, Messier C (2009) The role of cultural models in local perceptions of SFM—differences and similarities of interest groups from three boreal regions. J Environ Manage 90:740–751

Drescher M, Perera AH (2010a) Comparing two sets of forest cover change knowledge used in forest landscape management planning. J Environ Plan Manage 53(5):591–613

Drescher M, Perera AH (2010b) A network approach for evaluating and communicating forest change models. J Appl Ecol 47(1):57–66

Drescher M, Perera AH, Buse LJ et al (2006) Identifying uncertainty in practitioner knowledge of boreal forest succession in Ontario through a workshop approach. Ontario Ministry of Natural Resources, Ontario Forest Research Institute, Sault Ste Marie, Forest Research Report 165

Drescher M, Perera AH, Buse LJ et al (2008) Uncertainty in expert knowledge of forest succession: A case study from boreal Ontario. For Chron 84:194–209

Elson LT, Simon NPP, Kneeshaw D (2007) Regeneration differences between fire and clearcut logging in southeastern Labrador: a multiple spatial scale analysis. Can J For Res 37:472–480

Ford DN, Sterman JD (1998) Expert knowledge elicitation to improve formal and mental models. Syst Dynam Rev 14(4):309–340

Forestry Canada Fire Danger Group (1992) Development and structure of the Canadian Forest Fire Behavior Prediction System. Information Report ST-X-3. Forestry Canada, Science and Sustainable Development Directorate, Ottawa

Forsyth J, Innes L, Deering K, Moores L (2003) Forest Ecosystem Strategy Plan for Forest Management District 19 Labrador/Nitassinan. Innu Nation and Newfoundland and Labrador Department of Forest Resources and Agrifoods, Northwest River

Foster DR (1983) The history and pattern of fire in the boreal forest of southeastern Labrador. Can J Bot 61:2459–2471

Foster DR (1984) Phytosociological description of the forest vegetation of southeastern Labrador. Can J Bot 62:899–906

Foster DR (1985) Vegetation development following fire in *Picea mariana* (black spruce)–*Pleurozium* forests of southeastern Labrador, Canada. J Ecol 73:517–534

Foster DR, King GA (1986) Vegetation pattern and diversity in S.E. Labrador, Canada. *Betula papyrifera* (birch) forest development in relation to fire history and physiography. J Ecol 74:465–483

Gauthier S, Vaillancourt MA, Leduc A et al (2008) Aménagement écosystémique en foret boréale. Presse de l'Université du Québec, Québec

Gunderson LH, Holling C (2002) Panarchy: understanding transformations in human and natural systems. Island Press, Washington

Gustafson EJ, Shifley SR, Mladenoff DJ et al (2000) Spatial simulation of forest succession and timber harvesting using LANDIS. Can J For Res 30:32–43

Gustafson EJ, Sturtevant BR, Fall A (2006) A collaborative, iterative approach to transfer modeling technology to land managers. In: Perera AH, Buse LJ, Crow TR (eds) Forest landscape ecology: Transferring knowledge to practice. Springer Science & Business Media, New York, pp 123–134

Johnson CJ, Gillingham MP (2005) An evaluation of mapped species distribution models used for conservation planning. Environ Conserv 32(2):117–128

Landres PB, Morgan P, Swanson FJ (1999) Overview of the use of natural variability concepts in managing ecological systems. Ecol Appl 9:1179–1188

Mallik AU (2003) Conifer regeneration problems in boreal and temperate forests with ericaceous understory: role of disturbance, seedbed limitation, and keystone species change. Crit Rev Plant Sci 22:341–366

Messier C, Fortin M-J, Smiegelow F et al (2003) Modelling tools to assess the sustainability of forest management scenarios. In: Burton PJ, Messier C, Smith DW, Adamoviicz WL (eds) Towards sustainable management of the boreal forest. NRC Research Press, Ottawa, pp 531–580

Mladenoff DJ (2004) LANDIS and forest landscape models. Ecol Modell 180(1):7–19

Murray JV, Goldizen AW, O'Leary RA et al (2009) How useful is expert opinion for predicting the distribution of a species within and beyond the region of expertise? A case study using brush-tailed rock-wallabies *Petrogale penicillata*. J Appl Ecol 46(4):842–851

Newfoundland Forest Service (1995) Temporary sample plot program. Newfoundland Forest Service, St. John's, Draft 28 February 2002

Roberts BA, Simon NPP, Derring KW (2006) The forests and woodlands of Labrador, Canada: ecology, distribution and future management. Ecol Res 21:868–880

Robertson MP, Peter CI, Villet MH, Ripley BS (2003) Comparing models for predicting species' potential distributions: a case study using correlative and mechanistic predictive modelling techniques. Ecol Modell 164(2–3):153–167

Scheller RM, Domingo JB, Sturtevant BR et al (2007) Design, development, and application of LANDIS-II, a spatial landscape simulation model with flexible temporal and spatial resolution. Ecol Modell 201:409–419

Scheller RM, Mladenoff DJ (2004) A forest growth and biomass module for a landscape simulation model, LANDIS: Design, validation, and application. Ecol Modell 180:211–229

Shannon CE (1948) A mathematical theory of communications. Bell Syst Tech J 27:379–423

Simon NPP, Schwab FE (2005a) The response of conifer and broadleaved trees and shrubs to wildfire and clearcut logging in the boreal forests of central Labrador. Northern J Appl For 22:35–41

Simon NPP, Schwab FE (2005b) Plant community structure following wildfire in the subarctic forests of Labrador. Northern J Appl For 22:229–235

SPSS Inc (2009) PASW Statistics 18, Version 18.0.0 for Windows. SPSS, Chicago

Sturtevant BR, Fall A, Kneeshaw DD et al (2007) A toolkit modeling approach for sustainable forest management planning: achieving balance between science and local needs. Ecol Soc 12:7. Available from: http://www.ecologyandsociety.org/vol12/iss2/art7/ES-2007-2102.pdf (accessed February 2011)

Sturtevant BR, Scheller RM, Miranda BR et al (2009) Simulating dynamic and mixed-severity fire regimes: A process-based fire extension for LANDIS-II. Ecol Modell 220:3380–3393

Ward BC, Mladenoff DJ, Scheller RM (2005) Simulating landscape-level effects of constraints to public forest regeneration harvests due to adjacent residential development in northern Wisconsin. For Sci 51:616–632

Wiken EB (1986) Ecological Land Classification Series, No. 19, Lands Directorate, Environment Canada, Hull

Chapter 11
Use of Expert Knowledge to Develop Fuel Maps for Wildland Fire Management

Robert E. Keane and Matt Reeves

Contents

11.1 Introduction

Fuel maps are becoming an essential tool in fire management because they describe, in a spatial context, the one factor that fire managers can control over many scales – surface and canopy fuel characteristics. Coarse-resolution fuel maps are useful in global, national, and regional fire danger assessments because they help fire managers effectively plan, allocate, and mobilize suppression resources (Burgan et al. 1998).

R.E. Keane (✉)
Missoula Fire Sciences Laboratory, USDA Forest Service, Rocky Mountain Research Station, 5775 Highway 10 West, Missoula, MT 59808 USA
e-mail: rkeane@fs.fed.us

M. Reeves
RMRS Forestry Sciences Laboratory, USDA Forest Service, Rocky Mountain Research Station, 800 E. Beckwith Ave., Missoula, MT 59801 USA

A.H. Perera et al. (eds.), *Expert Knowledge and Its Application in Landscape Ecology*, 211
DOI 10.1007/978-1-4614-1034-8_11, © Springer Science+Business Media, LLC 2012

Regional fuel maps are useful as inputs for simulating carbon dynamics, smoke scenarios, and biogeochemical cycles, as well as for describing fire hazards to support prioritization of firefighting resources (Leenhouts 1998; Lenihan et al. 1998). Intermediate- and fine-resolution digital fuel maps are important for rating ecosystem health, targeting and evaluating tactical fuel treatments, computing fire hazard and risk (the potential damage and likelihood of that damage, respectively), and aiding in environmental assessments and fire danger forecasting programs (Pala et al. 1990; Hawkes et al. 1995; Gonzalez et al. 2007). However, landscape-level fuel maps have seen the most use in fire management because they provide the critical inputs for the spatially explicit fire behavior and growth models used to simulate fires so that they can be more effectively managed and fought (Keane et al. 1998, 2006).

Expert knowledge has been involved in the development of most fuel maps currently used in fire management (Keane et al. 2001). Experts in wildfire suppression, fuel management, fire modeling, and prescribed burning have provided the background information needed to create, refine, and validate the primary fuel maps required by fire behavior and growth models. This heavy reliance on expert knowledge is a result of many factors, including the high spatial and temporal variability of fuels, diversity of fuel beds, subjective nature of the fuel classifications, and lack of comprehensive fuel data across forests and rangelands. In this chapter, we summarize past, present, and potential future use of expert knowledge in the mapping of fuels to support fire management, primarily for the USA, but also including knowledge from other countries. We present a detailed example of how expert knowledge was used in the national landscape fire and resource management planning tools (LANDFIRE) mapping project, which created a set of national fuel maps. We also discuss the challenges involved in mapping fuel, review mapping approaches that have integrated expert knowledge in their design, and describe technologies and protocols needed to facilitate the development of accurate digital fuel maps.

11.1.1 Fuel Mapping Background

Wildland fuels comprise all the organic matter available to permit fire ignition and sustain combustion (Albini 1976; Sandberg et al. 2001). Specifically, fuel components are the live and dead surface and canopy biomass that foster the spread of wildland fire. Surface fuel is often divided into duff and litter, downed and dead woody biomass in a range of diameter classes, and live and dead standing vegetation (Fosberg 1970; DeBano et al. 1998). Canopy fuel is aerial biomass (typically >2 m above ground), primarily composed of branches and foliage and also included arboreal mosses, lichens, dead ladder fuels, and other hanging dead material such as needles and dead branches (Reinhardt et al. 2006). The amount (mass or volume) of biomass per unit area is often referred to as the "fuel load." Most fuels in fire-prone ecosystems accumulate in the absence of fire. Surface fuels usually increase until the decomposition rate equals the deposition rate, and canopy fuels tend to increase as shade-tolerant tree species become established in the understory and overstory (Keane et al. 2002).

Wildland fuels can be mapped using many approaches (Keane et al. 2001; Arroyo et al. 2008). Most efforts have mapped important fuel characteristics such as the fuel model (an abstract classification of fuel used as an input to fire behavior models), fuel bed depth, and canopy bulk density as a function of vegetation type (Agee et al. 1985; Menakis et al. 2000), ecosystem (Grupe 1998), and topography (Rollins and Yool 2002) to create spatial layers in a geographical information system (GIS). Some researchers have qualitatively or quantitatively related fuel information to various forms of remote-sensing data at multiple scales, including digital photographs (Oswald et al. 1999), LANDSAT images (Wilson et al. 1994), ASTER images (Falkowski et al. 2005), AVIRIS images (Roberts et al. 1998), AVHRR images (McKinley et al. 1985; Burgan et al. 1998), microwave-radar images (Arroyo et al. 2008), and LIDAR data (Mutlu et al. 2008). Others have mapped fuels using complex statistical modeling techniques coupled with comprehensive field data (Rollins et al. 2004) and knowledge-based systems (Goulstone et al. 1994; de Vasconcelos et al. 1998). Most efforts have combined two or more of these approaches into an integrated analysis, with the goal of developing more accurate and consistent fuel maps. One resource that has been used as the foundation for most of these mapping efforts was expert knowledge.

For many ecological reasons, it is difficult to map wildland fuels (Keane et al. 2001). The most notable factor that confounds mapping is the high temporal and spatial variability of fuel components (Brown and See 1981; Keane 2008). The components, fuel loads, and properties of the fuels are also highly diverse and vary across multiple scales; a fuel bed, for example, can consist of many fuel components, including litter, duff, logs, and coniferous cones, and the properties of each component, such as its heat content, moisture content, and size, can be highly variable even within a single type of fuel. The variability of fuel loads within a stand, for example, can be as high as the variability across a landscape, and this variability can be different for each fuel component and property (Brown and Bevins 1986). A single wind storm or wet snow can rapidly increase woody fuel load at the surface and change the entire structure of the fuel bed (Keane 2008).

There are also many methodological and technological factors that complicate fuel mapping. First, much of the remotely sensed data used in fuel mapping is derived using technologies that cannot detect surface fuels because the ground is often obscured by the forest canopy (Lachowski et al. 1995). Even if the canopy were removed, it is doubtful that today's coarse-resolution imagery could distinguish subtle differences in the characteristics of all fuel components. High fuel diversity and variability also preclude an accurate standardized measurement and mapping protocol; it is difficult to sample fine fuels (e.g., duff, litter, and fine woody material) and large fuels (e.g., logs) at the same scale, degree of rigor, and accuracy (Sikkink and Keane 2008). Fuel components can vary across different scales (e.g., logs vary over a larger area than fine fuels), and few of these scales match the resolution of the remote-sensing data, the sampling methods, or the available GIS data layers. Moreover, many fuel parameters required by current fire behavior models lack standardized measurement techniques. For example, some fire behavior models use fuel model classifications that were subjectively created to represent expected fire behavior (Anderson 1982).

11.1.2 Fuel Classifications

Because of the abovementioned factors, fire management has turned to the use of fuel classifications to simplify the collection of input data for fire modeling applications. Most fire models use fuel classifications to simplify the inputs for fuel characteristics, but the diversity of these inputs makes accurate, comprehensive, and consistent fuel classification difficult (Sandberg et al. 2001; Riccardi et al. 2007; Lutes et al. 2009). Some fuel classifications are designed to include subjective components and categories that are based on the objective of a given mapping project. For example, fire behavior prediction requires mapping of the fuel loads of downed and dead fine woody materials stratified into the size classes that are required by the fire behavior model (Burgan and Rothermel 1984).

Fuel classifications can be divided into those that were developed to simulate the effects of fire and those that were developed to predict fire behavior. The former fuel classifications summarize actual fuel characteristics (most often fuel load) for diverse fuel components based on vegetation type, biophysical setting, or fuel bed characteristics. Few of these classifications were developed to support unique identification of the classes in the field; most rely on the expertise of the fuel sampler and their ability to match the observed fuel bed conditions to the classification categories. The exception is the fuel loading model (FLM) classification (Lutes et al. 2009), which contains a comprehensive field key (Sikkink et al. 2009).

In contrast, fuel classifications designed to predict fire behavior have categories referred to as fire behavior fuel models (FBFMs), which are a set of summarized fuel characteristics (e.g., fuel load, ratios of surface area to volume, mineral content, heat content) for each fuel component that is required by the fire behavior model (Burgan and Rothermel 1984). The most commonly used FBFM classifications are the 13 models of Anderson (1982) and the 40+ models of Scott and Burgan (2005), all of which are used as inputs to the Rothermel (1972) fire-spread model that is implemented in the BEHAVE and FARSITE fire prediction systems, and the 26 fire danger models of Deeming et al. (1977) that are used in the US National Fire Danger Rating System. FBFMs are not a quantitative description of fuel characteristics, but rather a set of fuel inputs designed to compute an "expected" fire behavior; this is because the inherent complexity of the mechanistic fire behavior models of Rothermel (1972) and Albini (1976) makes it difficult to realistically predict fire behavior from the actual fuel load (Burgan 1987). As a result, a complicated procedure must be followed to develop FBFMs in which fuel loads and other characteristics are adjusted to match fire characteristics that have been observed in the field (Burgan 1987).

Therefore, without prior knowledge of fire behavior in local fuel types, it is nearly impossible to accurately and consistently use and interpret most FBFM classifications (Hardwick et al. 1998), and the identification of fuel models in the field is highly subjective because it is based on an individual's perception of fire behavior rather than on actual measurements of fuel loads. Because classifications based on fire behavior and fire effects form the backbone of most fire management analyses, and because these classifications are inherently subjective and difficult to use, most fuel mapping must rely on expert knowledge and experience during all phases of the mapping process.

11.2 The Use of Expert Knowledge in Fuel Mapping

11.2.1 Who Are the Experts?

The best experts to use when creating wildland fuel maps are people who are actually involved in the management of wildland fire (Table 11.1). Local and regional fire behavior analysts who have extensive experience in predicting fire behavior and effects for both wildfires and prescribed fires are probably the most desirable experts because they can provide integrated knowledge of the influence of topography, vegetation, disturbance, and climate on fuel bed characteristics and the consequences for fire behavior (Keane et al. 2000). Fuel specialists and fire management personnel are also important because they have extensive knowledge of how to implement a fuel model within a fire model and understand the temporal and spatial scales of various fuel characteristics. Any expert who assists in mapping fuels must understand both the conditions and properties of wildland fuels and the expected fire behaviors if these fuel complexes are burned. Experts can be selected from diverse pools; Keane et al. (1998) used fire managers and wildfire suppression specialists; Nadeau and Englefield (2006) used fire scientists; and Reeves et al. (2009) used scientists, managers, and any other fire resource professionals who were available.

Since the quality of the fuel maps used to predict fire behavior is nearly impossible to assess because of the subjective nature of the FBFM (Keane et al. 2001), it is essential that those who use the fuel maps approve of their utility. The complexity, resolution, and detail involved in the mapping procedures, such as whether the latest statistical techniques and state-of-the-art images are used, are less important than producing a map that fire managers trust enough to use. As a result, experts in GIS, digital mapping, analysis of satellite images, fire ecology, and spatial statistical analysis play a lesser role than fire managers in providing expert knowledge. Complex and novel mapping techniques may yield fuel map layers that fire managers may never use, whereas fuel maps developed from simplistic qualitative techniques may be easier for fire managers to understand and employ. This means that fuel mapping, especially for fire behavior prediction, should incorporate knowledge from fire management experts during map development to increase the likelihood

Table 11.1 A summary of the potential experts whose knowledge can be used to more effectively map wildland fuels

Title	Main job	Potential knowledge	Potential mapping tasks
Fire behavior analyst	Predicting fire behavior	FBFM sampling; fire behavior simulation, collecting fuel information as inputs	FBFM assignment and calibration; map validation and verification
Fuel specialist	Sampling, estimating, and treating wildland fuels	FBFM identification; fuel sampling; defining the biophysical context for fuels; prediction of fire effects	Collection of reference field data, estimation and verification of fuel loads
Fire manager	Managing fire in specific areas using fuel treatments, prescribed burning, and controlled wildfires	Local knowledge of wildland fuel characteristics; prediction of fire behavior and effects	Calibration, validation, and verification of local area references
Fire suppression specialist	Suppression of fires	FBFM identification and use; prediction of fire behavior	FBFM calibration; map validation
Fire scientist	Conducting fire and fuel research	Depends on the scientist and their field of study	Fuel collection, sampling, and identification; map validation and calibration
Fire prevention specialist	Fire danger warnings, public information, preventing unwanted ignitions	General fuel information	Map validation and verification

These titles vary among countries and government agencies, and many of these experts have multiple titles and perform multiple duties. *FBFM* fire behavior fuel model

that the resulting maps will be used. This is a somewhat subjective and self-affirming process, but one that is necessary until advances in fire behavior prediction use fuel inputs that can be readily measured, validated, and verified, while also being understood and accepted by fire managers.

11.2.2 How Is Expert Knowledge Used?

There are four general ways that expert knowledge can be integrated into the fuel mapping process. First, expert knowledge can be used in the field to estimate or measure the fuels to provide information that will be used for ground-truthing or as a reference in the mapping process (i.e., for *reference*). FBFMs, for example, must be estimated at a plot level by experts who have been trained to predict fire behavior when assessing fuel (Burgan and Rothermel 1984). Hornby (1935) used a team of experts who traversed the landscapes of the western United States to evaluate fire behavior characteristics from fuel and vegetation attributes, and who used these attributes to delineate differences in fire spread and intensity. Experts in visually assessing fuel loads could choose the most appropriate fuel characteristics classification system (FCCS) category that best describes a sample area (Riccardi et al. 2007). Keane et al. (1998) trained field crews to properly use Anderson's (1982) US National Forest Fire Laboratory fuel models in sample plots. These experts can also build local keys for identifying appropriate fuel models in the field to help other crews to consistently collect useful fuel data. Agee et al. (1985), for example, used local fire and fuel experts to construct and refine FBFM fuel keys for use in the field.

Second, fuel information can be assigned to the categories or values of other GIS layers, such as vegetation or topography, using expert knowledge to create the fuel maps (i.e., for *calibration*). In this approach, experts assign fuel characteristics, such as an FBFM, FLM, or FCCS category, to each combination of mapped categories across selected data layers (Keane et al. 1998). Vegetation maps are most often used in fuel mapping projects (Menakis et al. 2000), and experts have assigned fuel classification categories to combinations of potential vegetation (i.e., biophysical setting), cover type, and structural stage (Keane et al. 1998, 2000; Schmidt et al. 2002; Reeves et al. 2009). In Canada, Hawkes et al. (1995) used experts to assign fuel types based on tree height, canopy closure, crown type, and cover type, and Nadeau and Englefield (2006) integrated the opinions of forest fire scientists using a fuzzy-logic engine to combine spatial data layers of land cover, biomass, and leaf area to create a map of Canadian Forest Fire Danger Rating System fuel types for Alberta. Keane et al. (2000) used experts to select the most appropriate FBFM for the combination of categories across three vegetation maps in New Mexico.

A third approach is to review fuel maps to refine mapping methods and update the input databases using expert knowledge (i.e., for *validation*). Reeves et al. (2009) asked fire management experts to evaluate portions of the preliminary LANDFIRE fuel maps to refine the mapping protocols so that they accounted for local conditions. This approach is commonly referred to as the "sniff test," because fuel and fire experts use their local knowledge to determine whether things "make sense"; that is, based on their experience, they critique the value and reliability of the fuel map with respect to their management objectives, and suggest ways to improve map quality (Keane et al. 2006). Keane et al. (2000) conducted workshops on the Gila National Forest in which fire managers refined the mapping of surface fire behavior model assignments to vegetation map categories based on their knowledge of the fuels in the mapped areas.

Last, fire and fuel experts can be used to create the spatial reference or ground-truthing information needed to assess the accuracy and precision of fuel maps (i.e., for *verification*). Local fuel experts can delineate important fuel types on maps that can then be used as a ground-truthing reference for the development and evaluation of the spatial fuel information. Previous fuel maps or mapping efforts developed based on expert experience can also be used as a validation tool, as can vegetation and stand maps that can be correlated with fuel properties.

11.2.3 How Is Knowledge Obtained from the Experts?

Perhaps the most common vehicle for obtaining expert knowledge is a workshop in which experts participate in a focused meeting to build the background knowledge that will be used for developing the fuel maps. These workshops can be attended in person, by telephone, or using videoconferencing technology. Extensive preparation is critical so that the experts can efficiently and effectively summarize their knowledge while staying focused on the specific mapping objective. For example, Keane et al. (1998) prepared detailed worksheets for combinations of vegetation classification categories in the Selway-Bitterroot Wilderness area so that fuel specialists could more easily assign a surface fire behavior model (Anderson 1982) to each vegetation category for their area. To improve workshop efficiency, it is sometimes beneficial to provide default knowledge or a "straw man" for workshop participants to critique and improve. For example, Keane et al. (2000) assigned fuel models to New Mexico vegetation types and then asked fire managers to review and update these assignments.

The workshop participants should agree beforehand on the process and parameters that will be used for the fuel model assignments and map development, and they should attempt to reach consensus on the assignments to create more consistent maps. For example, some fuel specialists may select a fuel model based on severe drought conditions at the height of a wildfire season, but that may be inappropriate if the fuel map will be used to predict the spread of prescribed fires during less dangerous portions of the fire season. Therefore, it is important that the group work together, based on a clear understanding of the map's objectives, to permit calibration and increase consistency.

Other means of obtaining expert knowledge include surveys and interviews. Though these avenues can be easier to implement, they are less desirable because they fail to provide a process by which the experts can calibrate their expertise relative to the mapping objective and the knowledge of others. In contrast, Hirsch et al. (2004) interviewed 141 fire managers to obtain their knowledge about fireline efficiency and initial attack productivity, because the context for this information was the same for each expert and the goal was to create statistical distributions that described this body of information. Although many researchers believe that the interview process should be relaxed and confidential, Keane et al. (1998) found that a more active dialog that included challenges to statements and assignments was needed to ensure that the information was consistent across respondents. Sometimes experts have little knowledge of a specific fuel characteristic, but have considerable experience in assessing fire behavior when fuel with this characteristic is burned. In these cases, surveys and workshops can let researchers infer information about the fuel from the expert's assessment of the fire behavior characteristics in that fuel. For example, experts can be shown photos of vegetation types with known fuel loads and asked to estimate potential flame lengths; if fuel loads are unknown, fire behavior models can be used to work backward to approximate the fuel loads needed to achieve the estimated flame lengths (Reeves et al. 2009). The expert knowledge collected from workshops, interviews, and surveys can be synthesized using many types of technology. For example, Nadeau and Englefield (2006) used fuzzy logic to summarize the opinions of fire scientists, whereas others have stored expert assignments and estimates in databases (Keane et al. 1998; Reeves et al. 2009).

11.3 The LANDFIRE Fuel Mapping Effort

The LANDFIRE project mapped wildland fuels, vegetation, and fire regime characteristics across the USA to support multiagency, multiscale fire management (Rollins 2009; Rollins and Frame 2006). This project was unique because of its national scope and its creation of an integrated suite of spatial data at 30-m spatial resolution, with complete coverage of all lands within the lower 48 states in the USA, comprising 64 mapping zones. The LANDFIRE fuel maps were created to support the use of critical fire behavior models such as FARSITE (Finney 1998) and FLAMMAP (Finney 2006). LANDFIRE was the first project of its kind to offer high-resolution, wall-to-wall wildland fuel spatial data for the USA (eight fuel data layers were mapped by LANDFIRE).

11.3.1 Surface Fuel Mapping

Two LANDFIRE surface fire behavior fuel model layers [FBFM13 for the 13 fuel models of Anderson (1982) and FBFM40 for the 40 fuel models of Scott and Burgan (2005)] and the two surface fuel load classifications [FCCSM for the FCCS models of Riccardi et al. (2007) and FLM of Lutes et al. (2009)] were used in this project. They were mapped by linking unique combinations of categories from several vegetation classifications that described the existing vegetation, plant height, canopy cover, and biophysical setting (Reeves et al. 2009) to the categories in the two FBFM classifications (Keane et al. 2001) and to the categories in the two fuel loading classifications (FCCSM, FLM; Fig. 11.1). Assignments for the FBFMs assumed environmental conditions that typify the fire weather that is normally encountered during the peak of the burning season for each geographic region being evaluated.

Very few agencies have sufficient georeferenced field data on fuels to permit fuel mapping, so plot-level data were mostly unavailable to facilitate the assignment of surface FBFMs, FCCSMs, or FLM to the LANDFIRE vegetation data products for all regions (Caratti 2006). Therefore, all fuel mapping rules (assignments) were accomplished using a qualitative approach based on the experience and knowledge

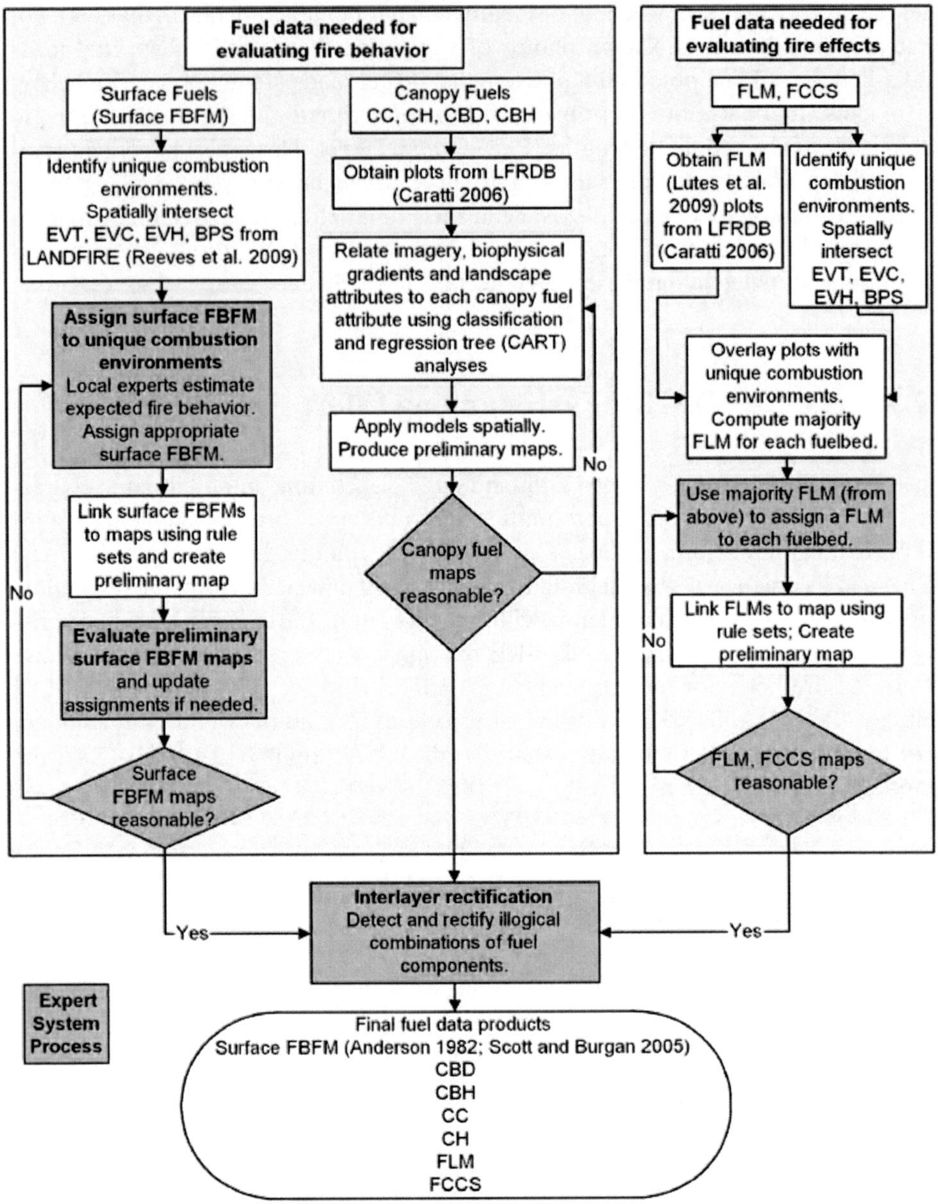

Fig. 11.1 A flowchart showing the procedures used to map surface and canopy fuel characteristics in the LANDFIRE national mapping project. The dark gray boxes indicate when expert knowledge was used to create, validate, and refine fuel maps. *BPS* biophysical setting, *CC* canopy cover (%), *CBD* canopy bulk density (kg m^{-3}); *CBH* canopy base height (m), *CH* canopy height (m), *EVC* existing vegetation cover (%), *EVH* existing vegetation height (m), *EVT* existing vegetation type, *FBFM* fire behavior fuel model, *FCCS* surface fire characteristics classification system, *FLM* fuel loading model, *LFRDB* LANDFIRE Reference DataBase

of the fire and fuel experts. These experts were usually fire behavior specialists who understood the fire behavior typically associated with the area being evaluated, but included other people associated with fire management (see Table 11.1), because many regions had no local fire behavior experts. In a series of integrated workshops, the fire experts evaluated each unique combination of vegetation classifications and predicted the fire behavior based on their experience. When experts were not available for a given LANDFIRE mapping zone, assignments from an adjacent mapping zone were used.

Initial review maps were created for each of the 64 LANDFIRE mapping zones once all unique combinations of the vegetation layers had been assigned surface fuel models. Each review map was then evaluated by a separate group of local fire and fuel specialists to detect areas where the surface FBFMs were obviously mischaracterized. During this intensive review period, approximately 5–20 local specialists updated, refined, and improved the review maps, and all disagreements between participants were resolved through majority vote when consensus could not be reached. When few experts were available for a mapping zone, experts from adjacent mapping zones were used. If obvious errors were detected, only the rule sets used to compare the surface FBFMs to the vegetation components were revised instead of subjectively updating individual pixels in the surface FBFM map.

11.3.2 Canopy Fuel Mapping

Four canopy map layers were created to describe canopy fuels – canopy bulk density (CBD, kg m^{-3}), canopy base height (CBH, m), canopy cover (CC, %), and canopy height (CH, m) – using regression-tree statistical modeling, in which field-referenced estimates were related to satellite imagery, biophysical gradients, stand structure, and vegetation composition data. The regression-tree models were formulated using the algorithm as implemented in the Cubist software (Rulequest Research, St. Ives, Australia). Canopy cover and height (CC, CH) were mapped using the methods of Zhu et al. (2006), and the CBH and CBD layers were mapped using the methods of Keane et al. (2006). The CBH and CBD canopy characteristics were estimated for each field plot using the FuelCalc software, which uses the algorithms of Reinhardt et al. (2006). Regression-tree models for CBH and CBD were developed using the spatially explicit predictor variables available in the LANDFIRE system (Keane et al. 2006), such as satellite reflectance, biophysical gradients, and vegetation structure and composition data (Reeves et al. 2009). Each regression tree was then applied across each mapping zone to produce preliminary maps of CBH and CBD. A gamma log-link generalized linear model (McCullagh and Nelder 1983) was then used to refine the CBD map by ensuring that the CBD predictions made sense in relation to CC (e.g., to eliminate high CBD values in areas of low CC; Reeves et al. 2009).

The resultant CBD, CH, CBH, and CC maps were evaluated by local experts in a series of workshops to eliminate illogical combinations. These experts determined thresholds for acceptable canopy fuel behavior stratified by other LANDFIRE mapping categories. This critical expert analysis ultimately improved the efficacy and accuracy of the canopy fuel maps. For example, during the interlayer rectification, experts assigned a CBH of 10 m and a CBD of 0.01 kg m^{-3} to deciduous stands to ensure that crown fires would not be simulated in this forest type.

A tenfold cross-validation procedure was used to assess the accuracy of the CBD and CBH regression-tree models by comparing plot-level estimates with mapped predictions at the same locations. No accuracy assessment was performed for the surface FBFMs because there were few independent datasets available and because different evaluators tend to estimate surface FBFMs differently, though consistent estimates between observers can sometimes be achieved (Burgan and Rothermel 1984). Despite this lack of an accuracy assessment for surface fuels, the abovementioned qualitative evaluation was performed during rule set development, expert review, and the annual postfire-season reviews (Fig. 11.1). Annual postfire-season reviews offered users of the LANDFIRE fuel data products a chance to discuss any issues with the data. Most of the maps derived from expert knowledge had low accuracies (<50%) and contained inconsistencies. Future improvements to this process must therefore include georeferenced field data to guide, evaluate, and eventually replace expert opinions. One advantage of using expert assignments of fuel attributes to the LANDFIRE vegetation map categories is that the successional models developed for LANDFIRE contain development pathways that can eventually be used to update the fuel maps for changes in the vegetation.

11.4 The Future of Expert Knowledge in Fuel Mapping

Expert knowledge has been indispensable in building contemporary fuel maps, and without this input, it is doubtful that today's fuel maps would be useful to fire managers. However, the goal of any mapping effort should be to minimize subjective bias by replacing qualitative expert knowledge with empirically driven, quantitative, and objective approaches. Although input from experts will continue to play an important role in the development of fuel maps, tomorrow's fuel layers should be designed so that the methods are repeatable, quantitative, and unbiased, and the maps are constructed using a combination of detailed georeferenced field data, high-resolution remote-sensing data, complex ecosystem simulations, novel GIS techniques, knowledge-based systems, and advanced statistical analyses (Keane et al. 2001). The first step is to develop new fire behavior prediction models that use inputs that can be easily defined, measured, and summarized in the field. Then, new surface fuel and canopy fuel sampling methods must be developed and adopted by land management agencies to allow the development of extensive, standardized, spatially explicit databases of fuel conditions that can be used for map development, testing, and validation (Krasnow et al. 2009; Lutes et al. 2009).

A variety of remote-sensing technologies, such as LIDAR (Koetz et al. 2008), radar (Bergen and Dobson 1999), digital photography (Bailey and Mickler 2007), hyperspectral imagery (Jia et al. 2006), high-resolution images (Lasaponara and Lanorte 2007), and a melding of various images (Keramitsoglou et al. 2008; Koetz et al. 2008), will vastly improve fuel mapping compared with current methods that rely heavily on the use of LANDSAT images. Ecosystem simulation models can be used in combination with climate, soil, and topographic information to spatially describe the biophysical environment and thereby improve fuel mapping (Rollins et al. 2004). Progressive GIS techniques can be used to integrate spatial data layers in such a way as to predict the most appropriate fuel model (Hawkes et al. 1995; Chuvieco and Salas 1996). In addition, new statistical analysis techniques, such as regression trees, gradient nearest-neighbor analysis, fuzzy logic, and hierarchical modeling, are needed to integrate the biophysical controls on fuel properties within fuel maps (Ohmann 1996; Ohmann and Spies 1998; Nadeau and Englefield 2006).

In the meantime, innovative analytical techniques must integrate expert knowledge into a repeatable, quantitative map-building process, based on the best expert-systems technologies (Goulstone et al. 1994; de Vasconcelos et al. 1998). For example, CBH can be indirectly estimated by mathematically solving an empirical equation for the CBH required to allow a fire to transition from a surface fire into a crown fire assuming various fire behavior parameters (Reeves et al. 2009). The expert contribution to this technique involves panels of local fire behavior prediction experts who collectively determine the conditions under which a stand will likely transition from a surface fire into a crown fire. This approach combines empirical modeling with expert knowledge and will consistently estimate crown fire activity if the assumed environmental conditions are realized.

Fuel classifications such as those discussed in this chapter could also be improved so that the resulting fuel maps provide higher quality inputs for fire behavior and fire effects models. The first step in this process would be to build fire behavior models that are more sensitive to realistic (i.e., field-based) estimates of fuel inputs (Arroyo et al. 2008). For example, most fire behavior models are implemented in only one dimension (point models), but wildland fire behavior occurs across three dimensions (3D). Thus, 3D fire behavior and fire effects models must be built to account for fire processes that are influenced by the vertical, longitudinal, and horizontal distributions of fuels. For example, radiation and convection are important heat-transfer processes that must be simulated in 3D to fully describe complex fire behaviors in complex fuels (Linn 1997). Once the necessary 3D models are developed, they will require innovative fuel classifications that not only describe fuel properties such as fuel load across components, but also describe how these properties are distributed both spatially (Reich et al. 2004) and temporally (Keane 2008). Future fuel classifications should be based on extensive (regional to continental) and comprehensive (all fuel components) field data (McKenzie et al. 2007), and they should be designed to emphasize differences in fuel bed properties, not only the vegetation type, structure, and topographic setting (Lutes et al. 2009). Last, there should be considerable expert knowledge built into these classifications to ensure that they will be useful to the fire managers who will use them.

We also believe that fuel experts, fire behavior analysts, and fire managers may need to rethink their paradigms for fuel description to allow for the development of higher quality fuel maps in the future. Fuel classifications, though popular, efficient, and easy to use, may be inappropriate in the future because fuel properties are not correlated across fuel components, their properties vary across different scales, and the classification categories are limited, restrictive, and subjective. Fuel components may need to be mapped independently at the most appropriate scale for a given management task to ensure accurate fire behavior prediction. For example, a digital map of coarse woody fuels could be created at a 30-m pixel resolution, whereas a fine-fuel map might require a pixel resolution of 1–5 m. Fuel component definitions should also be investigated to develop more flexible and comprehensive methods for describing the fuels and providing model inputs. For example, the size class distributions for coarse woody debris could be quantified for a fuel bed so that the fuel load can be computed for woody fuels of any size instead of using the four restrictive size classes that are currently used. Designing woody fuel size classes based on the drying time (Fosberg 1970) is probably inappropriate for accurately estimating fuel loadings and carbon pools. Tomorrow's fuel experts must be willing to modify their view of fuel complexes to permit the development of innovative wildland fuel maps. And, for these experts to modify their approach, researchers must present them with a new approach that they can understand, trust, and learn how to apply in their daily work.

11.5 Summary

Expert knowledge is an indispensable tool in the development of wildland fuel maps, and most mapping efforts have extensively used information gained from experts to support many phases of the fuel mapping process. However, the high variability of fuels, coupled with the subjective nature of expert knowledge, will require a stronger reliance on empirical data and statistical analysis to generate effective fuel maps in the future. Although expert knowledge will continue to play a critical complementary and supplementary role in future fuel mapping efforts, fuel mapping must incorporate a less-subjective means of map development. This will be difficult because it will require a complete overhaul of how fire managers think about fuel and the development of new fire behavior models and leads to the design of new fuel classifications and new fuel sampling protocols. For this change in thought to be possible, researchers must find ways to understand the real-world challenges faced by fire managers so that it is possible to communicate the advances in fire science in a way fire managers can understand and accept. Only in this way will the new science be adopted and incorporated into future fire management.

References

Agee JK, Pickford SG, Kertis J et al (1985) Vegetation and fuel mapping of North Cascades National Park Service complex. National Park Service Cooperative Park Studies Unit, College of Forest Resources, University of Washington, Seattle, Final Report Contract CX-9000-3-E029

Albini FA (1976) Estimating wildfire behavior and effects. USDA Forest Service, Intermountain Research Station, Ogden. General Technical Report INT-30

Anderson HE (1982) Aids to determining fuel models for estimating fire behavior. USDA Forest Service Intermountain Research Station, Ogden, General Technical Report INT-122

Arroyo LA, Pascual C, Manzanera JA (2008) Fire models and methods to map fuel types: the role of remote sensing. Ecol Manage 256:1239–1252

Bailey AD, Mickler R (2007) Fine scale vegetation classification and fuel load mapping for prescribed burning. In: Butler BW, Cook W (eds) The fire environment – innovations, management, and policy. USDA Forest Service, Rocky Mountain Research Station, Fort Collins, Proceedings RMRS-P-46CD, pp 261–270

Bergen KM, Dobson MC (1999) Integration of remotely sensed radar imagery in modeling and mapping of forest biomass and net primary production. Ecol Modell 122:257–274

Brown JK, Bevins CD 1986. Surface fuel loadings and predicted fire behavior for vegetation types in the northern Rocky Mountains. USDA Forest Service, Intermountain Forest and Range Experiment Station, Ogden, Research Note INT-358

Brown JK, See TE (1981) Downed dead woody fuel and biomass in the northern Rocky Mountains. USDA Forest Service, Intermountain Forest and Range Experiment Station, Ogden, General Technical Report INT-117

Burgan RE (1987) Concepts and interpreted examples in advanced fuel modeling. USDA Forest Service, Intermountain Research Station, Ogden, General Technical Report INT-238

Burgan RE, Klaver RW, Klaver JM (1998) Fuel models and fire potential from satellite and surface observations. Internat J Wild Fire 8:159–170

Burgan RE, Rothermel RC (1984) BEHAVE: fire behavior prediction and fuel modeling system – FUEL subsystem. USDA Forest Service, Intermountain Research Station, Ogden, General Technical Report INT-167

Caratti J (2006) The LANDFIRE prototype project reference database. In: Rollins M, Frame C (eds) The LANDFIRE prototype project: nationally consistent and locally relevant geospatial data for wildland fire management. USDA Forest Service, Rocky Mountain Research Station, Ogden, RMRS-GTR-175, pp 367–396

Chuvieco E, Salas J (1996) Mapping of spatial distribution of forest fire danger using GIS. Internat J Geogr Inf Syst 10:333–345

DeBano LF, Neary DG, Ffolliott PF (1998) Fire's effect on ecosystems. John Wiley and Sons, New York

De Vasconcelos MJP, Paul JCU, Silva S et al (1998) Regional fuel mapping using a knowledge based system approach. In: Viegas DX (ed) 3rd International Conference on Forest Fire Research and 14th Conference on Fire and Forest Meteorology, Luso,University of Coimbra, pp 2111–2123

Deeming JE, Burgan RE, Cohen JD (1977) The National Fire Danger Rating System – 1978. USDA Forest Service, Intermountain Forest and Range Experiment Station, Ogden, General Technical Report INT-39

Falkowski MJ, Gessler PE, Morgan P et al (2005) Characterizing and mapping forest fire fuels using ASTER imagery and remote sensing. For Ecol Manage 217:129–146

Finney MA (1998) FARSITE: Fire Area Simulator – model development and evaluation. USDA Forest Service, Rocky Mountain Research Station, Ft. Collins, Research Paper RMRS-RP-4

Finney, MA (2006) An overview of FlamMap fire modeling capabilities. In: Andrews P, Butler B (eds) Fuels Management – How to Measure Success. USDA Forest Service, Rocky Mountain Research Station, Fort Collins, Proceedings RMRS-P-41, pp 213–220

Fosberg MA (1970) Drying rates of heartwood below fiber saturation. For Sci 16:57–63
Gonzalez JR, Kolehmainen O, Pukkala T (2007) Using expert knowledge to model forest stand vulnerability to fire. Comp Elec Agric 55:107–114
Goulstone AG, Xiang WN, Sox J (1994) GIS, expert system technologies improve forest fire management techniques. GIS World 7:32–36
Grupe MA (1998) Assessing the applicability of the terrestrial ecosystem survey for FARSITE. University of New Mexico, Albuquerque, Master's Thesis
Hardwick PE, Lachowski H, Forbes J et al (1998) Fuel loading and risk assessment Lassen National Forest. In: Greer JD (ed) Proceedings of the Seventh Forest Service Remote Sensing Applications Conference. American Society for Photogrammetry and Remote Sensing, Bethesda, Maryland, pp 328–339
Hawkes B, Niemann O, Goodenough D et al (1995) Forest fire fuel type mapping using GIS and remote sensing in British Columbia. In: Proceedings of the Symposium GIS Applications in Natural Resources 2–9th symposium on Geographic Information Systems, Vancouver. GIS World, Fort Collins, pp 290–299
Hirsch KG, Podur JJ, Janser RF et al (2004) Productivity of Ontario initial-attack fire crews: results of an expert-judgement elicitation study. Can J For Res 34:705–715
Hornby LG (1935) Fuel type mapping in Region One. J For 33:67–72
Jia GJ, Burke IC, Goetz AFH et al (2006) Assessing spatial patterns of forest fuels using AVIRIS data. Remote Sens Environ 102:318–327
Keane RE (2008) Surface fuel litterfall and decomposition in the northern Rocky Mountains, USA. USDA Forest Service, Rocky Mountain Research Station, Fort Collins, Research Paper RMRS-RP-70
Keane RE, Burgan RE, Wagtendonk JV (2001) Mapping wildland fuels for fire management across multiple scales: Integrating remote sensing, GIS, and biophysical modeling. Internat J Wild Fire 10:301–319
Keane RE, Frescino TL, Reeves MC, Long J (2006) Mapping wildland fuels across large regions for the LANDFIRE prototype project. In: Rollins M, Frame C (eds) The LANDFIRE prototype project: nationally consistent and locally relevant geospatial data for wildland fire management. USDA Forest Service, Rocky Mountain Research Station, Ogden, RMRS-GTR-175, pp 367–396
Keane RE, Garner JL, Schmidt KM et al (1998) Development of input spatial data layers for the FARSITE fire growth model for the Selway-Bitterroot Wilderness complex, USA. USDA Forest Service, Rocky Mountain Research Station, Fort Collins, General Technical Report RMRS-GTR-3
Keane RE, Mincemoyer SE, Schmidt KM et al (2000) Mapping vegetation and fuels for fire management on the Gila National Forest Complex. USDA Forest Service, Rocky Mountain Research Station, Ogden, General Technical Report RMRS-GTR-46-CD
Keane RE, Veblen T, Ryan KC et al (2002) The cascading effects of fire exclusion in the Rocky Mountains. In: Baron J, Hauer R, Fagre D (eds) Rocky Mountain Futures: An Ecological Perspective. Island Press, Washington, pp 133–153
Keramitsoglou I, Kontoes C, Sykioti O et al (2008) Reliable, accurate, and timely forest mapping for wildfire management using ASTER and Hyperion satellite imagery. For Ecol Manage 255:3556–3562
Koetz B, Morsdorf F, van der Linden S et al (2008) Multi-source land cover classification for forest fire management based on imaging spectrometry and LiDAR data. For Ecol Manage 256:263–271
Krasnow K, Schoennagel T, Veblen TT (2009) Forest fuel mapping and evaluation of the LANDFIRE fuel maps of Boulder County, Colorado, USA. For Ecol Manage 257:1603–1612
Lachowski H, Maus, P, Golden, M et al (1995) Guidelines for the use of digital imagery for vegetation mapping. USDA Forest Service, Engineering Staff, Ogden, EM-7140–25
Lasaponara R, Lanorte A (2007) On the capability of satellite VHR Quickbird data for fuel type characterization in fragmented landscapes. Ecol Modell 204:79–84
Leenhouts B (1998) Assessment of biomass burning in the conterminous United States. Conserv Ecol 2:1–23

Lenihan JM, Daly C, Bachelet D, Neilson RP (1998) Simulating broad scale fire severity in a dynamic global vegetation model. Northwest Sci 72:91–103

Linn RR (1997) A transport model for prediction of wildfire behavior. New Mexico State University, Las Cruces, Ph.D. thesis

Lutes DC, Keane RE, Caratti JF (2009) A surface fuels classification for estimating fire effects. Internat J Wild Fire 18:802–814

McCullagh P, Nelder JA (Eds) (1983) Generalized Linear Models. Chapman and Hall, London

McKenzie D, Raymond CL, Kellogg L et al (2007) Mapping fuels at multiple scales: landscape application of the Fuel Characteristic Classification System. Can J For Res 37:2421–2437

McKinley RA, Chine EP, Werth LF (1985) Operational fire fuels mapping with NOAA-AVHRR data. In: Pecora X Symposium Proceedings. American Society for Photogrammetry and Remote Sensing, Bethesda, pp 295–304

Menakis JP, Keane RE, Long DG (2000) Mapping ecological attributes using an integrated vegetation classification system approach. J Sust For 11:245–265

Mutlu M, Popescu SC, Zhao K (2008) Sensitivity analysis of fire behavior modeling with LIDAR-derived surface fuel maps. For Ecol Manage 256:289–294

Nadeau LB, Englefield P (2006) Fine-resolution mapping of wildfire fuel types for Canada: Fuzzy logic modeling for an Alberta pilot area. Environ Mon Assess 120(1–3):127–152

Ohmann JL (1996) Linking plot data, models, and maps in regional ecological analysis. In: Society of American Foresters 1995 Convention. Society of American Foresters, Bethesda, pp 99–103

Ohmann JL, Spies TA (1998) Regional gradient analysis and spatial pattern of woody plant communities of Oregon forests. Ecol Monogr 68:151–182

Oswald BP, Fancher JT, Kulhavy DL, Reeves HC (1999) Classifying fuels with aerial photography in East Texas. Internat J Wild Fire 9:109–113

Pala S, Taylor D, Holder G (1990) Integrating satellite-derived forest fuel data into fire management decision support models. In: Proceedings, Second National GIS Conference, GIS World, Fort Collins, pp 345–356

Reeves MC, Ryan, KC, Rollins, MC et al (2009) Spatial fuel data products of the LANDFIRE project. Internat J Wild Fire 18:250–267

Reich RM, Lundquist JE, Bravo VA (2004) Spatial models for estimating fuel loads in the Black Hills, South Dakota, USA. Internat J Wild Fire 13:119–129

Reinhardt E, Scott JH, Gray KL, Keane RE (2006) Estimating canopy fuel characteristics in five conifer stands in the western United States using tree and stand measurements. Can J For Res 36:1–12

Riccardi CL, Prichard SJ, Sandberg DV, Ottmar RD (2007) Quantifying physical characteristics of wildland fuels using the Fuel Characteristic Classification System. Can J For Res 37:2413–2420

Roberts D, Gardner M, Regelbrugge J et al (1998) Mapping the distribution of wildfire fuels using AVIRIS in the Santa Monica Mountains. In: Proc. 7th AVIRIS Earth Science Workshop, NASA, Pasadena, JPL 98–21, pp 345–352

Rollins AM, Yool SR (2002) Characterizing fuel load and topographic relationships in a montane canyon of Southern Arizona. MS Thesis, University of Arizona, Tucson

Rollins MG (2009) LANDFIRE: a nationally consistent vegetation, wildland fire, and fuel assessment. Internat J Wild Fire 18:235–249

Rollins MG, Frame C (2006) The LANDFIRE Prototype Project: nationally consistent and locally relevant geospatial data for wildland fire management. USDA Forest Service, Rocky Mountain Research Station, Fort Collins, General Technical Report RMRS-GTR-175

Rollins MG, Keane RE, Parsons RP (2004) Mapping ecological attributes using gradient analysis and remote sensing. Ecol Appl 14:75–95

Rothermel RC (1972) A mathematical model for predicting fire spread in wildland fuels. USDA Forest Service, Intermountain Forest and Range Experiment Station, Ogden, Research Paper INT-115

Sandberg DV, Ottmar RD, Cushon GH (2001) Characterizing fuels in the 21st century. Internat J Wild Fire 10:381–387

Schmidt, KM, Menakis JP, Hardy CC, Hann WJ, Bunnell, DL (2002) Development of coarse-scale spatial data for wildland fire and fuel management. USDA Forest Service Rocky Mountain Research Station, Fort Collins, General Technical Report RMRS-GTR-87

Scott J, Burgan RE (2005) A new set of standard fire behavior fuel models for use with Rothermel's surface fire spread model. USDA Forest Service Rocky Mountain Research Station, Fort Collins, General Technical Report RMRS-GTR-153

Sikkink P, Keane RE (2008) A comparison of five sampling techniques to estimate surface fuel loading in montane forests. Internat J Wild Fire 17:363–379

Sikkink P, Keane RE, Lutes DC (2009) Field guide for identifying fuel loading models. USDA Forest Service, Rocky Mountain Research Station, Fort Collins, General Technical Report RMRS-GTR-225

Wilson BA, Ow CFY, Heathcott M et al (1994) Landsat MSS classification of fire fuel types in Wood Buffalo National Park, Northern Canada. Global Ecol Biogeogr Lett 4:33–39

Zhu Z, Vogelmann J, Ohlen D et al (2006) Mapping existing vegetation composition and structure. In: Rollins M, Frame C (eds) The LANDFIRE prototype project: nationally consistent and locally relevant geospatial data for wildland fire management. USDA Forest Service, Rocky Mountain Research Station, Fort Collins, General Technical Report RMRS-GTR-175, pp 195–215

Chapter 12
Using Bayesian Mixture Models That Combine Expert Knowledge and GIS Data to Define Ecoregions

Kristen J. Williams, Samantha Low-Choy, Wayne Rochester, and Clair Alston

Contents

K.J. Williams (✉)
CSIRO Ecosystem Sciences, G.P.O. Box 1700, Canberra, ACT 2601, Australia
e-mail: kristen.williams@csiro.au

S. Low-Choy
Cooperative Research Centre for National Plant Biosecurity and Faculty of Science and Technology, Queensland University of Technology, 2 George Street, Brisbane, QLD 4001, Australia

W. Rochester
CSIRO Marine and Atmospheric Research, G.P.O. Box 2583, Brisbane, QLD 001, Australia

C. Alston
Faculty of Science and Technology, Queensland University of Technology, 2 George Street, Brisbane, QLD 4001, Australia

A.H. Perera et al. (eds.), *Expert Knowledge and Its Application in Landscape Ecology*, DOI 10.1007/978-1-4614-1034-8_12, © Springer Science+Business Media, LLC 2012

12.1 Introduction

Conservation planning and management programs typically assume relatively homogeneous ecological landscapes. Such "ecoregions" serve multiple purposes: they support assessments of competing environmental values, reveal priorities for allocating scarce resources, and guide effective on-ground actions such as the acquisition of a protected area and habitat restoration. Ecoregions have evolved from a history of organism–environment interactions, and are delineated at the scale or level of detail required to support planning. Depending on the delineation method, scale, or purpose, they have been described as provinces, zones, systems, land units, classes, facets, domains, subregions, and ecological, biological, biogeographical, or environmental regions. In each case, they are essential to the development of conservation strategies and are embedded in government policies at multiple scales.

For simplicity, we have chosen "ecoregion" to imply a relatively homogeneous landscape planning unit defined based on ecological principles that will guide the design of conservation practices to produce specific outcomes. To guide conservation policy, a hierarchy of these units must provide a context for biological patterns that may be poorly understood (Leathwick et al. 2011), including future biodiversity compositions arising from dynamic ecological processes and drivers of environmental change (Beier and Brost 2010).

Researchers generally use a spatially nested or hierarchical structure for ecoregions, with sizes depending on the study's purpose and boundaries based on distinct landscape components (Hardman-Mountford et al. 2008). At higher hierarchical levels, ecoregions resemble and broadly correspond to climatic and landform systems; at lower levels, they define increasingly homogeneous climatic, physiographic, and biotic characteristics (Bailey 2004). Though some boundaries are clearly defined by sharp gradients such as geological changes (e.g., mountains) that are reflected in structural changes in the vegetation, other boundaries are arbitrary divisions along gradually changing environmental and biotic gradients. In these latter situations, boundary locations are imprecise and experts choose a position appropriate for the study goals (McMahon et al. 2001).

Because ecoregion boundaries are often fuzzy, and because the degree of heterogeneity varies, different jurisdictions have adopted different methods of combining data and expert knowledge (McMahon et al. 2004). To be implemented in policies and plans, ecoregions must be widely accepted by land managers from various backgrounds. Therefore, ecoregions are commonly derived first using data-driven methodologies and second using expert judgments (Loveland and Merchant 2004). The data-driven ecoregions are reviewed by experts and practitioners to ensure boundaries are realistic and useful. For example, experts may adjust how they have weighted particular aspects to build a mental model of the landscape that better reflects their perceptions (Hargrove and Hoffman 2004). Agencies that have invested significant time and resources in developing a regionalization won't readily switch systems, so as new datasets, models, and classification methods are developed, new ways to combine them with existing systems become necessary.

In this chapter, we present a case study on validating and refining subregion boundaries in Australia's Southeast Queensland ecoregion. Specifically, we discuss how best to create subregions that match subjective expert assessments while integrating relevant empirical data. To do so, we discuss appropriate statistical methods and the respective roles of expert knowledge and empirical data. Based on this discussion, we demonstrate the use of Bayesian mixture models and their interpretation, thereby providing a rigorous and repeatable way to create or revise ecoregions that integrate expert knowledge with empirical data. We conclude by identifying the strengths and weaknesses of the approach, challenges to its implementation, and recommendations for better capturing expert knowledge generated by existing management processes.

12.2 Case Study

Australia's terrestrial ecoregions ("bioregions") provide a guiding framework for biodiversity conservation strategies (Thackway and Cresswell 1995; Environment Australia 2000). They provide crucial inputs for several planning mechanisms, including selection of representative conservation areas (Natural Resource Management Ministerial Council 2004), assessment of landscape health (Morgan 2000), and State of the Environment reporting (Cork et al. 2006). These assessments allow quantitative reporting and monitoring of trends, and therefore influence future planning for sustainable development.

Our case study focuses on developing a subregional characterization of the Southeast Queensland ecoregion (Fig. 12.1). This region is recognized for its high biodiversity, which includes many unique endemic species, many near their northern or southern distribution limits. It contains 151 regional ecosystems including numerous significant wetlands (Sattler and Creighton 2002).

The Southeast Queensland ecoregion was first delineated by Stanton and Morgan (1977), who described it as a place of great physical and biological diversity, characterized by a subtropical climate. Landforms and vegetation, distinguished by airphoto interpretation, were the main attributes used to delineate ecoregions at this time, but were supplemented by published information. The ecoregion extends as far north as Curtis Island, near the 900-mm rainfall isohyet, and is bounded by the watershed of the ranges that form part of the Queensland state border with New South Wales to the south (Fig. 12.1).

The rapid assessment by Stanton and Morgan (1977) (see their Appendix A) proposed arbitrary contour lines to differentiate lowland and highland landforms and landscapes, but could not define Southeast Queensland's complex subregions. Young and Dillewaard (1999) subsequently delineated ten subregions from the 13 "environmental provinces" proposed by Young and Cotterell (1993) (Fig. 12.1). These subregions were principally defined based on expert knowledge guided by maps of environmental zones following the approach of Morgan and Terrey (1990), who emphasized structural geology (major geological formations) and local climates. Where suitable land system maps (Christian and Stewart 1968) existed, they also informed the delineation of subregion boundaries.

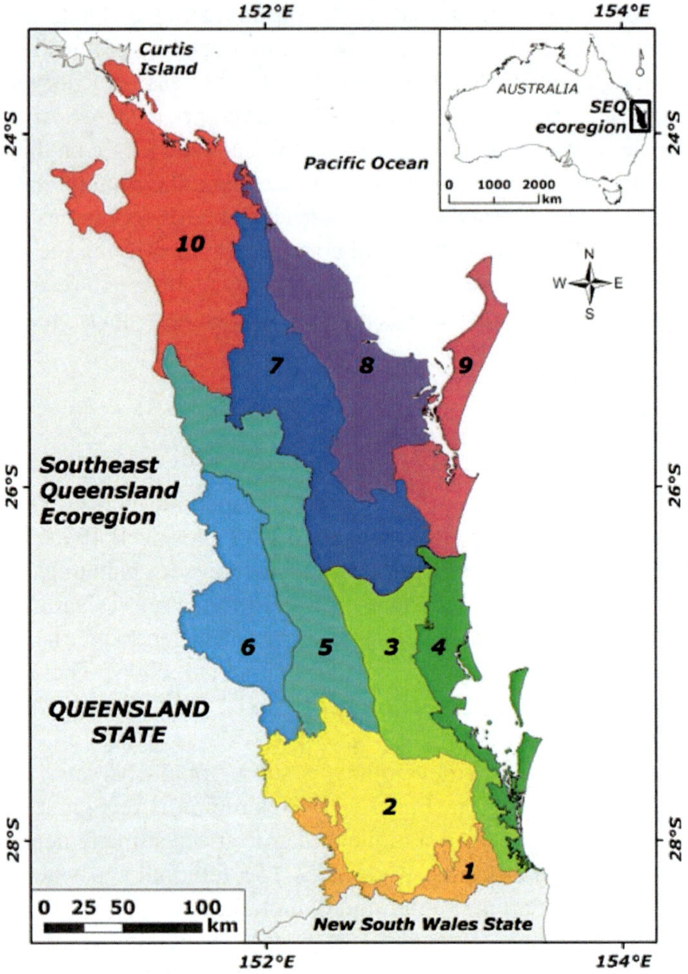

Fig. 12.1 Location of the ten subregions of the Southeast Queensland ecoregion. Subregions: 1, Scenic Rim; 2, Moreton Basin; 3, Southeast Hills and Ranges; 4, Southern Coastal Lowlands; 5, Brisbane–Barambah Volcanics; 6, South Burnett; 7, Gympie Block; 8, Burnett-Curtis Coastal Lowlands; 9, Great Sandy; 10, Burnett-Curtis Hills and Ranges. Data sources: ecoregions – Queensland Herbarium BIOPROV v4.2 2001; coastline and state borders – Geoscience Australia GEODATA Coast 100K 2004

12.3 Methodology for Developing Subregions

We chose Austin's (2002) framework for statistical analysis in ecology to guide our subregion development. Austin's ecology–statistics–data (ESD) framework is an approach explicitly grounded in ecological theory, with statistical methods and data chosen based on that theory. By using an explicit multivariate normal mixture model (MNMM; Fraley and Raftery 2006; Frühwirth-Schnatter 2006), we provide a foundation for statistical and therefore ecological inference. Here, we describe the ESD framework's three components for ecoregion delineation: the underlying (ecological) conceptual model; the choice and scale of attributes (in the data) used to develop the ecoregions; and the quantitative (statistical) method that accepts these attributes as inputs and provides ecoregion boundaries and descriptions as outputs.

12.3.1 The Underlying Ecoregion Conceptual Model

Any set of hypothetical ecoregions is based on attributes that capture environmental, topographic, bioclimatic, or biodiversity themes. Selection of attributes depends on the conceptual ecological model (the framework) chosen. We therefore start by discussing how these attributes relate to a conceptual ecological model (the "E" in ESD). Such models start with the main drivers of biophysical patterns then match them to attributes that can be adequately measured. For ecoregions, these are the ecological and geomorphological processes that drive the formation and distribution of biota and whose degree of homogeneity characterizes each ecoregion at a given scale.

Terrestrial ecoregions are defined by factors that determine variability at landscape to regional scales, primarily topography, annual bioclimatic indices (temperature, rainfall), and geology. In contrast, subregions are defined by factors that determine variability at local to landscape scales, primarily fine-scale topography, seasonal variation in bioclimatic indices (temperature, rainfall extremes), and soil characteristics that influence resource availability to plants. These ecosystems form readily identifiable land or habitat types that capture the mosaic of variation associated with soil sequences and disturbance regimes.

These principles are well-understood and embedded within both environmental science and regulations, such as when regional ecosystems are defined within an ecoregion-based planning framework (Sattler and Williams 1999). The greatest certainty in ecoregion boundary delineation occurs at either end of a spectrum that recognizes the fine-scale diversity among regional ecosystems, as in the example of the Queensland vegetation types (Neldner et al. 2005), or that recognizes broad-scale differences, as in Australia's ecoregions (Thackway and Cresswell 1995). Our Southeast Queensland study focuses on an intermediate scale, with high uncertainty in boundary delineation. Subregion definition, therefore, lets us test a model-based approach that can incorporate information from both experts and empirical sources.

12.3.2 Information (Attributes)

We used two information sources in this case study: expert delineations of ten subregions in the interim biogeographic regionalization for Australia (Environment Australia 2000) and geographical information system (GIS) spatial layers for environmental attributes (Rochester et al. 2004).

12.3.2.1 Expert-Delineated Boundaries

All of Queensland's ecoregions and subregions were defined subjectively using expert knowledge supported by maps of the geology, soils, and land systems and by LANDSAT images (Sattler and Williams 1999; Environment Australia 2000). The resulting boundaries quantify the experts' beliefs and initial assessments of relatively homogeneous and spatially contiguous environmental groupings. Typically, boundaries were attached to mapped topographical features (e.g., ridgelines, waterways), boundaries from finer-scale vegetation or soil mapping, or administrative borders (e.g., between states). Anchoring expert boundaries to identifiable attributes facilitated spatial alignment with the GIS layers in the preceding data-driven component.

Queensland's ecoregions and subregions are formally reviewed by expert panels who successively refine the boundaries used to guide environmental policies and legislation (i.e., the *Vegetation Management Act 1999*; the *Integrated Planning Act 1997*). Experts have experience in vegetation mapping, soil science, or conservation planning in the ecoregion under review. Key information is provided (including recent publications), particularly in relation to locations where there is ecological evidence for boundary shifts. Maps, including GIS visualizations, facilitate exploration, discussion, and definition of boundaries. Consensus is achieved through facilitated discussions, and both conclusions and the underlying rationale are recorded. Revision opportunities arise at intervals determined by the availability of new information or by unresolved issues.

Although the resulting ecoregions are sound, and major changes are not desired, minor changes are made through regional ecosystem reviews (Neldner et al. 2005). However, there is less consensus on subregion boundaries because the higher levels of environmental variation have confounded expert interpretation. Thus, formal integration of expert knowledge with empirical data to refine subregion boundaries would be desirable (e.g., Accad et al. 2005).

We used the expert-delineated Southeast Queensland subregion boundaries, derived following a process similar to that described above, to infer the mean values and ranges for each subregion's key environmental factors (see Sect. 12.3.2.2). Though the uncertainty in boundary positions was not explicitly captured by the expert panels, it should be possible to elicit this information. Our model compares this expert classification with one based on relatively independent environmental factors using proxies (empirical data) drawn from available GIS layers (Rochester et al. 2004).

12.3.2.2 GIS Data

We inferred ecological hypotheses from the bioregional framework descriptions of Stanton and Morgan (1977), Sattler and Williams (1999), and Thackway and Cresswell (1995) corresponding to expert-defined subregion boundaries and their associated geological, landform, annual rainfall, and elevation descriptions by Young and Dillewaard (1999) to guide our selection of GIS data (Table 12.1). In this study, we

Table 12.1 Relationships between the GIS datasets and the ecological hypotheses tied to the proxies for environmental factors (geology, landform, elevation, rainfall, and vegetation) that were used by experts to describe the ten subregions of the Southeast Queensland ecoregion. Proxy environmental factors were interpreted from Table 12.1 in Young and Dillewaard (1999)

Ecological hypothesis	Environmental proxy	GIS dataset
Rock attributes influence soil formation processes, thereby affecting the soil conditions, nutrients, and water availability that affect plant growth, resulting in characteristic vegetation types	Geology (surface lithology) was grouped into 12 provinces that represent the variation from basic to acid igneous rocks, fine- to coarse-grained sediments, sedimentary sources and ages, the degree of consolidation, metamorphosis, and development of duricrusts	Clay content in the A horizon (Clay) soil water-holding capacity of the A and B horizons (Moisture)
Landform type, topographic position, and landform heterogeneity interact with geology and climate to generate localized variations in soil formation, thereby influencing the development of characteristic vegetation types	Landforms were grouped into seven types that characterize the topographic variation associated with mountain ranges, plateaus, narrow to broad valleys, rolling hills, coastal plains, and sand dunes	Compound topographic index (CTI)
Annual rainfall is an important determinant of plant productivity and biomass. When balanced against evaporation, it represents water availability, which may be seasonally limiting, thereby influencing the vegetation types that can be supported by an area	Annual rainfall was defined in 50-mm increments up to 900 mm and then in 100-mm increments, representing moisture limitations during the dry season that influence vegetation type	Annual rainfall (Rain) total rainfall during the warmest quarter (Wettest)
Elevation is topographically correlated with climatic factors such as temperature and rainfall. Higher elevations experience different weather conditions than lower elevations due to topographic funneling of wind, greater exposure, and adiabatic cooling of air, leading to increased precipitation. Lower elevations are generally found in coastal regions, where air temperatures are moderated by proximity to the ocean	Mean elevation was defined in 25-m increments up to 50 m, 50-m increments to 300 m, and then in 100-m increments to reflect the steep gradient from coastal lowlands to inland mountains	Mean annual temperature (Heat) mean minimum temperature during the coolest month (Coldest) mean maximum temperature during the warmest month (Hottest)

defined each 1,250-m pixel in the GIS data layer as a single site, and used the environmental characteristics of that site in our subsequent analysis of the empirical data. Four groups of environmental variables were selected from Rochester et al.'s (2004) data: for geology, the clay content and water-holding capacity of the soils, which represent the texture and depth of surficial deposits; for landforms, a topographic wetness index derived from a digital elevation model; for rainfall, the annual total and total during the warmest quarter; and for elevation, mean annual temperature and the mean maximum and minimum temperatures during the warmest and coolest month, respectively. Although these factors are relatively independent proxies for subregional ecological patterns, they are interrelated through the physical processes of climate–soil–landscape systems. We selected them because they are more directly correlated with the resources and conditions that influence vegetation patterns than indirect factors such as altitude and geology. To estimate the degree of interdependence among these variables, we calculated pairwise correlations among them.

12.3.3 Quantitative (Statistical) Model

Bayesian statistics provides an intuitive framework for examining the spectrum of results from an expert-dominated to a data-dominated analysis, but requires the definition of an explicit and quantitative model (Low-Choy et al. 2009). To achieve this, it helps to view ecoregion classification methods as model-based or model-free. The model-based approaches provide a well-specified statistical model whose components can be directly interpreted using the conceptual ecological model (i.e., strongly linking the "S" and "E" of ESD). Since the model is Bayesian and statistical it uses the data (i.e., the "D" in ESD) to update initial estimates of parameters that define the model. In some ecological applications, such as mapping natural hazards or rainfall-runoff, classification could also be achieved via a deterministic model (Goswami et al. 2007), which relies entirely on the ecological conceptual model. Although data may be used to evaluate the predictive performance of such deterministic ecoregions, data are not used to update parameter estimates.

Model-free approaches range from expert specification of ecoregion boundaries to the use of data-mining algorithms to group similar sites; a spectrum documented by Loveland and Merchant (2004) for mapping American ecoregions. In contrast to model-based approaches, model-free approaches utilize an algorithm that focuses on accurate prediction of boundaries rather than on providing a single explanatory model to describe ecoregions (e.g., Hargrove and Hoffman 2004; Snelder et al. 2010).

12.3.3.1 Bayesian Multivariate Normal Mixture Model

Here, we consider MNMMs, a model-based approach that has been used to detect multispecies assemblages (Georgescu et al. 2009). Model-based approaches have the advantage of being able to support all stages of clustering, description, and site allocation common to ecoregion definition. The Bayesian framework provides a clear link between statistical and ecological inference in combining expert and empirical data and addresses uncertainty in model parameters. Indeed, many clustering algorithms represent particular MNMM forms (Cucala et al. 2009). Fitting MNMMs within the Bayesian framework naturally allows us to consider a spectrum of models with more or less emphasis on expert knowledge compared to empirical data. This framework is characterized by a learning cycle that starts with a prior distribution for model parameters, and then incorporates the empirical data to generate a *posterior* (updated) distribution. This updating is achieved via an MNMM likelihood function that describes the likelihood of the observed data under each parameterization of the model.

The *prior* in a Bayesian statistical model represents the experts' prior assessment of the most plausible classification. For example, a weakly informative *prior* stipulates that subregions have a mean and variance similar to that of the overall study area; that is, subregions do not exist and the ecoregion is homogeneous at the model's scale. A strongly informative *prior* encodes expert-delineated boundaries, as detailed in case study B by Low-Choy et al. (2009); this implies that subregions exist, and that the ecoregion is highly heterogeneous at the model's scale. The Southeast Queensland expert-specified subregions tend to follow boundaries defined by different combinations of topography, climate, and structural geology that were modified to obtain contiguous areas. This reflects the expert's tacit assessment of trade-offs among key drivers for a given set of boundaries (Table 12.1) and highlights the potential, through elicitation or modeling, to reveal the underlying conceptual model. Advances in computers and statistical algorithms have made it feasible to define regions using Bayesian MNMMs, the method chosen for our case study.

12.3.3.2 Model Specification and Implementation

Within each ecoregion, a multivariate normal density function is used to describe the joint statistical distribution of all attributes (occurring across sites within the ecoregion), defined in terms of two sets of parameters:

1. A mean value for each attribute in the ecoregion.
2. A covariance matrix describing how each pair of attributes is related within the ecoregion.

The MNMM is formed as a mixture of these multivariate normal distributions, where each site has a probability of being allocated to each ecoregion (defined by its multivariate distribution). The core modeling "trick" underlying mixture models creates a third set of parameters to capture this allocation:

3. Indicator variables that stipulate which ecoregion each site is allocated to.

The probability of allocating each site to an ecoregion is simply calculated as a weighted average of the (multivariate normal) likelihood of the site's attribute values across an ecoregion. These weights provide the fourth set of parameters for the MNMM.

4. The overall probability that an arbitrary pixel is allocated to a specific ecoregion, which therefore reflects the size of the ecoregion.

Hence, all three phases of ecoregion definition (clustering, description, and site allocation) are embraced within the same model: the site allocations determine how training sites are clustered together as well as predicting allocation of new sites, while the other three sets of parameters (means, covariance matrices, and size, represented by the number of sites) provide the ecological description of the ecoregions. Accad et al. (2005) and Low-Choy et al. (2009) provide more details of model development, including a mathematical specification.

The Southeast Queensland ecoregion expert panel identified subregion boundaries (Young and Dillewaard 1999) that effectively allocate each site to a subregion. From these allocations, each subregion's mean, variance, and size can be imputed (Accad et al. 2005). This combined information defines an expert-informed *prior*. We express this information as a hierarchical *prior* (multivariate normal–inverse Wishart), as described by Frühwirth-Schnatter (2006).

We compare three *prior* models with these subregional means and variances:

Model A. The *prior* is set to the overall ecoregion values of the empirical data.
Model B. The *prior* is imputed from the expert boundaries, but given a moderate degree of belief, equivalent to five times more than that of the empirical data.
Model C. The *prior* is defined as in B, but with a high weight for expert knowledge, effectively ten times more than the weight applied in Model B, and 50 times more than the weight assigned to the empirical data.

The *posterior* distribution provides a range of plausible models, which are different mixtures of multivariate normal distributions that fit well to the measured attributes. For each plausible mixture model, each site may be allocated to the most likely subregion. This provides a *posterior* predictive distribution of all plausible groupings of sites. The *posterior* distribution of these allocations shows how many sites are "stable" (i.e., consistently allocated to the same subregion).

Site allocations are an important parameter in the *posterior* MNMM distribution. For mapping, each site is allocated to the ecoregion with the highest *posterior* probability. Point estimates of other *posterior* parameters are used to interpret the ecoregions. These are the sizes of each ecoregion, and the means and covariances of the environmental attributes within each ecoregion. The Bayesian implementation of the MNMM also provides *posterior* measures of uncertainty for all of these parameters. Predictive uncertainty is given in terms of the alternative allocations for sites that are not completely stable in their allocation to one ecoregion. Then 95% credible intervals for ecoregion size and attribute means and covariances contain the parameter with a 95% chance (a more intuitive interpretation but the logical reverse of confidence intervals). This rich information on uncertainty differs from classical inference for MNMMs (e.g., Fraley and Raftery 2006) and data-mining algorithms (Hastie et al. 2008), in which bootstrapping (or another resampling method) is required to provide a specific measure of uncertainty that reflects random permutations of the input data.

We fit an MNMM to the empirical data with hierarchical *priors* informed by the expert-defined boundaries using a Markov-chain Monte Carlo (MCMC) algorithm, as described by Frühwirth-Schnatter (2006). The assumption that data can be adequately modeled can be assessed via *posterior predictive checks* (Gelman et al. 2004). This procedure checks whether the data fall within the range of many simulations from the model. It uses the plausible parameter values generated from the *posterior* distribution, and is a by-product of the MCMC computations. Then, attributes of hypothetical sites are generated for each plausible parameter value. These hypothetical sites can be compared to the actual sites to reveal whether the actual data fall within the bounds of the model predictions. We performed these computations using the R software (Ihaka and Gentleman 1996).

12.4 Case Study Results

The expert-delineated subregions (Fig. 12.1) were based on five biophysical themes, but two or three of the themes dominated the choice of boundaries – the geological province and the landform-altitude zones, which varied in importance across the subregions. Models B and C used these delineations as the basis for the *prior*, effectively "anchoring" the analysis of the GIS data, whereas Model A was purely data-driven. Model B was intermediate between the data-driven Model A and the expert-driven Model C.

12.4.1 Interpretation of the Mixture Models for Model A

The essential ideas underlying an MNMM are accessible to anyone who understands normal distributions. To make these concepts concrete, we have illustrated the results from fitting a data-driven Bayesian MNMM (Model A, Sect. 12.3.3.2), using a weakly informative *prior*, for the ten subregions in the Southeast Queensland ecoregion (Fig. 12.2). Though still data-driven, this form of MNMM is model-based and also integrates uncertainty (from grouping to description to prediction) and provides richer information than clustering alone.

In each subregion, each attribute follows a normal distribution described by its mean and variance (Fig. 12.2). To facilitate comparisons, each attribute has been standardized by subtracting its ecoregional mean (equivalent to zero on the y-axis) and dividing the result by its ecoregional standard deviation; so that the horizontal black lines through the center of the box-plots then represent the average relative contribution of each attribute. Values on the y-axis indicate the number of standard deviations away from the ecoregional average. Thus, attributes that are predominately positive or negative respectively fall above or below this overall average.

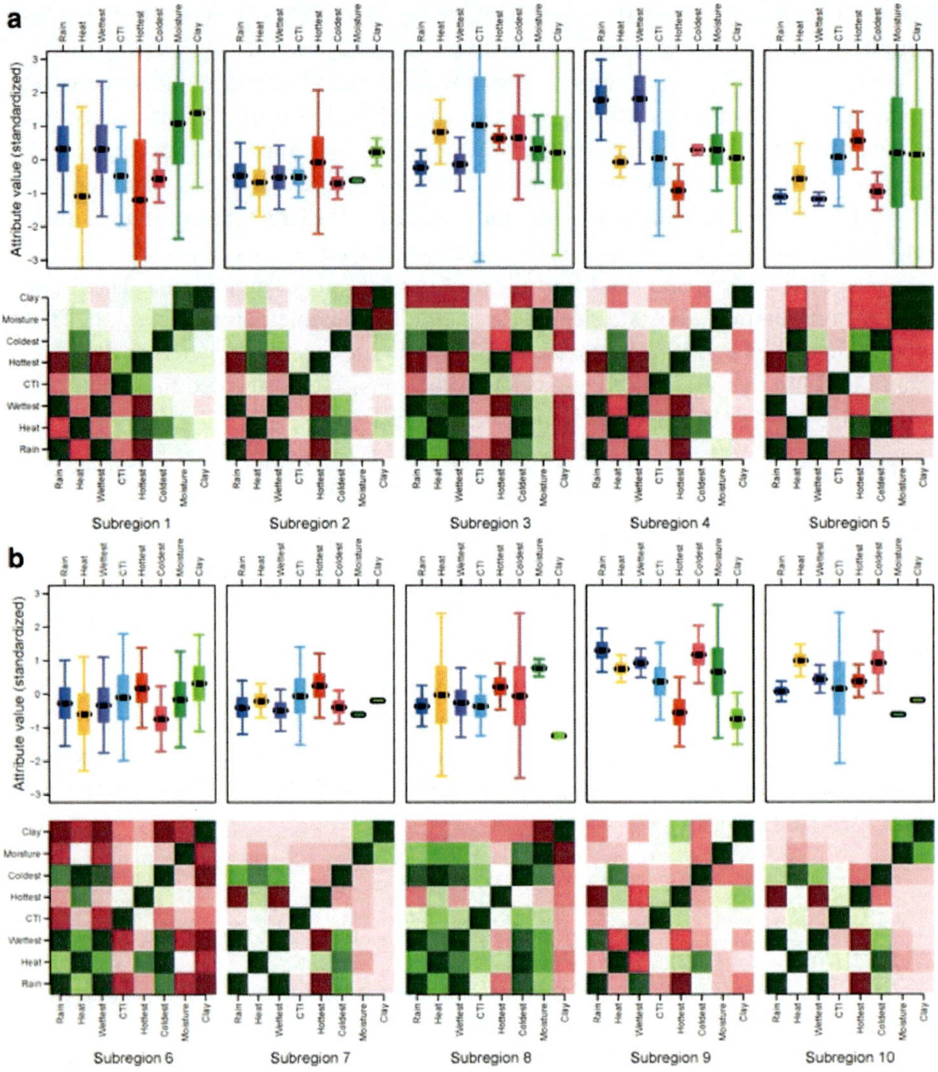

Fig. 12.2 Results from fitting a data-driven Bayesian MNMM (Model A) for ten subregions (numbered to match Fig. 12.1); (**a**) subregions 1–5, (**b**) subregions 6–10. Boxplots (*top row*) summarize the normal distribution fit to each attribute within each subregion. All attributes have been standardized so that the *y*-axis reflects the number of standard deviations from the ecoregion's mean. *Black line*, mean; box, middle 50% of the distribution; whiskers enclose values between the 2.5th and 97.5th quantile of the modeled distribution based on the mean *posterior* estimate of the parameters for the mean, variance and correlations. Image plots (*bottom row*) show the correlations among variables within a subregion: *green*, highly positive; *white*, negligible; *red*, highly negative. Closely related variables (either negatively or positively correlated for the ecoregion) are located close to each other. Environmental variables: Clay (A horizon clay content, %); Coldest (mean minimum temperature during the coolest month, °C); CTI (compound topographic index, dimensionless); Heat (mean annual temperature, °C); Hottest (mean maximum temperature during the warmest month, °C); Moisture (soil water-holding capacity, mm); Rain (annual rainfall, mm); Wettest (total rainfall during the warmest quarter, mm). For subregion descriptions, see Fig. 12.1 caption

The box and whiskers enclose 50 and 95%, respectively, of the modeled (standardized) values for each attribute. Hence, the size of the box is proportional to the variance. For example, annual rainfall in the Brisbane–Barambah Volcanics (subregion 5), which closely corresponds with the expert's delineation, shows little variance (a small box and small whiskers) and an average that is one standard deviation (y-axis) below the ecoregion's average; this means that annual rainfall is consistent and lower than average. In contrast, soil moisture and clay content are close to the ecoregion's average, but vary widely (more than three standard deviations above and below the ecoregion's average).

The relationships among attributes are reflected by the correlation matrix (Fig. 12.2, image plots below the box-plots). In the matrix, each cell represents the Pearson's correlation coefficient for each pair of attributes, when all attributes are considered simultaneously. For example, soil moisture and clay content in subregion 5 are highly positively correlated, since this area is dominated by heavy clay. In contrast, the hottest temperature is moderately to highly positively correlated with annual temperature, but highly negatively correlated with annual rainfall.

This example illustrates the normal and multivariate aspects of the model. We used *posterior predictive checks* to test the assumption of multivariate normality, and found no major discrepancies between the actual sites and the hypothetical sites in the *posterior* model (Low-Choy et al., unpublished data). The "mixture" aspect arises because we also modeled the probability that each site belongs to each subregion. This gives a measure of the "size" (number of sites) of each subregion. The subregion with the greatest proportion of sites either wholly or partially allocated was the Gympie Block (subregion 7), which comprised nearly 20% of the ecoregion (top, Fig. 12.3a), followed by the Burnett-Curtis Hills and Ranges (subregion 10), which comprised nearly 15% of the ecoregion. Most of the remaining subregions accounted for between 5 and 10% of the ecoregion.

Some sites were stable, and were consistently allocated to the same subregion (bottom, Fig. 12.3a). In the Burnett-Curtis Coastal Lowlands (subregion 8) and subregion 10, more than 90% of the sites were always allocated to the respective subregion, whereas in the South Burnett (subregion 6), only 60% of sites were stably allocated to that subregion. In subregion 8, the approximately 10% portion of unstable sites (Fig. 12.3b) were allocated to the Southeast Hills and Ranges (subregion 3) and subregion 6, whereas nearly all unstable sites in subregion 10 were allocated to subregion 7. For the larger number (40%) of unstable sites in subregion 6, the second choice for allocation was spread among several options: most often the Scenic Rim (subregion 1), followed by subregions 2, 4, or 8, then 3, 5, or 9.

12.4.2 Comparison with Models B and C

Figure 12.4 shows that some expert-delineated subregions were not clearly defined by the data-driven analysis in Model A. For example, subregions 3, 6, 5, 7, and 1, in that order, are reduced in size, displaced, or fragmented compared with their

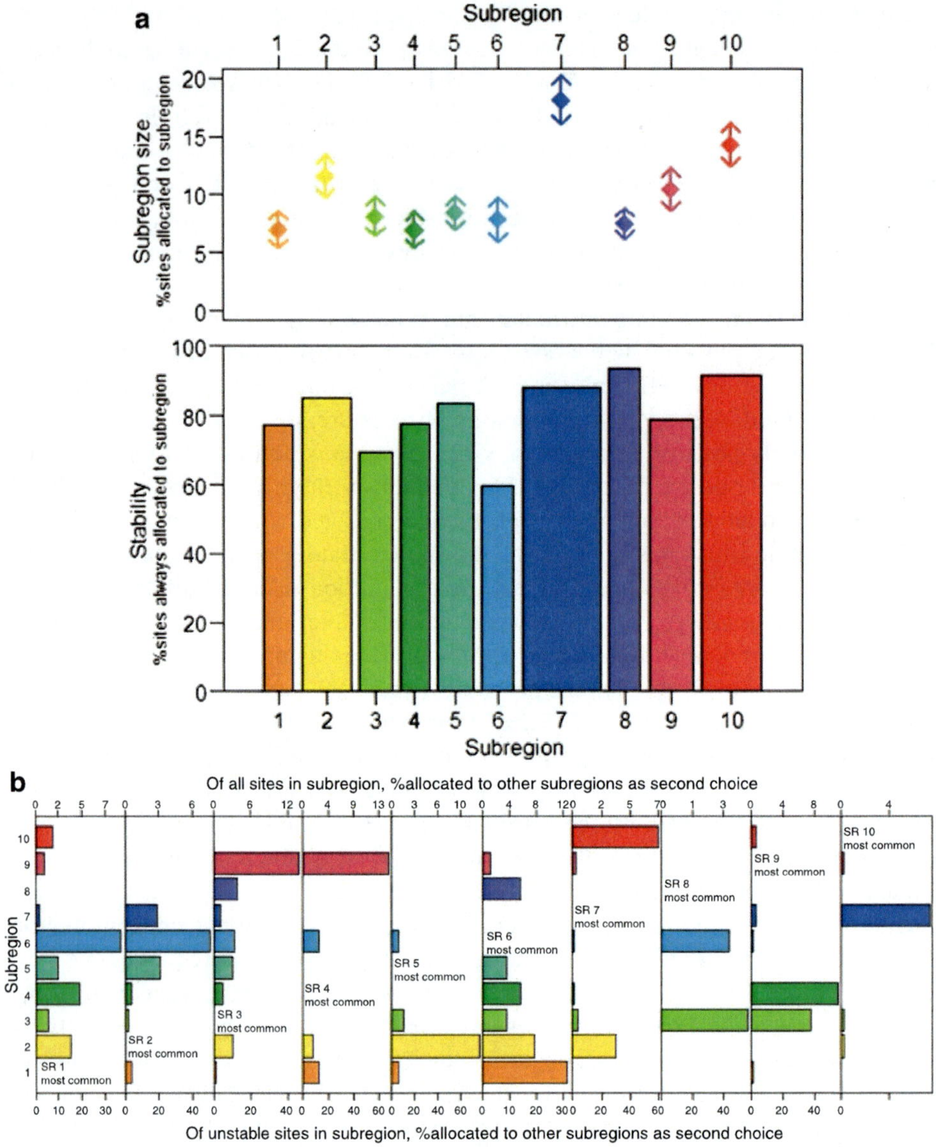

Fig. 12.3 Subregion size and stability for data-driven Model A. (**a**) *Top*: proportion of sites allocated (*y*-axis, *dots*) to each subregion (*x*-axis). Error bars ending in arrows show the 95% credible intervals for these proportions, reflecting the model's degree of uncertainty. (**a**) *Bottom*: stability of site allocation based on the proportion of sites (*y*-axis) that were consistently (across all *posterior* simulations) allocated to one subregion (*x*-axis). The width of each bar reflects the relative size of each subregion (i.e., reflects the position of the dots in the *top-left graph*). (**b**) instability plot for each subregion showing how often unstable sites are reallocated to another subregion as the second choice (shown by the colored bars), calculated over all *posterior* simulations. This is calculated as 100 minus the stability in the graph (**a**) *Bottom*. For subregion descriptions, see Fig. 12.1 caption

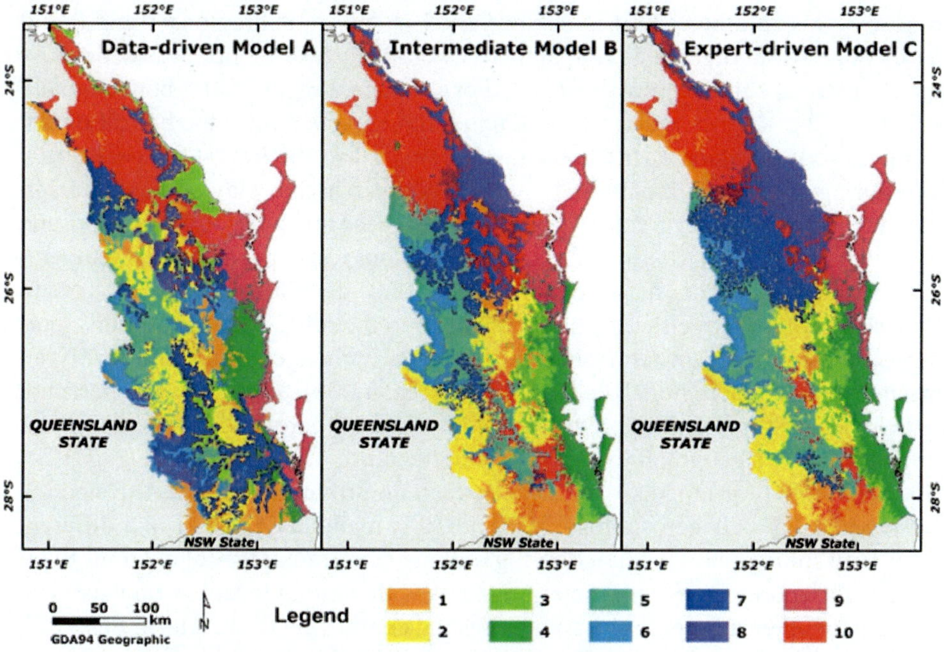

Fig. 12.4 Maps showing predicted allocation of sites ranging from data-driven Model A given no prior boundaries (*left*), to expert-driven Model C with high weight on expert boundaries (*right*), and the intermediate Model B with some weight on expert boundaries (*middle*). For subregion descriptions, see Fig. 12.1 caption

delineation in Fig. 12.1. Some subregions were robust and did not vary greatly between Model A (data-dominated) and Model C (expert-dominated); these include subregion 10 in the north and the Great Sandy (subregion 9) in the mid-east. Though Model C closely resembled the expert-delineated subregions (Fig. 12.1), it showed segmentation among some subregions (e.g., subregions 1, 2, 7, and 10), which suggests that some environmental attributes may not have been explicitly accounted for in the expert delineations. Alternatively, additional subregions could be delineated in the western part of the Moreton Basin (subregion 2) and the northern part of subregion 1, or the western and southern margins of the ecoregion boundary could be refined. The intermediate Model (B) illustrates a gradual evolution from the data-driven to expert-driven delineations, such as the increasing dissolution of subregion 7 and its redistribution to the south and west with decreasing emphasis on the expert knowledge.

The results from the data-driven model (Model A; Fig. 12.2) reveal the pivotal role of soil moisture (water content at field capacity) in defining subregions. The standardized value of soil moisture was consistently low in the data-dominated model of subregions 2, 7, and 10; higher than average in subregions 1, 8, and 9; and average or slightly above average for the other subregions (Model A; Fig. 12.2). These differences are consistent with location-based variation in soil types. Heavier clays and deeper weathered soils are predominant in some subregions and shallow or sandy soils in others. Soil moisture varied widely in subregions 1 and 5, but hardly at all in subregions 2, 7, 8, and 10. Overall, each subregion was defined by relative extremes in different attributes. For instance, low rainfall – both annually and during the warmest quarter – and high temperatures during the hottest month characterized subregion 5. In contrast, subregion 1 had the lowest temperatures – both during the hottest month and annually – and a higher clay content and soil moisture potential, with greater variability in all these attributes. Subregions 7 and 10 had similar profiles (both means and correlations), except for coldest temperatures, which were higher in subregion 10 (Fig. 12.2). In addition, sites allocated to these subregions were either stable or changed to the other subregion in this pair; that is, two-thirds of unstable sites in subregion 7 changed to 10 and nearly all unstable sites in subregion 10 changed to 7 (Fig. 12.3b). This indicates potential for combining these two subregions (not considered here), which share similar longitude and are geographical neighbors in the north of the ecoregion (Fig. 12.4, Model A). Even for models with higher expert contribution (Fig. 12.4, B and C), this relationship between subregions 7 and 10 is maintained, albeit in a different form: both subregions have neighboring outliers scattered in the inland mid- to far south of the ecoregion. Similar interpretations with varying levels of emphasis can be made for other subregions by comparing and contrasting the results in Figs. 12.2 and 12.3.

Interestingly, three subregions (3, 4, and 9) were closely related (Model A, Fig. 12.3b) in terms of unstable sites being allocated with high probability to one or the other two. These do not share many similar attributes, the only common points are higher than average coldest temperatures and soil characteristics that are neither highly variable (as in subregions 1 and 5) nor highly consistent (as in subregions 2, 7, 8, and 10). Geographically, these regions are located closest to one another in the expert-dominated Model C (Fig. 12.4). To a lesser extent, three subregions (1, 2, and 6) were related by sharing unstable sites (Model A, Fig. 12.3b). These subregions are located in the south and west of the ecoregion.

The *posterior* uncertainty in all means and variances (80 attribute–subregion pairs) and the correlations was investigated (Low-Choy et al., unpublished data). All subregional parameters (means, variances, and correlations) were fairly stable because their *posterior* standard deviations were quite narrow. This information on uncertainty (a useful feature of the Bayesian MNMM) provided confidence that the model was relatively robust and useful for ecoregion definition.

12.5 Discussion

12.5.1 Value of Expert Knowledge in Ecoregion Studies

McMahon et al. (2004) recognized the complementary value of expert- and data-driven approaches to ecoregion delineation and recommended collaborative exploration of these methods by researchers with quantitative and qualitative expertise.

Mackey et al. (2008), in their review of ecoregion approaches, noted the conflict between qualitative versus quantitative approaches to biogeographic regionalizations. They recommended an approach that formalized the conceptual framework and objectives of the study. Loveland and Merchant (2004) also recognized the confluence of expert- and data-driven approaches in ecoregional mapping and the need for expertise in combining these approaches. Hargrove and Hoffman (2004) suggested that existing maps, with their implied weights, could become direct inputs for a fully quantitative model, but that increasingly fine ecoregion subdivisions quickly surpassed the ability of experts to resolve. However, the ability of Bayesian statistical approaches to reconcile the differences between data- and expert-driven approaches by formalizing the role of expert (subjective) information has not been noted in previous ecoregion studies. Our applications (Rochester et al. 2004; Accad et al. 2005; Low-Choy et al. 2009) are the first trials of this novel approach.

Although expert contributions have been acknowledged in classical statistical analyses which are purely data-driven, an explicit record of the current state of knowledge is generally not required prior to the analysis. Data-mining techniques are also susceptible to *post hoc* manipulation to fit *a priori* assumptions (Hastie et al. 2008). Although both empirical and expert data are accommodated by increasingly popular conditional probability network models (also known as Bayesian Belief Networks), the two information sources are treated as though interchangeable. This contrasts with the explicit directionality afforded by the Bayesian paradigm, in which expert knowledge forms the basis for a learning cycle that uses empirical evidence to update the *prior* results (Low-Choy et al. 2009; Kuhnert et al. 2010). This is made possible by the Bayesian broad definition of probability, which embraces uncertainty in expert knowledge as well as in the empirical data.

Our investigations worked with existing expert knowledge, which was expressed as existing delineations of subregional boundaries. Since these boundaries were derived through a well-established environmental management process, they distilled the experts' "best" judgment at the time. The indirect approach to expert elicitation of site allocation, rather than using means and covariance matrices of attributes within ecoregions, also contributes to the accuracy of the expert's assessments. This provides benefits similar to indirect elicitation of expert knowledge in a regression context (Chap. 3). By asking experts about observable quantities (here, boundaries) we avoid several potential sources of inaccuracy (Kynn 2008; Chap. 3). Scales are easily misjudged (Kuhnert et al. 2010) and this kind of misjudgment could occur here if experts were asked to estimate means and variances instead. Nevertheless, further work could explore effective ways of determining whether eliciting some of these parameters may help reveal the experts' reasoning behind the boundaries. Site allocations reflect qualitative expert knowledge about boundaries, and correspond to parameters within the MNMM, so that inaccuracies in translation do not arise (Kuhnert et al. 2010). A sensitivity analysis (Accad et al. 2005) allows us to adjust the level of belief in the experts' boundary delineations (Fig. 12.4), which is another important component of accuracy assessment (Low-Choy et al. 2009; Kuhnert et al. 2010).

12.5.2 Bayesian Ecoregion Analysis

Thackway and Cresswell (1995) identified a number of assumptions and limitations of the Australian ecoregions that remain unresolved (Environment Australia 2000). For example, an analysis of environmental variability and heterogeneity within- and between-ecoregions was desired to validate the choice of boundaries defining contiguous areas and their implications for conservation planning. However, the diversity in the way jurisdictions derived boundaries prevented comparison. Rigorous testing of boundaries and heterogeneity assessment using regional and continental datasets and analytical tools was therefore recommended. For these reasons, Australian ecoregions continue to be termed "interim." These concerns can now be addressed using Bayesian MNMMs and GIS data.

Bayesian MNMMs can address both boundary-specific questions (e.g., Accad et al. 2005), such as whether a boundary is at the correct position, and broader questions about how to balance boundaries across a larger area (e.g., the present case study). A particular concern involves minimizing the revision of boundaries unless strong evidence indicates the need for change. Our intermediate model (B) illustrates how the boundaries move or dissolve as the model becomes increasingly data-dominated (like model A) or expert-dominated (like model C). For instance, as expert influence increased from models A to B to C, the Burnett-Curtis Hills and Ranges (subregion 10) in the north was increasingly allocated to islands of sites occurring within those mapped to two southern subregions – Scenic Rim (1) and parts of the Moreton Basin (2). The MNMMs revealed that annual and coldest temperatures underlie the similarity, and that the hottest temperatures underlie the discrimination among these subregions (particularly 2 and 10) in their attribute means and correlation structure (Fig. 12.2).

We also noted that several subregions – South Burnett (6), Gympie Block (7), and Burnett-Curtis Coastal Lowlands (8) – were not well defined by the data-driven analysis (Model A). This suggests two things. First, these subregions are prime candidates for additional expert panel consultations to refine the underlying drivers, attribute selection, and boundaries. Second, our use of GIS data as proxies could be refined to better accord with the expert-delineated boundaries. These proxies were inferred from the underlying ecological rationales, which were not documented systematically, rather than from descriptions imputed from the boundaries (Young and Dillewaard 1999). These descriptions assisted the choice of attributes used in the Bayesian models (Table 12.1) and may have given greater emphasis to some facets of the environment than the experts had intended. Interestingly, the Bayesian uncertainty assessment revealed that in the data-dominated analysis subregion 7 was stably defined (Fig. 12.3a) and shared no sites with its closest geographic neighbor, subregion 8 (Fig. 12.3b), whereas subregion 6 was the least consistently defined subregion, sharing sites with seven other subregions. More closely aligning the choice of GIS data with the expert-delineation framework might increase the stability of site allocation to subregions.

12.5.3 *The Bayesian Ecoregion Analysis Identifies Boundary Uncertainty*

The Bayesian approach identifies the plausible range of ecoregion clusters based on available data. Alternatively, a data-driven (classical MNMM or data mining) classification supplemented by bootstrapping would choose the ecoregions that maximizes the likelihood of the observed data. This latter approach provides less information on uncertainty of parameters defining the explanatory model, and can suffer from numerical and interpretative problems if competing models lead to similarly low likelihood values. However, Bayesian inference requires a deeper understanding of the model through more explicit or formal definition of its conceptual underpinnings and purpose, as required by the ESD framework in ecology (Austin 2002), and more intensive computation. An important step in utilizing MCMC analysis, as in this study, is to assess whether simulations have indeed converged on the desired *posterior* distribution (Low-Choy et al. 2009).

In this context, utilizing expert-delineated boundaries in a *prior* model allows the use of holistic expert knowledge rather than burdening experts with the difficult task of expressing their qualitative knowledge in terms of explicit quantitative ecological propositions and proxy GIS attributes or with the need to specify variable weights and transformations. The MNMM framework also provides a basis for integrating new types of information such as "fuzzy" boundaries, thereby placing varying degrees of emphasis on boundaries with different levels of expert certainty.

The use of a model that accommodates both description and delineation of ecoregions provides a basis for focusing further dialogue with experts on specific factors (e.g., temperature) as well as holistic issues (e.g., similarity between two subregions). Our case study provided robust model-based support for the usual informal and internal estimates provided by experts and showed how to modify data-driven ecoregion delineations based on expert input. The continuum of models also revealed the likely path of reasoning followed by experts as they increasingly modify their decisions to account for empirical data.

12.5.4 *Future Directions*

Bayesian MNMMs have been promoted for use in ecoregion delineations in software such as Autoclass (Cheeseman and Stutz 1996), in which a data-driven approach was implemented using strictly noninformative *priors*, which can lead to computational problems (Frühwirth-Schnatter 2006). More recently, Bayesian MNMMs with weakly informative or expert-informed *priors* have been implemented (Rochester et al. 2004) using approaches such as that of Bensmail et al. (1997). The same method was used to integrate expert knowledge with GIS data (using different attributes) for subregional delineation of Southeast Queensland

(Rochester et al. 2004) and terrestrial ecoregions in northern Queensland (Accad et al. 2005). Here, we implemented an improved formulation of the Bayesian MNMM using a hierarchical rather than independent *prior* formulation, leading to better numerical stability (Frühwirth-Schnatter 2006). Moreover, to strengthen the explanatory role of the subregions under the ESD framework, attributes were chosen to encompass more direct measures, such as soil moisture and clay content, rather than indirect measures, such as elevation and classes of structural geology.

Since the introduction of expert-informed Bayesian MNMMs (Rochester et al. 2004), several technological changes have occurred. These include improved capabilities of GIS and statistical software and of computer hardware, and a better understanding of the properties of Bayesian MNMMs (Frühwirth-Schnatter 2006). This has considerably simplified model implementation. Current research on mixture models, especially for large datasets, aims to provide model-based solutions to the complex and highly computation-intensive issues of covariate selection, choosing the number of regions, and allowing non-normal distributions within regions.

It is of ecological interest to reveal the expert's conceptual model underlying their proposed ecoregion boundaries. As discussed in Sect. 12.5.3, a sensitivity analysis from this modeling exercise provides some indication of how experts supplement incomplete information provided by GIS attributes. Instead of comparing two end-points that contrast expert- and data-driven regionalizations, the sensitivity analysis highlighted the "evolutionary path" that boundaries follow, as more or less emphasis is placed on expert knowledge. This is highly informative, since this path is just one of many possible paths. These ideas could be further developed and tested to reveal the attributes and weights underlying the experts' choice of boundaries. This would require additional elicitation from experts, for example, using indirect methods tailored to classification via mixture models, similar to an approach recently developed for regression (Chap. 3).

12.6 Conclusions

Given the necessary use of both expert knowledge and empirical data to define ecoregion boundaries for use in conservation planning and management, and the controversy that accompanies the development of such boundaries, Bayesian approaches can facilitate efforts to achieve consensus. Existing ecoregion delineations can be modified or validated using Bayesian methods, permitting continuous improvement of the delineations in response to new data or changed objectives (e.g., to achieve ecological resilience under a variable climate).

Our case study demonstrated a Bayesian framework that leverages all available information (both empirical data and expert knowledge) to delineate ecoregions. We showed the potential of combining expert knowledge within an empirical analysis, with examples of how the derivation can be weighted toward expert knowledge, empirical data, or a combination of the two. Ecoregions with intermediate weights can provide useful insights into how this trade-off evolves. Our results justify seeking ways to combine expert knowledge with empirical data. For example, uncertainty such as diversity of opinions could be defined more explicitly during environmental management discussions. Bayesian MNMMs can be used to highlight boundary uncertainty and provide support for the robustness and acceptability of existing ecoregion delineations.

This approach provides an efficient mechanism for incorporating all information sources and addressing the inherent limitations of each. Environmental GIS layers, which are constructed from many smaller datasets, generally suffer from varying levels of consistency in quality and resolution. Experts typically have most confidence in their knowledge of particular areas, yet are often asked to extrapolate that knowledge to similar areas. The final analysis should reflect a balance in which empirical data or expert knowledge dominates when appropriate, while identifying knowledge gaps where neither source is superior so that stronger evidence can be sought, and increasing confidence where the two sources concur.

Bayesian MNMMs provide ecologically meaningful insights about subregions, including their heterogeneity (variances), key ecological drivers and their interactions (means, covariances), and the location and stability of boundaries (site allocation). The computational elegance of a single model that encompasses all the necessary steps of ecoregion delineation, from classification to ecoregion description and boundary delineation, is superior to existing model-free approaches. The rich information on uncertainty, both in parameter estimation and in prediction of site allocation, provides an integrated and intuitive basis for evaluating a set of ecoregions based on current information and for targeting future collection of both expert and empirical data that has not previously been possible.

References

Accad A, Low-Choy S, Pullar D, Rochester W (2005) Bioregion classification using model-based clustering: a case study in north eastern Queensland. In: Zerger A, Argent RM (eds), MODSIM 2005 International Congress on Modelling and Simulation. Modelling and Simulation Society of Australia and New Zealand, Melbourne, pp 1326–1332

Austin MP (2002) Spatial prediction of species distribution: an interface between ecological theory and statistical modelling. Ecol Modell 157(2–3):101–118

Bailey RG (2004) Identifying ecoregion boundaries. Environ Manage 34:S14-S26

Beier P, Brost B (2010) Use of land facets to plan for climate change: conserving the arenas, not the actors. Conserv Biol 24(3):701–710

Bensmail H, Celeux G, Raftery AE, Robert CP (1997) Inference in model-based cluster analysis. Statistics Comp 7(1):1–10

Cheeseman P, Stutz J (1996) Bayesian classification (AutoClass): theory and results. In: Fayyad UM, Piatetsky-Shapiro G, Smyth P, Uthurusamy R (eds), Advances in knowledge discovery and data mining. American Association for Artificial Intelligence Press/MIT Press, Menlo Park, CA, pp 153–180

Christian CS, Stewart GA (1968) Methodology of integrated surveys. In: UNESCO (ed), Aerial survey and integrated studies. Conference Proceedings. UNESCO, Toulouse, pp 233–280

Cork S, Sattler P, Alexandra J (2006) Biodiversity theme commentary prepared for the 2006 Australian State of the Environment Committee. Department of the Environment and Heritage, Australian Government, Canberra

Cucala L, Marin J-M, Robert CP, Titterington DM (2009) A Bayesian reassessment of nearest-neighbor classification. J Am Stat Assoc 104(485):263–273

Environment Australia (2000) Revision of the interim biogeographic regionalisation for Australia (IBRA) and development of version 5.1 - summary report. Australian Government, Canberra

Fraley C, Raftery AE (2006) MCLUST version 3 for R: normal mixture modeling and model-based clustering. Technical Report No. 504. Department of Statistics, University of Washington, Seattle

Frühwirth-Schnatter S (2006) Finite mixture and Markov switching models. Springer, New York

Gelman A, Carlin JB, Stern HS, Rubin DB (2004) Bayesian data analysis. Chapman and Hall/CRC, New York

Georgescu V, Soubeyrand S, Kretzschmar A, Laine AL (2009) Exploring spatial and multitype assemblages of species abundances. Biom J 51(6):979–995

Goswami M, O'Connor KM, Bhattarai KP (2007) Development of regionalisation procedures using a multi-model approach for flow simulation in an ungauged catchment. J Hydrol 333(2–4):517–531

Hardman-Mountford NJ, Hirata T, Richardson KA, Aiken J (2008) An objective methodology for the classification of ecological pattern into biomes and provinces for the pelagic ocean. Remote Sens Environ 112(8):3341–3352

Hargrove WW, Hoffman FM (2004) Potential of multivariate quantitative methods for delineation and visualization of ecoregions. Environ Manage 34(S1):39–60

Hastie T, Tibshirani R, Friedman JH (2008) The elements of statistical learning (2nd ed). Springer Verlag, New York

Ihaka R, Gentleman R (1996) R: a language for data analysis and graphics. J Comput Graph Stat 5(3):299–314

Kuhnert PM, Martin TG, Griffiths SP (2010) A guide to eliciting and using expert knowledge in Bayesian ecological models. Ecol Lett 13(7):900–914

Kynn M (2008) The 'heuristics and biases' bias in expert elicitation. J Roy Stat Soc Ser A (Stat Soc) 171:239–264

Leathwick JR, Snelder T, Chadderton WL et al (2011) Use of generalised dissimilarity modelling to improve the biological discrimination of river and stream classifications. Freshw Biol 56(1):21–38

Loveland TR, Merchant JM (2004) Ecoregions and ecoregionalization: geographical and ecological perspectives. Environ Manage 34(S1):1–13

Low-Choy S, O'Leary R, Mengersen K (2009) Elicitation by design in ecology: using expert opinion to inform priors for Bayesian statistical models. Ecology 90(1):265–277

Mackey BG, Berry SL, Brown T (2008) Reconciling approaches to biogeographical regionalization: a systematic and generic framework examined with a case study of the Australian continent. J Biogeogr 35(2):213–229

McMahon G, Gregonis SM, Waltman SW et al (2001) Developing a spatial framework of common ecological regions for the conterminous United States. Environ Manage 28(3):293–316

McMahon G, Wiken EB, Gauthier DA (2004) Toward a scientifically rigorous basis for developing mapped ecological regions. Environ Manage 34:S111-S124

Morgan MG, Terrey J (1990) Natural regions of western New South Wales and their use for environmental management. Proc Ecol Soc Australia 16:67–73

Morgan G (2000) Landscape health in Australia: a rapid assessment of the relative condition of Australia's bioregions and subregions. Environment Australia, Australian Government, Canberra

Natural Resource Management Ministerial Council (2004) Directions for the national reserve system – a partnership approach. Department of the Environment and Heritage, Australian Government, Canberra

Neldner VJ, Wilson BA, Thompson EJ, Dillewaard HA (2005) Methodology for survey and mapping of regional ecosystems and vegetation communities in Queensland, version 3.0. Environmental Protection Agency, Queensland Government, Brisbane

Rochester W, Accad A, Low-Choy SJ et al (2004) Final report UQ-EPA subregion classification project. The University of Queensland, Brisbane

Sattler P, Creighton C (2002) Australian terrestrial biodiversity assessment 2002. National Land and Water Resources Audit, Australian Government, Canberra

Sattler P, Williams R (eds) (1999) The conservation status of Queensland's bioregional ecosystems. Environmental Protection Agency, Queensland Government, Brisbane

Snelder T, Lehmann A, Lamouroux N et al (2010) Effect of classification procedure on the performance of numerically defined ecological regions. Environ Manage 45(5):939–952

Stanton JP, Morgan G (1977) The rapid selection and appraisal of key endangered sites: the Queensland case study. School of Natural Resources, University of New England, Armidale

Thackway R, Cresswell ID (1995) An interim biogeographic regionalisation for Australia: a framework for setting priorities in the national reserves system cooperative program, version 4.0. Australian Nature Conservation Agency, Canberra

Young PAR, Cotterell MA (1993) A conservation assessment of the South-eastern Queensland 2001 region (draft report). Department of Environment, Queensland Government, Brisbane

Young PAR, Dillewaard HA (1999) Chapter 12: Southeast Queensland. In: Sattler PS, Williams RD (eds), The conservation status of Queensland's bioregional ecosystems. Environmental Protection Agency, Queensland Government, Brisbane, pp 12.1-12.75

Chapter 13
Eliciting Expert Knowledge of Ecosystem Vulnerability to Human Stressors to Support Comprehensive Ocean Management

Carrie V. Kappel, Benjamin S. Halpern, Kimberly A. Selkoe, and Roger M. Cooke

Contents

C.V. Kappel (✉) • B.S. Halpern • K.A. Selkoe
National Center for Ecological Analysis and Synthesis, 735 State Street,
Santa Barbara, CA 93101, USA
e-mail: kappel@nceas.ucsb.edu

R.M. Cooke
Resources for the Future, 1616 P Street NW, Washington, DC 20036-1400, USA

A.H. Perera et al. (eds.), *Expert Knowledge and Its Application in Landscape Ecology*,
DOI 10.1007/978-1-4614-1034-8_13,

13.1 Introduction

More than 38% of the world's population lives within 100 km of the coast, and the coastal zone is becoming more heavily populated each year (Small and Cohen 2004). More and more human activities depend upon and compete for coastal and marine ecosystem goods and services. This intensification of use is necessitating a shift toward more comprehensive and integrated approaches to management – a shift that is already underway via approaches such as ecosystem-based management and ocean zoning (Day 2002; McLeod et al. 2005; Crowder et al. 2006; Douvere et al. 2007; Douvere 2008; Office of the President 2010). Given the diversity of human uses and natural resources that converge in coastal waters, understanding the potential independent and cumulative impacts of those uses and associated stressors on marine ecosystems can be very challenging. Little empirical data is available to weigh the relative vulnerability of the range of ecosystem types to the full set of human stressors (Halpern et al. 2007). Nevertheless, decision-makers require scientific input to support the setting of priorities, spatial planning, and zoning of the marine environment that they are increasingly being called to undertake (Leslie and McLeod 2007; Ehler and Douvere 2009).

For example, the Commonwealth of Massachusetts (USA) passed the Massachusetts Oceans Act in 2008, which required the Executive Office of Energy and Environmental Affairs (EOEEA) to draft and implement a comprehensive ocean management plan for most of the Massachusetts state waters (0.3–3 nautical miles from shore), the first plan of its kind in the USA. The plan must address ocean uses and development that are incompatible with each other or with sustainable use of natural resources and must address the overall balance among use, protection, and development. Among other directives, the Act requires that EOEEA's plan "value biodiversity and ecosystem health" and "identify and protect special, sensitive, or unique estuarine and marine life and habitats." Accordingly, the plan identifies special, sensitive, and unique marine resources, establishes marine resource management areas and management measures for those areas that both protect their resources and allow for appropriate ocean uses within designated areas.

To inform this effort, we collaborated with the Massachusetts Ocean Partnership to conduct a survey of regional (New England) experts in each of 15 marine ecosystem types to gauge the relative vulnerability of each ecosystem to each of 58 current and emerging anthropogenic stressors. The resulting matrix of vulnerability scores is one of several tools being used by EOEEA to identify vulnerable ecosystem types and the risks from particular human activities that occur in or have been proposed for Massachusetts' coastal waters. The vulnerability assessment helped to identify special, sensitive, and unique marine resources and to inform an analysis of compatibility between traditional and emerging ocean uses and these resources, which is critical to the development and implementation of the plan. In a separate effort, we also used the scores in combination with spatial information on the distribution of marine ecosystems and the distribution and intensity of human activities to map cumulative impacts in the marine environment (unpublished data; the approach was similar to that of Halpern et al. 2008b, 2009). These maps are also being used to support the Massachusetts ocean management planning process.

Expert knowledge was critical to assess ecosystem vulnerability to human stressors within the short timeframe of the Massachusetts planning process (12 months for plan development, 6 months for review of the draft plan). Managers were faced with making decisions about where to allow certain human uses and how to protect vulnerable ecosystems within multiuse areas. They needed a way to rank overall vulnerability and to understand the vulnerability of specific ecosystem types to particular stressors, yet they did not have in-house expertise about many of the stressors and ecosystems, a robust method for quantifying vulnerability, or the time and resources necessary to empirically evaluate the entire set of stressor–ecosystem combinations. With 15 different ecosystem types, ranging from tidal flats to deep-water soft sediment habitats, and 58 different human uses and associated stressors, there were 870 stressor–ecosystem combinations. Most of these combinations have not been investigated scientifically, so empirical data with which to quantify the relative vulnerability of the ecosystems to different stressors were inadequate or nonexistent. Furthermore, we felt it was preferable to seek the advice of regional experts in each of these ecosystem types rather than relying solely on a review of the scientific literature, which was less likely to provide comprehensive information specific to the region.

We had previously conducted a similar survey in cooperation with experts in the California Current region (from the border between Canada and the United States south to Baja California Sur, Mexico). For that study, we used techniques from decision theory, described in Sect. 13.2, to create a mathematical model that represented how experts judge the relative vulnerability of marine ecosystems to different stressors (Neslo et al. 2008; Teck et al. 2010). In Massachusetts, we replicated this process to address two primary objectives. Our first objective was to see whether the mathematical model developed from the California Current expert knowledge elicitation process could be generalized to other experts in other regions by comparing the Massachusetts model results with those from the California Current analysis. Our hypothesis was that experts from both regions would base their judgments on the same vulnerability criteria in the same way, and that model results should therefore be similar for both regions. If our analysis supported the hypothesis, it would potentially be unnecessary to develop a new model for each new region or application, thereby increasing the flexibility of the expert knowledge elicitation framework that we developed and the ease with which it can be deployed to collect expert knowledge in new regions and for novel human stressors. Our second objective was to use the knowledge gathered from the New England experts to provide a robust and comparable assessment of ecosystem vulnerability across a broad suite of human stressors and marine ecosystem types, thereby providing a more scientific basis for ocean management in Massachusetts and the broader New England region.

13.1.1 Vulnerability Assessment

There is a long history of assessing threats to species and the environment and prioritizing actions to mitigate them (e.g., Mace and Lande 1991; Master 1991; Bryant et al. 1998; Roberts 2002; Burke et al. 2002; Wilson et al. 2005; Vié et al. 2009). However, many of these efforts have focused on single species, habitats, or stressors. They have seldom examined threats at the ecosystem scale, nor have they integrated multiple stressors or multiple ecosystem types, although exceptions include the community-level methods reviewed in Nicholson et al. (2009) and the integrated ecosystem assessments of Wickham (1999), Noss et al. (2002), and Tran et al. (2002). Though some studies have attempted to compare and rank the severity of a suite of stressors (e.g., Wilcove et al. 1998; Kappel 2005), it is difficult to compare the impacts of very different stressors (e.g., biomass harvest versus sea surface temperature rise), especially at the ecosystem level. In addition, setting of conservation priorities has often relied on expert judgments collected "behind closed doors" (e.g., Noss et al. 2002), leading to public perception of the process as a "black box" (Regan et al. 2004; Beazley et al. 2010). In addition to the growing need for transparency in the current politicized decision environment, it is also necessary to ensure that the elicitation is performed in a way that meets scientific standards for rigor. If the elicitation of expert knowledge is not transparent and clearly documented, the results are unlikely to be repeatable (Tversky and Kahneman 1982; Plous 1993; Keith 1998; Burgman 2001; Rush and Roy 2001; Regan et al. 2004; Aspinall 2010).

13.1.2 Expert Knowledge Elicitation Framework

Given increasing resource constraints and more frequent human use conflicts in many coastal zones around the world, there is a growing need for a way to compare the effects of multiple stressors across ecosystems using the same measurement scale. To address this need, we developed a flexible way to elicit expert assessments of ecosystem vulnerability to human stressors that accounts for ecological context, acknowledging that the same activity may have different effects in different ecosystems (Halpern et al. 2007; Neslo et al. 2008; Teck et al. 2010). This structured survey tool provides for transparency and repeatability. Furthermore, the quantitative approach collects expert knowledge as independent samples rather than a group consensus, thereby preserving information on the level of disagreement (i.e., uncertainty) among experts and reducing the bias that can arise in efforts to develop consensus opinions as a result of dominant personalities and idiosyncratic group dynamics (Chap. 8). The main output of the survey is a matrix of vulnerability scores across all ecosystems and human stressors. Scores can be used to rank stressors or ecosystems, and the rankings can then guide management decisions and setting of priorities.

We used the elicitation framework to develop a matrix of vulnerability scores for New England marine and coastal ecosystems (or habitats), such as beaches and dunes, seagrass beds, rocky reefs, and pelagic waters. We asked experts with experience studying each ecosystem to assess its vulnerability to a list of different stressors associated with human activities. We defined a stressor as anything that can perturb an ecosystem beyond its natural limits of variation. For example, rising sea surface temperature is a stressor associated with anthropogenic climate change, and the destructive demersal fishing that occurs at or just above the ocean floor perturbs marine ecosystems by damaging or destroying seafloor habitat, removing biomass of target species, and catching nontarget species as bycatch. Whether a stressor affects a particular ecosystem depends on that ecosystem's *vulnerability*. Vulnerability is dictated by *exposure*, which represents the chance that an ecosystem will encounter a given stressor; by *sensitivity*, which represents the degree to which the ecosystem will be affected by the stressor; and by *resilience*, which represents the ability of the affected ecosystem components to recover from disturbance caused by the stressor (Millennium Ecosystem Assessment 2005).

To capture these different aspects of vulnerability, we used five vulnerability criteria, each anchored to specific units of measurement: spatial scale (km^2) for a single occurrence of the stressor, frequency (days per year) that the stressor occurs at a given location, trophic impact (from a single species up to the entire community), percent change in biomass of the affected ecosystem component (%), and recovery time (years) required for the ecosystem to return to natural conditions (Table 13.1). The first two criteria assess the degree of exposure to a given stressor. The trophic impact and percent change in biomass parameters address which components of the ecosystem are sensitive to a given activity or stressor and how sensitive they are. The final criterion, recovery time, measures an aspect of ecosystem resilience by asking how long it would take for the system to recover following the disturbance. Each expert considers their ecosystem of expertise and scores the five criteria for each stressor. The results are combined (as described in Sect. 13.2) into a single vulnerability score after transformation to account for the different ranges of values (i.e., to make the parameter values more directly comparable).

Structuring the collection of expert knowledge along these five axes aids experts in formalizing their knowledge of ecosystem vulnerability (Regan et al. 2004). It improves transparency and repeatability by requiring all experts to make decisions based on the same set of criteria and the same scales. This quantification of expert opinion within a standardized framework makes it possible to compare estimates of vulnerability across different stressors and different ecosystem types. A separate task in the survey allowed us to quantify the *weight* (importance) that a given pool of experts gives to each of the five criteria, so we were not limited to equal weighting when we summed the criteria into a single vulnerability score. For example, experts may feel that the trophic impact is more important than a stressor's frequency of occurrence when assessing vulnerability. Techniques from the field of decision theory let us determine the relative weight of each criterion (its importance) in the expert judgments of vulnerability and let us test how these weights varied among experts.

Table 13.1 Description of the five vulnerability criteria used to assess ecosystem vulnerability to each stressor and the values provided to survey respondents as choices when scoring each criterion in the second section of the survey

Spatial scale	Frequency	Trophic impact	Percentage change in biomass	Recovery time
The spatial scale (km^2) at which a single instance of an activity or stressor impacts the ecosystem, both directly and indirectly	The mean annual frequency (days per year) of the stressor at a particular location within a given region	The primary extent of marine life affected by a stressor within a given ecosystem and region (i.e., single or multiple species, single or multiple trophic levels, or the entire ecosystem)	The degree to which the species, trophic level or levels, or entire ecosystem's "natural" state is affected by the stressor	The mean time (in years) required for the affected species, trophic level or levels, or entire ecosystem to return to its former, "natural" state following disturbance by a particular stressor
0	0	0	0%	0
<1 km^2	Once every 100 years	1 (1 or more species)	1–10%	<6 months
1–10 km^2	Once every 50 years	2 (1 trophic level)	11–20%	6 months to 1 year
10–100 km^2	Once every 20 years	3 (>1 trophic level)	21–30%	1–5 years
100–1,000 km^2	Once every 10 years	4 (entire community)	31–40%	6–10 years
1,000–10,000 km^2	Once every 5 years	Don't know	41–50%	11–25 years
>10,000 km^2	Once every 2 years		51–60%	26–50 years
Don't know	1–7 days per year		61–70%	51–75 years
	7–14 days per year		71–80%	76–100 years
	15–31 days per year		81–90%	>100 years
	31–90 days per year		91–100%	Don't know
	91–180 days per year		Don't know	
	181–364 days per year			
	365 days per year			
	Don't know			

Values for vulnerability scenarios in the first section of the survey were chosen to span these same ranges. Values for these five criteria are weighted and combined (as described in Sect. 13.2) to calculate a single vulnerability score for each combination of stressor and ecosystem

Using these techniques, we tested the generalizability of our model of ecosystem vulnerability by comparing the similarity in model weights from two different regions: New England and the California Current. We also report the expert-assessed marine ecosystem vulnerability across the full suite of human stressors and marine ecosystems in the New England region and discuss how these results can inform ocean management in Massachusetts and beyond.

13.2 Methods

13.2.1 Experts

Experts were defined as academic, government agency, nongovernmental organization (NGO), or private sector scientists and managers with expertise in the ecology, conservation, or management of the ecosystems of interest and experience with some or all of the 58 human stressors. All experts had at least 2 years of experience working in these ecosystems within the waters of New England. Many experts were identified through Google Scholar searches using the ecosystem types in combination with different human stressors. Authors of published, peer-reviewed papers that addressed one or more stressors within relevant marine ecosystems and who had significant experience in the region (e.g., working at a New England institution, publishing multiple relevant papers, participating in major research projects in the area) based on their curriculum vitae or Web sites were considered potential experts. Others were identified through our contacts in the Massachusetts EOEEA and Massachusetts Ocean Partnership, especially for government experts, and we supplemented our pool of experts using snowball sampling, in which the experts we selected identified other potential experts (Goodman 1961; Meyer 2001).

13.2.2 Survey Instrument and Data Collection

The list of 15 ecosystems was derived based on input from the Massachusetts Office of Coastal Zone Management and EOEEA (Table 13.2). The 58 current and emerging human stressors were identified based on our previous work (Teck et al. 2010), and refined based on input from the Massachusetts Ocean Partnership, Office of Coastal Zone Management, and EOEEA. The five vulnerability criteria were developed previously in a workshop at the National Center for Ecological Analysis and Synthesis that brought together conservation scientists and ecologists (Halpern et al. 2007).

The Massachusetts marine ecosystem vulnerability survey was based on and refined from previous instruments that have been deployed globally (Halpern et al. 2007),

Table 13.2 Coastal and marine ecosystem types of New England, with brief descriptions

	Ecosystem	Description
Intertidal	Beach	Sandy shoreline habitat within the tidal zone
	Barrier beach	Sandy intertidal habitat parallel to and separated from shore by a body of water
	Rocky intertidal	Rocky shoreline habitat within the tidal zone
	Salt marsh	Vegetated marine or estuarine habitat within the tidal zone
	Tidal flat	Unvegetated sand or mud habitat within the tidal zone
Subtidal coastal	Eelgrass	Near-shore subtidal habitat dominated by *Zostera marina*
	Algal zone	Near-shore subtidal habitat <10 m deep dominated by algal cover
	Near-shore soft bottom	Near-shore subtidal habitat 10–60 m deep with silt, mud, or sand substrate
	Near-shore hard bottom	Near-shore subtidal habitat 10–60 m deep with cobble, boulder, or bedrock substrate
Offshore	Hard bottom shelf	Subtidal habitat 60–200 m deep with cobble, boulder, or bedrock substrate
	Soft bottom shelf	Subtidal habitat 60–200 m deep with silt, mud, or sand substrate
	Hard bottom bathyal	Subtidal habitat >200 m deep with cobble, boulder, or bedrock substrate
	Soft bottom bathyal	Subtidal habitat >200 m deep with silt, mud, or sand substrate
	Shallow pelagic	Water column above 200 m in all areas >30 m deep[a]
	Deep pelagic	Water column below 200 m in all areas >200 m deep

These represent the categories of ecosystem for which we identified the experts who would be consulted in our study

[a]Pelagic habitat in waters <30 m deep was considered as part of the benthic habitat, i.e., fully coupled

in the Northwest Hawaiian Islands (Selkoe et al. 2009), and in the California Current region (Neslo et al. 2008; Teck et al. 2010). A preliminary draft of the survey instrument was tested and revised based on input from a sample group of seven experts, none of whom participated in the final survey. Potential experts were then contacted and invited to participate. Those who agreed received surveys, and were reminded up to three times until their surveys were returned or they were classified as a nonrespondent. Expert knowledge was collected through a spreadsheet-based survey instrument with pull-down menus and accompanying documentation (including a video tutorial), which were distributed via the Web and returned anonymously. Respondents filled out one survey for each ecosystem in which they had expertise.

The survey was divided into two sections. The first section focused on deriving the vulnerability weights needed for model formulation. These results also addressed our first objective, of determining how these weights would differ from those in the

California Current study. This section of the survey was added to the survey instrument and deployed during a second round of sampling and so was filled out by a subset of all survey respondents. For this section, experts had a discrete-choice task, in which they were asked to rank a set of 30 fictitious human stressor scenarios to determine which one represented the greatest vulnerability for a hypothetical ecosystem based on given values for the vulnerability criteria, as well as the second-, third-, fourth-, and fifth-highest vulnerabilities. Respondents were asked to evaluate either a set of coastal scenarios or a set of offshore scenarios, depending on their area of expertise. We analyzed the results for the coastal and offshore groups separately to assess whether there was a division in perspectives within the expert pool.

The set of 30 human stressor scenarios we created comprised different combinations of values for the five vulnerability criteria. These scenarios were designed to sample the full range of possible combinations of criteria values (see Table 13.1 for the ranges of values for each criterion). We emphasized to respondents that the scenarios were hypothetical, as the intent of this section of the survey was to elicit the perceived importance of each criterion for determining vulnerability. For example, we asked experts to compare two scenarios: one in which a stressor is known to occur at a spatial scale of 10 km^2 with a frequency of once per year, with impacts on the entire community (trophic level = 4) that included a 25% biomass change and a recovery time of 1 year, and another in which the spatial scale was 1 km^2 and the frequency was 365 days/year, with impacts on only a single species (trophic level = 1), a 80% decline in biomass, and a 6-month recovery time. We asked experts to choose the five "worst" stressor scenarios for the hypothetical ecosystem and rank them from 1 to 5 in decreasing order of ecosystem vulnerability. We then used this information to determine the relative weights that experts implicitly place on the five criteria. For example, if trophic impact was most important to experts in making their rankings, then scenarios with high values for trophic impact ought to appear more often in their top-ranked scenarios than those with low values for this criterion. In Sect. 13.2.3, we describe in detail the mathematical technique used to derive the vulnerability criteria weights.

The second section of the survey was treated separately and addressed our second objective, namely to gather quantitative estimates of vulnerability for every combination of ecosystem and stressor. However, because experts could only be expected to comment in detail on ecosystems for which they had knowledge, each survey addressed only a single ecosystem type; experts could fill out more than one survey if they believed they had sufficient expertise in more than one ecosystem type. We later combined the results for each ecosystem to allow comparisons among ecosystem types. For each of the 58 human stressors, the experts estimated a value for each of the five vulnerability criteria from a pull-down menu of ranges (see Table 13.1) or stated that they did not know what the value should be. These values were averaged across the surveys received for each ecosystem (e.g., all beach surveys were combined to obtain vulnerability criteria values for beach habitat for each stressor). In the next section, we explain how these average criteria values were combined into a single score.

13.2.3 Multicriteria Decision Model

We treated vulnerability as the weighted sum of the five vulnerability criteria:

$$\text{Vulnerability} = \sum_{k=1}^{5} W_k S_{ijk}, \tag{13.1}$$

where S_{ijk} is the value of criterion k for stressor i in ecosystem j (e.g., the trophic impact of commercial shipping in shallow pelagic waters), and W_k is the weight assigned to criterion k, such that $\sum_{k=1}^{5} W_k = 1$. Because we expected vulnerability to be monotonic with respect to all five criteria (i.e., vulnerability of the ecosystem should increase with increasing values of each criterion), we chose an additive linear model with positive weights on the criteria as the simplest and most easily interpreted model form. Although using a multiplicative function or another formula to combine the criteria is possible and may be conceptually justifiable in some situations, we lacked sufficient empirical information about the interactions among the criteria to justify such a formulation. In addition, relative rankings will often be fairly insensitive to the exact formulation of the function. Furthermore, a multiplicative model tends to produce a skewed distribution of values that is hard to interpret. Using a simple multiple linear regression let us compute scores for new scenarios that were not included in the first section of the survey by simply supplying values for the five criteria, multiplying them by the model weights, and summing them. We assumed that the weight values were consistent across all stressors and ecosystems, which let us use a single model for all ecosystem–stressor combinations and let us compare them directly.

We used a multicriteria decision model (MCDM), which is a type of random-utility model (Keeney and Raiffa 1993), to derive the model weights. The statistical joint distribution of the model weights (i.e., their means and variances) represents the preferences of a population of experts, in this case their relative preferences for (or perceived importance of) the five different vulnerability criteria, in judging an ecosystem's vulnerability to a particular stressor. We then calculated 95% confidence intervals for the weights to represent the level of disagreement among the experts. MCDMs have been used in a wide range of fields to represent expert judgments of perceived risk or vulnerability (Kraan and Bedford 2005; Kurowicka and Cooke 2006; Neslo et al. 2008; Cooke 2009; Kurowicka et al. 2010), including a limited but growing use in ecology (e.g., Tran et al. 2002). Teck et al. (2010) provide more details on our use of MCDM in this context.

Rather than asking the experts to directly evaluate the criterion weights, we instead inferred values of these weights from their relative importance in the expert rankings of the hypothetical scenarios in the first section of the survey (the discrete-choice task). This inference was based on probabilistic inversion, a mathematical technique analogous to the maximum-likelihood approach that determines the joint distribution of the weights that best reproduces the observed distribution of ranks assigned by the experts (Cooke and Goossens 2004). For example, if 20% of the experts ranked scenario 15 first and 45% ranked scenario 30 second, the algorithm searches for a distribution of weights that realizes these probabilities. We assumed that experts use the information provided in the hypothetical scenarios consistently and that other types of information not provided in the scenario were not important to their ranking decisions. These assumptions are both testable by examining inconsistencies in the scenario rankings (e.g., when a scenario with high values for all five criteria is ranked below a scenario with lower values).

Prior to all analyses, we transformed spatial scale and frequency values with different units of measurement to produce a similar range of values for all the vulnerability criteria. To do so, we used the following transformations: scale = ln[scale × 100] and frequency = ln[frequency × 360]. This was necessary to prevent certain criteria from unduly biasing the results simply because they had larger ranges of values. Models were fit using the top two scenarios ranked by the experts. For each hypothetical set of vulnerability criteria values, we calculated the percentage of experts who ranked that scenario first or second, and used probabilistic inversion to derive the set of model weights that most closely reproduced the observed percentages.

Note that the weights calculated using probabilistic inversion are specific to the exact transformations that were used and to the actual ranges of the data. If these weights are used to calculate vulnerability scores from new sets of criteria values (e.g., those elicited from a new group of experts or for a new set of stressors), the same transformations must be applied.

13.2.4 Analyses

Once we had determined weights for each of the criteria, we multiplied them by the average values for each criterion, derived by averaging the survey responses in the second section across the population of surveys for each ecosystem type, and summed across the five criteria as shown in (13.1). This let us produce a vulnerability score for each ecosystem–stressor combination and a matrix of scores for all possible combinations. From this matrix, we calculated average scores for each stressor (across all ecosystems) and for each ecosystem (across all stressors), thereby allowing us to compare and prioritize stressors and ecosystems and to identify important knowledge gaps.

We used *G*-tests (a maximum-likelihood method analogous to a chi-square test) to test for potential bias in gender or affiliation between respondent and nonrespondent pools. Within the respondent pool, we averaged vulnerability across all criteria and all stressors to test for differences among responses associated with gender (using the *t*-test), affiliation (using ANOVA for the groups academic, federal, state, NGO, or private sector), and years of experience (by means of least-squares linear regression).

13.2.5 Comparison of the Massachusetts and California Current Models

By comparing model weights from the Massachusetts pool of experts with those from the California Current experts, we sought to assess whether our model of ecosystem vulnerability was robust with respect to the pool of experts; that is, we tested whether it could be generalized to other regions with other expert pools. We interpreted a good correspondence between the model weights from these two pools of experts as evidence that the experts had a shared conception of ecosystem vulnerability and of the relative importance of the five vulnerability criteria. Finding similar weights across disparate expert groups would suggest that the model can be used in other regions and applied to other stressors.

13.3 Results

13.3.1 Survey Response Rate

We invited 332 potential experts to participate in the survey. We removed 21 who self-identified as nonexperts, and 112 potential experts did not respond. Of the remaining 199 potential experts, 57 agreed to participate, yielding a participation rate of 28.6%. Some experts filled out surveys for more than one ecosystem, resulting in a total of 87 completed surveys for the ecosystem vulnerability section of the survey. One survey was discarded because the expert did not specify an ecosystem type and only provided information on a single stressor. One survey for the algal zone habitat was identified as an outlier based on extremely high criteria values (more than four standard deviations from the mean for that ecosystem) and was removed from subsequent analyses. We believe that this individual may have misunderstood the scoring task. As no expert evaluated the hard bottom bathyal habitat, we eliminated this ecosystem from further analyses. Though our target was three to five survey responses per ecosystems type, in some cases, we received fewer than three responses. We included results for these ecosystems for comparison's sake, but caution the reader that our confidence in these results is lower than for better-sampled ecosystems.

As the ranking section of our survey was added during a second round of sampling, only a subset of experts completed this portion of the survey: we received 35 rankings (26 from coastal habitat experts and 9 from offshore habitat experts).

13.3.2 Demographics of the Survey Respondents

We received completed surveys from 37 men and 20 women. Respondents had an average age of 47.1 ± 1.6 years (mean ± standard error henceforth), with an average of 18.9 ± 1.4 years of experience working in the region and an average of 19.1 ± 1.6 years

Table 13.3 Compositions of the respondent and nonrespondent pools (only confirmed experts) in terms of their institutional affiliation and gender

	Affiliation					Gender		
Category	Academic	State agency	Federal agency	NGO	Private	Male	Female	Total
Respondents	20	13	19	4	1	39	18	57
Nonrespondents	69	8	51	13	1	106	36	142
Total	89	21	70	17	2	145	54	199

There was significant bias in the respondent pool based on affiliation but not gender (*G*-test; see Sect. 13.3 for details)

of experience in their ecosystem of expertise. Most respondents (65%) were PhD-level scientists, but 25% had a Master's degree and 10% had a B.S. or other degrees. The majority of respondents were employed by academic institutions (35%) or by government agencies (33% federal, 23% state). The remainder of respondents came from NGOs (7%) and private firms (2%). The distribution of regional expertise was balanced: 63% of respondents indicated that their answers applied to the entire New England region, versus 21% who said their answers applied to the Acadian biogeographic province north of Cape Cod, and 16% who said their answers applied to the Virginian province, south of the Cape.

13.3.3 Potential Survey Bias

Log-likelihood-ratio *G*-tests with Yates' correction applied (Table 13.3) showed a significant difference in the composition of the respondent ($n=58$) and nonrespondent ($n=199$) pools with respect to affiliation ($G=12.4$, df=4, $p=0.014$), but not with respect to gender ($G=1.97$, df=1, $p=0.161$). Affiliation was significant because the respondent pool was biased toward state agency participants and the nonrespondent pool was slightly biased toward academic scientists. However, we found no significant difference among the average criteria scores of respondents from different types of organization (ANOVA, $F_{4,52}=1.31, p=0.280$). We also found no effect of gender (t-test, $t=-1.35, p=0.184$) or years of experience (least-squares linear regression, $R^2=0.001, p=0.796$) on the average scores.

13.3.4 Model Results

We used the 35 responses to the first section of the survey (26 coastal, 9 offshore) to fit a linear MCDM composed of the weighted sums of the five vulnerability criteria. Only 7 of the 30 scenarios could be ranked first, over other scenarios, without violating the assumptions of a monotonic, linear model. These are the only scenarios that were not *dominated* by another scenario in the list (i.e., one with equal or higher values across all five vulnerability criteria). The remaining 23 scenarios were dominated and therefore could have produced inconsistent rankings if the experts ranked them first. We discarded 13 responses from experts who chose such inconsistent rankings. This left 16 consistent first rankings for coastal experts and 6 for offshore experts.

Coastal and offshore experts ranked the scenarios similarly. Both sets of experts relied most heavily on the trophic-level impact and the biomass change of the affected ecosystem component or components in making their ranking decisions (weights of 0.466 ± 0.074 and 0.345 ± 0.059, respectively, for coastal experts and 0.542 ± 0.103 and 0.283 ± 0.077, respectively, for offshore experts). The emphasis on these two criteria was similar to that observed with the California Current experts (Table 13.4). Trophic impact was negatively correlated with frequency and recovery time; the frequency and recovery time were slightly positively correlated with spatial scale and percent change in biomass, respectively.

13.3.5 Comparison of the Massachusetts and California Current Models

Given the large number of dominated scenarios in the survey design, we interpreted the number of consistent ranks (16/26 for coastal and 6/9 for offshore) as evidence that the preferences of the New England experts were broadly, but not wholly, consistent with a linear monotonic model of vulnerability based on our five criteria. Vulnerability was perceived by the experts to increase with an increasing number of species and trophic levels affected, with the magnitude of the biomass change in these species or trophic levels, and with the length of time required for recovery, as well as with the spatial extent and the frequency of stressor events. Inconsistent rankings suggest that the experts used other information (e.g., scenario names) beyond the five vulnerability criteria to make their ranking decisions or that there was some degree of misunderstanding of the task. The New England results support the generalizability of the model developed in the California Current project. In both settings, experts placed a combined weight of 81–89% on the trophic impact and the percent change in biomass when determining ecosystem vulnerability (Table 13.4).

13.3.6 Ecosystem Vulnerability

Experts judged the coastal subtidal and offshore benthic habitats, and specifically the hard bottom shelf, near-shore soft bottom, soft bottom shelf, algal zone, near-shore hard bottom, and tidal flat habitats to be most vulnerable to human impacts in this region (Table 13.5). Most intertidal and pelagic habitats received lower vulnerability scores. These results should be treated with caution, however, as the overall and individual sample sizes were quite small. Of particular concern is the fact that one of the most vulnerable habitats (the algal zone) and one of the least vulnerable

Table 13.4 Mean weights for the vulnerability criteria based on results of the ranking section of the survey from both the Massachusetts project and the California Current project (Teck et al. 2010)

Model	No. top ranks included	Spatial extent (km^2)	Frequency (days/year)	Trophic impact (levels 1–4)	Percent change (%)	Recovery time (year)
Coastal (MA)	2	0.0709 ± 0.012	0.0964 ± 0.023	0.466 ± 0.074	0.345 ± 0.059	0.0214 ± 0.006
Offshore (MA)	2	0.0644 ± 0.021	0.0848 ± 0.029	0.542 ± 0.103	0.283 ± 0.078	0.0257 ± 0.009
Combined (CA Current)	4	0.061 ± 0.008	0.046 ± 0.007	0.221 ± 0.022	0.665 ± 0.029	0.008 ± 0.001

The second column gives the number of top stressor ranks used to fit the model. Values are means and standard errors. Weights for the coastal (MA) model were used for subsequent analyses

Table 13.5 Vulnerability scores for the 58 stressors in 14 of the 15 ecosystems (no responses were received for the hard bottom bathyal habitat)

	Coastal ecosystems									Offshore ecosystems					Mean
Human activity	b	bb	r int	salt	tide	eel	alg	nh	ns	h sh	s sh	bath	sp	dp	Score
Expert sample size (*n*)	5	5	8	15	6	10	2	6	6	4	7	1	7	3	6.1
Aquaculture: finfish (herbivores)	0.0	0.0	1.9	0.8	0.3	0.6	nd	2.2	1.9	1.5	0.6	0.0	1.7	0.0	0.9
Aquaculture: finfish (predators)	0.0	0.0	1.9	0.9	0.2	0.6	nd	2.1	1.9	3.0	0.6	0.0	1.7	0.0	1.0
Aquaculture: marine plants	0.0	0.0	2.0	0.8	0.8	0.7	nd	1.0	1.8	1.5	0.6	0.0	1.0	0.0	0.8
Aquaculture: shellfish	0.9	1.7	2.5	1.4	2.0	2.0	1.7	2.1	2.5	2.4	0.6	0.0	1.0	0.6	1.5
Benthic structures	0.0	1.3	2.8	1.9	3.3	1.2	3.7	2.6	3.3	3.8	3.2	3.6	1.8	1.9	2.5
Climate change: ocean acidification	1.8	3.2	4.0	3.7	4.3	2.5	nd	3.0	1.5	5.0	nd	nd	4.7	3.8	3.4
Climate change: sea level rise	4.4	5.0	4.8	4.6	4.8	3.5	nd	2.4	2.0	4.7	1.0	0.0	2.5	1.1	3.1
Climate change: sea temperature change	4.3	5.5	3.9	4.2	6.1	2.8	nd	4.6	4.8	5.9	3.5	4.3	5.1	4.6	4.6
Climate change: UV change	2.1	2.5	3.6	3.9	5.3	4.2	nd	3.1	3.8	5.3	nd	0.0	3.8	3.3	3.4
Coastal engineering: altered flow dynamics	3.6	4.2	4.0	3.7	3.6	2.8	nd	1.2	3.2	1.4	1.3	0.0	1.2	1.5	2.5
Coastal engineering: habitat alteration	3.7	4.0	3.7	3.4	3.6	2.9	nd	1.6	4.2	1.6	1.4	0.0	1.3	1.3	2.5
Direct human impact: trampling	3.0	2.9	2.0	2.2	2.7	1.6	0.0	0.6	0.8	0.0	0.5	0.0	0.4	0.9	1.3
Diseases and pathogens	1.5	1.6	2.9	1.8	3.2	2.3	nd	2.5	1.8	2.5	6.1	2.3	2.8	2.8	2.6
Dredging	2.8	3.9	1.3	1.8	3.3	2.6	3.4	3.2	3.5	4.1	3.2	0.0	1.3	1.8	2.6
Energy infrastructure: liquid natural gas	1.2	2.1	3.3	1.3	3.1	2.5	nd	2.1	4.6	3.3	3.4	nd	2.8	2.4	2.7
Energy infrastructure: tidal	1.4	2.6	1.0	1.5	3.0	2.7	nd	2.6	5.9	3.0	2.2	0.0	2.6	0.7	2.3

Human activity	b	bb	r int	salt	tide	eel	alg	nh	ns	h sh	s sh	bath	sp	dp	Mean
Energy infrastructure: wave	1.4	2.7	1.0	0.5	3.1	2.5	nd	2.6	5.5	3.0	3.1	0.0	3.3	0.7	2.3
Energy infrastructure: wind	1.3	2.4	4.0	0.6	3.1	2.4	nd	2.8	4.7	3.4	3.5	0.0	3.4	1.8	2.6
Fishing: aquarium	0.0	0.6	0.3	0.3	0.0	0.0	nd	0.9	0.4	1.9	0.6	0.0	0.7	0.7	0.5
Fishing: demersal destructive	0.0	0.0	0.9	0.5	3.1	2.6	nd	3.5	3.8	3.5	3.4	3.4	2.2	2.8	2.3
Fishing: demersal nondestructive, high bycatch	0.0	0.0	0.3	0.8	1.1	0.5	2.8	2.5	2.6	3.9	3.5	2.7	1.8	1.4	1.7
Fishing: demersal nondestructive, low bycatch	0.0	0.0	1.4	0.7	1.1	0.8	2.5	2.2	2.1	2.3	2.2	0.0	2.5	1.7	1.4
Fishing: destructive artisanal (subsistence)	0.2	0.0	0.0	0.5	1.1	0.0	nd	1.3	0.6	1.7	0.7	0.0	1.1	0.0	0.5

Fishing: nondestructive artisanal (subsistence)	0.6	0.0	0.2	0.7	1.0	0.3	nd	2.0	0.0	2.3	1.0	0.0	0.3	0.0	0.6
Fishing: pelagic, high bycatch	0.0	0.4	0.0	0.2	1.0	0.0	nd	1.6	1.5	3.0	2.9	0.0	3.2	2.9	1.3
Fishing: pelagic, low bycatch	0.0	0.3	0.0	0.2	1.0	0.0	nd	1.3	1.5	2.3	2.6	0.0	2.8	1.1	1.0
Fishing: recreational	1.1	1.7	1.2	1.6	2.1	1.1	1.6	2.5	1.5	2.8	2.0	0.0	2.2	0.9	1.6
Freshwater input: decrease	0.2	0.0	1.7	2.6	0.2	1.8	nd	3.0	0.0	3.4	2.4	0.0	1.2	0.8	1.3
Freshwater input: increase	0.5	1.1	2.6	3.5	2.2	2.4	nd	3.0	1.9	3.4	2.7	0.0	2.4	0.8	2.0
Invasive species (from ballast, etc.)	3.6	3.5	3.8	3.5	1.8	2.6	nd	3.5	4.0	4.0	3.3	nd	3.2	3.1	3.3
Military activity	0.0	0.8	0.5	0.2	3.1	0.0	nd	2.2	nd	1.8	nd	nd	1.7	nd	1.1
Nutrient input: causing harmful algal blooms	0.2	1.5	2.1	1.7	3.6	2.6	nd	2.7	1.5	3.5	2.9	0.0	3.1	1.2	2.0
Nutrient input: causing hypoxic zones	0.2	1.6	1.9	1.9	3.6	2.4	nd	3.2	3.7	3.2	3.0	0.0	1.5	1.1	2.1
Nutrient input: into eutrophic waters	0.2	1.8	1.3	2.9	4.0	2.9	nd	3.2	3.7	3.7	2.7	0.0	2.3	1.1	2.3
Nutrient input: into oligotrophic waters	0.2	1.7	0.5	2.5	2.4	2.3	2.9	1.6	nd	3.4	2.9	0.0	0.4	1.1	1.7
Ocean dumping: lost fishing gear	2.5	2.5	1.1	0.7	1.4	1.0	2.5	2.4	3.0	2.8	3.1	2.9	2.3	2.6	2.2
Ocean dumping: marine debris (trash, etc.)	3.6	4.2	2.2	1.6	2.7	0.8	3.1	2.4	3.3	2.8	3.7	nd	3.1	2.9	2.8
Ocean dumping: shipwrecks	0.5	2.7	1.6	0.3	1.2	0.5	nd	1.6	3.4	2.0	2.8	nd	1.3	1.5	1.6
Ocean dumping: toxic materials	1.9	2.2	2.6	1.5	3.4	0.7	3.7	2.2	2.2	3.9	4.3	5.2	2.3	2.6	2.8
Ocean mining (sand, minerals, etc.)	2.3	2.8	1.6	0.7	1.4	0.5	nd	2.3	3.8	3.7	3.3	0.0	2.0	2.9	2.1
Ocean pollution (from ships, ports, etc.)	3.4	3.3	3.3	2.9	3.2	1.8	nd	3.1	3.5	3.8	4.0	3.0	3.5	2.9	3.2
Pollution input: atmospheric	1.0	3.0	1.8	2.8	nd	1.8	nd	4.2	nd	4.0	4.7	0.0	3.4	2.8	2.7

Human activity	b	bb	r int	salt	tide	eel	alg	nh	ns	h sh	s sh	bath	sp	dp	Mean
Pollution input: inorganic	1.3	1.7	2.0	3.0	2.6	2.0	nd	3.5	nd	3.9	4.7	0.0	2.8	2.6	2.5
Pollution input: light	nd	2.1	1.0	0.7	0.0	0.0	nd	1.5	nd	1.8	nd	0.0	1.5	1.4	1.0
Pollution input: noise	nd	1.9	0.6	1.3	1.6	0.0	nd	1.7	nd	2.4	nd	0.0	2.7	2.9	1.5
Pollution input: organic	2.3	2.3	3.0	3.0	3.0	3.0	nd	3.4	3.8	3.6	3.8	0.0	3.5	3.2	2.9
Pollution input: trash, etc. (urban runoff)	3.1	3.3	2.1	2.7	2.7	1.9	3.6	3.1	4.1	2.4	3.2	0.0	2.8	2.1	2.7
Power plants and desalination plants	0.0	0.0	1.4	2.1	1.5	1.5	nd	2.6	3.1	3.2	1.3	0.0	1.5	1.1	1.5
Scientific research: collecting	0.6	1.4	1.7	2.1	1.3	0.4	0.5	0.9	1.3	1.5	1.8	0.0	0.9	1.6	1.1

(continued)

Table 13.5 (continued)

Human activity	Coastal ecosystems									Offshore ecosystems					Mean
	b	bb	r int	salt	tide	eel	alg	nh	ns	h sh	s sh	bath	sp	dp	Score
Scientific research: experiments/surveys	0.6	1.9	1.6	2.3	1.4	0.6	0.5	0.6	1.3	2.2	2.1	0.0	0.4	0.6	1.2
Sediment input: decrease	2.7	4.3	1.3	4.1	2.0	1.4	nd	1.9	0.7	3.2	1.5	0.0	1.0	0.8	1.9
Sediment input: increase	3.1	3.8	2.4	2.4	3.6	2.2	nd	2.2	2.3	3.4	2.8	0.0	1.8	0.8	2.4
Shipping (commercial, cruise, ferry)	0.9	3.1	1.2	0.8	0.9	1.7	1.8	2.8	nd	3.2	1.7	0.0	2.7	0.8	1.7
Tourism: kayaking	0.7	1.5	1.1	1.2	0.8	0.5	nd	0.6	0.0	0.0	0.0	0.0	0.5	0.0	0.5
Tourism: recreational boating	1.4	2.1	1.6	2.0	2.3	2.8	nd	2.0	1.7	1.2	0.7	0.0	1.2	0.8	1.5
Tourism: scuba diving	0.5	0.8	1.1	0.1	0.6	0.7	nd	1.2	1.5	0.0	0.0	0.0	0.5	0.6	0.6
Tourism: surfing	1.2	1.7	0.6	0.1	0.0	0.0	nd	0.0	0.0	0.0	0.0	0.0	0.4	0.0	0.3
Tourism: whale watching	nd	1.9	0.0	0.0	0.0	0.0	nd	0.6	0.0	1.9	0.0	0.0	2.3	1.1	0.7
Mean score	1.3	2.0	1.8	1.8	2.2	1.5	2.3	2.2	2.5	2.8	2.3	0.5	2.1	1.6	1.9
Standard error of the mean score	0.1	0.1	0.1	0.1	0.1	0.1	0.2	0.1	0.1	0.1	0.1	0.2	0.1	0.1	
Mean sample size	2.7	2.5	5.3	8.5	2.5	6.6	0.6	3.7	2.4	1.9	3.8	0.9	4.0	2.4	
Standard error of the mean sample size	0.1	0.1	0.2	0.3	0.1	0.2	0.1	0.2	0.2	0.1	0.2	0.0	0.1	0.1	

Ecosystem abbreviations: *b* beach, *bb* barrier beach, *r int* rocky intertidal, *salt* salt marsh, *tide* tidal flat, *eel* eelgrass, *alg* algal zone, *nh* near-shore hard bottom, *ns* near-shore soft bottom, *h sh*, hard bottom shelf, *s sh* soft bottom shelf, *bath* soft bottom bathyal, *sp* shallow pelagic, *dp* deep pelagic. Ecosystem–stressor vulnerability scores were derived from the average expert criteria values from the vulnerability portion of the survey, which were weighted and then summed using the MCDM. We used weights derived from the coastal experts' ranking given its greater sample size and its similarity to the values provided by the offshore experts. Scores are color-coded into four groups by dividing the range equally: 0–1.532 in *white*, 1.533–3.065 in *light gray*, 3.066–4.598 in *dark gray*, and 4.599–6.1 in *black* (with *white* numbers). Mean scores for each stressor across all ecosystems and for each ecosystem across all stressors are reported in the "Mean score" column and row, respectively. The mean sample sizes (±SE) were calculated by averaging across the number of responses provided for each vulnerability criterion (this number differs from the overall sample size by ecosystem given at the top of the table because not all experts filled out values for every criterion and every stressor) (soft bottom bathyal) had overall sample sizes of two or fewer experts, average score sample sizes below two, and missing vulnerability scores for many stressors (i.e., experts selected "don't know" for the criteria under these stressors). The top stressors included climate change impacts from rising ocean temperatures and ocean acidification, invasive species, increased ultraviolet radiation exposure, and ocean pollution from ships, ports, and spills.

13.4 Discussion

13.4.1 Model Results

As in the California Current project, the New England experts placed the most emphasis on the combination of trophic impacts and the percent change in biomass of the affected ecosystem components in their ranking of ecosystem vulnerability. However, the New England experts placed slightly greater emphasis on trophic impacts over percent change, whereas the California Current experts emphasized percent change in biomass in their rankings. As on the west coast (data not shown), the results for the two separate groups of experts (coastal and offshore) were quite similar. By weighting the trophic impact and percent change in biomass heavily, the experts were focusing primarily on the *sensitivity* of an ecosystem to particular stressors when determining its vulnerability. These results suggest that the model is generalizable both across regions and across groups of experts within a region.

The perceptions of ecosystem vulnerability by the New England experts, like those in the California Current project, were broadly consistent with a monotonic model of vulnerability based on spatial scale, frequency, trophic impact, percent change in biomass, and recovery time (Neslo et al. 2008). Given the large number of dominated scenarios in the survey design, inconsistent rankings are not unexpected. However, the fairly high number of inconsistent rankings suggests that other criteria or interactions among the criteria may also be important, that respondents did not understand the discrete-choice task, or that respondents were misidentified as experts. We cannot distinguish among these possibilities based on the data available to us, though our prevalidation of the survey instrument gives us confidence in the clarity of our instructions and its effectiveness as an elicitation tool. Sorting out the source of inconsistent rankings could be the subject of future investigations.

Prior investigations of this model, with greater sample sizes, have allowed for out-of-sample model validation (i.e., using a subset of the overall sample to develop the model and the remainder of the sample to validate the model) and have shown strong internal model validity (Neslo et al. 2008; Teck et al. 2010). Combined with the present results, these results suggest that our approach can provide a reliable model of expert perceptions of ecosystem vulnerability (within the global context of expert knowledge), and that the model can be used in a variety of settings, across a wide array of marine ecosystem types.

13.4.2 *Data Requirements for Determining Marine Ecosystem Vulnerability*

The approach described in this chapter fills a critical need for objective and quantitative ways to compare the relative vulnerability of a diversity of ecosystem types to a broad suite of human stressors. The resulting data fill many important data gaps for understudied ecosystems and stressors. The survey also served to identify particularly important data gaps and uncertainties for the region that even expert knowledge is hard-pressed to fill. For example, despite our efforts, few experts filled out surveys for several ecosystem types (only two for the algal zone, one for the soft bottom bathyal zone, and three for the deep pelagic zone), which leads to relatively high uncertainty about the average scores for these ecosystems. These and other ecosystems are also plagued by gaps for individual stressors, for which all experts answered "don't know." Even ecosystems with reasonable sample sizes (e.g., near-shore soft bottom and soft bottom shelf) sometimes had large numbers of understudied stressors for which no experts were able to score our vulnerability criteria ("nd" values in Table 13.4). The small sample sizes may have resulted from a dearth of experts working on individual human stressors or ecosystems in this region, reluctance on the part of such experts to participate, or a combination of these factors.

The New England project identified an extremely large candidate pool relative to other expert-based studies, which typically have sample sizes on the order of five to nine (Meyer 2001; see Table 14.1, Chap. 14), though sample sizes for expert panels in ecological assessments may be similar in magnitude to the size that we used (Noble 2004). Small sample sizes result primarily from the need to divide the expert pool among so many different ecosystem types. Knowledge gaps highlighted by this work may benefit from targeted funding to increase the understanding of the impact of certain stressors in certain marine environments. Small sample sizes may limit the willingness of decision-makers to use expert knowledge; targeted filling of the knowledge gap may increase the credibility, applicability, and utility of these results.

13.4.3 *Application to Management Decisions in Massachusetts*

The results of this survey have contributed to the development of the Massachusetts Ocean Management Plan by providing key information on ecosystem vulnerability and potential incompatibilities between particular ecosystems and human uses of the ocean. The Ocean Management Plan divides the planning region into three different types of management areas: prohibited, renewable energy, and multiuse areas. The majority of the planning area is designated as multiuse. Management of multiuse areas is determined by the vulnerability of the specific marine resources identified as being important within those areas. Compatibility analysis was a critical step in developing management measures for the multiuse areas, and the vulnerability scores matrix presented in this chapter was one of a small number of tools that managers had at their disposal to complete that analysis. In particular, this ecosystem-level analysis provided important information about the vulnerability of eelgrass, hard or complex seafloor, and intertidal flat habitats, all of which were identified as special, sensitive, and unique marine resources that required special protection.

The vulnerability scores were also an integral part of the development of cumulative-impact maps for the region, which show the spatial distribution and intensity of cumulative impacts based on both the distribution and intensity of activities and on the relative vulnerability of the underlying ecosystems to those uses. These maps and their input data layers will serve as important tools for management decisions as EOEEA moves forward with implementation and revision of the Ocean Management Plan. The Massachusetts Ocean Partnership is actively working with state managers on further ways to apply the survey results and cumulative impact maps. Potential uses of these tools that we are exploring with the state include providing guidance when evaluating permit and siting decisions, informing tradeoff analyses, and identifying future research needs.

13.4.4 Benefits of Using an Expert Knowledge Approach

Using expert knowledge to inform the ocean management planning process had other ancillary benefits. Seeking expert knowledge helped to broaden and strengthen the network of experts who the Massachusetts Ocean Partnership and the State can call upon for input into the process (N. Napoli, Massachusetts Ocean Partnership, personal communication). This study also increased the participation and engagement of regional scientists in the management planning process and increased knowledge within the research community of the work being done by the Massachusetts Ocean Partnership. Similar objectives of coordinating and eventually adopting conservation recommendations by expert participants are reported by Moody and Grand (Chap. 6).

13.4.5 Challenges of Using Expert Knowledge

There are some challenges associated with using expert knowledge in ocean management and planning, as we have done in this study. In practice, it may not always be feasible to conduct such a survey, as obtaining sufficient numbers of experts requires significant effort. Even with a relatively large respondent pool, the final sample sizes for each ecosystem–stressor combination will generally be small given the large number of ecosystem types into which the expert pool is split. The survey instrument itself is time-consuming and labor-intensive to complete, generally requiring from 45 min to 1 h, and identifying, communicating with, and cajoling appropriate experts to fill out the survey took several months. We found that some scientists, as trained empiricists, were reluctant to provide quantitative values in the form of "expert opinion," especially around topics they perceived to have high uncertainty or to be associated with controversy. Finally, with a limited pool of experts in the relevant fields, one soon runs into survey fatigue (i.e., experts become unwilling to repeatedly respond to survey requests), so repeat deployments of this or similar surveys become increasingly difficult. However, given the robustness of the model we developed and its validation across disparate groups of experts, it can be used in novel settings and to address emerging stressors without having to redeploy the ranking portion of the survey. Ecosystem vulnerability to new and emerging human stressors could be evaluated with a smaller group of experts and a shorter survey instrument (restricted to the new stressors and relevant ecosystems), so future elicitation of expert knowledge to improve our understanding of marine ecosystem vulnerability in this region could be significantly streamlined. Finally, given that the experts appeared to agree on the importance of trophic impact and the percent change in biomass as the driving factors that underlie ecosystem vulnerability, future elicitation of expert knowledge could potentially be made more efficient by focusing on these two vulnerability criteria.

13.5 Conclusions

Robust methods for comparing multiple ecosystem types and their vulnerabilities to a broad suite of human stressors are sorely needed in order to meet the requirements of ecosystem-based management and other forms of comprehensive ocean management (Leslie and McLeod 2007; Halpern et al. 2008a; Ehler and Douvere 2009). We demonstrated the utility and broad applicability of a structured framework for eliciting expert knowledge about the vulnerability of marine ecosystems to human stressors. The framework is ecologically grounded, flexible, transparent, and easily updated to accommodate emerging stressors. In addition, we provided clear and consistent instructions to all respondents, which have been archived for future use; this means that future research can follow the same approach and produce results that will be directly comparable to the results presented in this chapter. The framework is objective-neutral, meaning that it does not have the end goal of informing a particular kind of management action, such as the design of marine protected areas. Instead, the results of this survey approach can be used to inform a variety of conservation and management prioritization and planning exercises. Our results were applied in the development of the Massachusetts Ocean Management Plan, where they helped inform the assessment of compatibility between ocean uses and the protection of vulnerable marine resources. These results also contributed to the development of spatial maps of the cumulative impact of human uses of the waters off Massachusetts. The Massachusetts Ocean Partnership is actively working with state managers to use the results further in plan implementation and revision.

As the coastal zone becomes increasingly crowded, comprehensive spatial planning is emerging as a necessary management tool to sustain important marine ecosystem services, reduce conflicts, and protect vulnerable ecosystems (Douvere 2008; Ehler and Douvere 2009). Unfortunately, few tools exist to easily and robustly assess the impacts of the myriad human uses that compete for space in these waters. Even worse, only fragmentary data exist to help us understand the relative threat to marine ecosystems from the cumulative impacts of these various activities. Approaches that can aggregate the collective knowledge of the experts who know these ecosystems best will be necessary in order to move forward with spatial planning in the face of such pervasive and severe data gaps. In this chapter, we describe one framework for doing so – one that we believe is flexible enough to be applied in a variety of settings and easily transferred to new situations. Future work might expand on this approach to evaluating anthropogenic stressors by assessing management criteria such as the feasibility of addressing a particular stressor, the enforceability of any resulting guidelines, and the enforcement costs of management measures. Finally, marine spatial planning may benefit from expanding our definition of "expert." Our survey can and should be tailored to elicit information from other ecological knowledge holders like fishermen (Murray et al. 2006; St Martin et al. 2007; Johannes et al. 2008).

References

Aspinall W (2010) A route to more tractable expert advice. Nature 463:294–295

Beazley KF, Baldwin ED, Reining C (2010) Integrating expert judgment into systematic ecoregional conservation planning. In: Trombulak SC, Baldwin RF (eds) Landscape-scale conservation planning. Springer Science+Business Media, New York, pp 235–255

Bryant D, Burke L, McManus J, Spalding M (1998) Reefs at risk: A map-based indicator of threats to the world's coral reefs. World Resources Institute, Washington

Burgman MA (2001) Flaws in subjective assessments of ecological risks and means for correcting them. Austral J Environ Manage 8:219–226

Burke L, Selig E, Spalding M (2002) Reefs at risk in Southeast Asia. World Resources Institute, Washington

Cooke, RM (2009) Obtaining distributions from groups for decisions under uncertainty. In: Williams TM, Samset K, Sunnevag KJ (eds) Making essential choices with scant information: front-end decision making in major projects. Palgrave Macmillan, Basingstoke, pp 257–274

Cooke RM, Goossens LHJ (2004) Expert judgement elicitation for risk assessments of critical infrastructures. J Risk Res 7:643–656

Crowder LB, Osherenko G, Young OR et al (2006) Sustainability: resolving mismatches in U.S. ocean governance. Science 313:617–618

Day JC (2002) Zoning – lessons from the Great Barrier Reef Marine Park. Ocean Coast Manage 45:139–156

Douvere F (2008) The importance of marine spatial planning in advancing ecosystem-based sea use management. Mar Policy 32:762–771

Douvere F, Maes F, Vanhulle A, Schrijvers J (2007) The role of marine spatial planning in sea use management: the Belgian case. Mar Policy 31:182–191

Ehler C, Douvere F (2009) Marine spatial planning: A step-by-step approach toward ecosystem-based management. Intergovernmental Oceanographic Commission and Man and the Biosphere Programme, UNESCO, Paris. IOC Manual and Guides No. 53, ICAM Dossier No. 6

Goodman LA (1961) Snowball sampling. Ann Math Stat 32:148–170

Halpern BS, McLeod KL, Rosenberg AA, Crowder LB (2008a) Managing for cumulative impacts in ecosystem-based management through ocean zoning. Ocean Coast Manage 51:203–211

Halpern BS, Walbridge S, Selkoe KA, Kappel CV et al (2008b) A global map of human impact on marine ecosystems. Science 319:948–952
Halpern BS, Selkoe KA, Micheli F, Kappel CV (2007) Evaluating and ranking the vulnerability of global marine ecosystems to anthropogenic threats. Conserv Biol 21:1301–1315
Halpern BS, Kappel CV, Selkoe KA, Micheli F, Ebert C, Kontgis F, Crain C, Martone MRG, Shearer C, Teck S (2009) Mapping cumulative human impacts to California Current marine ecosystems. Conserv Lett 2:138–148
Johannes RE, Freeman MMR, Hamilton RJ (2008) Ignore fishers' knowledge and miss the boat. Fish Fisheries 1:257–271
Kappel CV (2005) Losing pieces of the puzzle: Threats to marine, estuarine, and diadromous species. Front Ecol Environ 3:275–282
Keeney RL, Raiffa H (1993) Decisions with multiple objectives: Preferences and value tradeoffs. Cambridge University Press, Cambridge
Keith DA (1998) An evaluation and modification of World Conservation Union Red List criteria for classification of extinction risk in vascular plants. Conserv Biol 12:1076–1090
Kraan, BCP, Bedford TJ (2005) Probabilistic inversion of expert judgements in the quantification of model uncertainty. Manage Sci 51:995–1006
Kurowicka D, Bucura C, Cooke RM, Havelaar A (2010) Probabilistic inversion in priority setting of emerging zoonoses. Risk Anal 30:715–723
Kurowicka D, Cooke RM (2006) Uncertainty analysis with high dimensional dependence modelling. Wiley, New York
Leslie HM, McLeod KL (2007) Confronting the challenges of implementing marine ecosystem-based management. Front Ecol Environ 5:540–548
Mace GM, Lande R (1991) Assessing extinction threats: toward a reevaluation of IUCN threatened species categories. Conserv Biol 5:148–157
Master LL (1991) Assessing threats and setting priorities for conservation. Conserv Biol 5:559–563
McLeod KL, Lubchenco J, Palumbi SR, Rosenberg AA (2005) Scientific consensus statement on marine ecosystem-based management. Communication Partnership for Science and the Sea, Washington. Available from http://www.compassonline.org/science/EBM_CMSP/EBMconsensus (accessed February 2011)
Meyer MA (2001) Eliciting and analyzing expert judgment: a practical guide. Society for Industrial and Applied Math, Philadelphia
Millennium Ecosystem Assessment (2005) Ecosystems and human well-being. Island Press, Washington
Murray G, Neis B, Johnsen JP (2006) Lessons learned from reconstructing interactions between local ecological knowledge, fisheries science, and fisheries management in the commercial fisheries of Newfoundland and Labrador, Canada. Hum Ecol 34:549–571
Neslo REJ, Micheli F, Kappel CV et al (2008) Modeling stakeholder preferences with probabilistic inversion: application to prioritizing marine ecosystem vulnerabilities. In: Linkov I, Ferguson E, Magar VS (eds) Real-time and deliberative decision making. Springer, Dordrecht, NATO Science for Peace and Security Series C: Environmental Security
Nicholson E, Keith DA, Wilcove DS (2009) Assessing the threat status of ecological communities. Conserv Biol 23:259–274
Noble BF (2004) Strategic environmental assessment quality assurance: evaluating and improving the consistency of judgments in assessment panels. Environ Impact Assess 24:3–25
Noss RF, Carroll C, Vance-Borland K et al (2002) A multicriteria assessment of the irreplaceability and vulnerability of sites in the Greater Yellowstone Ecosystem. Conserv Biol 16:895–908
Office of the President (2010) Stewardship of the oceans, our coasts, and the Great Lakes. 75 Federal Register 43023, FR Doc. 2010–18169 (Executive Order, 22 July 2010)
Plous S (1993) The psychology of judgment and decision making, McGraw-Hill Inc., New York
Regan TJ, Master LL, Hammerson GA (2004) Capturing expert knowledge for threatened species assessments: a case study using NatureServe conservation status ranks. Acta Oecol 26:95–107
Roberts CM (2002) Marine biodiversity hotspots and conservation priorities for tropical reefs. Science 295:1280–1284

Rush C, Roy R (2001) Expert judgement in cost estimating: modelling the reasoning approach. Concurrent Engin-Res A 9:271–284

Selkoe KA, Halpern BS, Ebert CM et al (2009) A map of human impacts to a "pristine" coral reef ecosystem, the Papahānaumokuākea Marine National Monument. Coral Reefs 28:635–650

Small C, Cohen JE (2004) Continental physiography, climate, and the global distribution of human population. Curr Anthropol 45:269–277

St Martin K, McCay BJ, Murray GD, Johnson TR (2007) Communities, knowledge and fisheries of the future. Int J Global Environ 7:221–239

Teck SJ, Halpern BS, Kappel CV et al (2010) Using expert judgment to estimate marine ecosystem vulnerability in the California Current. Ecol Appl 20:1402–1416

Tran LT, Knight CG, O'Neill RV et al (2002) Fuzzy decision analysis for integrated environmental vulnerability assessment of the Mid-Atlantic Region 1. Environ Manage 29:845–859

Tversky A, Kahneman D (1982) Judgment under uncertainty: heuristics and biases. In: Kahneman D, Slovic P, Tversky A (eds) Judgment under uncertainty. Cambridge University Press, Cambridge

Vié JC, Hilton-Taylor C, Stuart SN (2009) Wildlife in a changing world: An analysis of the 2008 IUCN Red List of threatened species. World Conservation Union, Gland

Wickham JD (1999) Environmental auditing: an integrated environmental assessment of the US Mid-Atlantic Region. Environ Manage 24:553–560

Wilcove DS, Rothstein D, Dubow J et al (1998) Quantifying threats to imperiled species in the United States. BioScience 48:607–615

Wilson K, Pressey RL, Newton A et al (2005) Measuring and incorporating vulnerability into conservation planning. Environ Manage 35:527–543

Chapter 14
Elicitation and Use of Expert Knowledge in Landscape Ecological Applications: A Synthesis

Chris J. Johnson, C. Ashton Drew, and Ajith H. Perera

Contents

14.1 Introduction

There are many pressing questions and challenges in landscape ecology that have important consequences for sustainable resource management and the conservation of biodiversity. Given the spatial and temporal scopes and the resulting complexity of these issues, many landscape ecologists struggle to provide evidence-based solutions.

C.J. Johnson (✉)
Ecosystem Science and Management Program, University of Northern British Columbia, 3333 University Way, Prince George, BC, Canada V2N 4Z9
e-mail: johnsoch@unbc.ca

C.A. Drew
Department of Biology, U.S. Geological Survey and North Carolina Cooperative Fish and Wildlife Research Unit, North Carolina State University, Campus Box 7617, Raleigh, NC 27695, USA

A.H. Perera
Ontario Forest Research Institute, 1235 Queen Street East, Sault Ste. Marie, ON, Canada P6A 2E5

A.H. Perera et al. (eds.), *Expert Knowledge and Its Application in Landscape Ecology*, DOI 10.1007/978-1-4614-1034-8_14, © Springer Science+Business Media, LLC 2012

This is especially apparent when we rely exclusively on the traditional approaches and data employed in the natural sciences to understand broad-scale phenomena that have interacting ecological and human elements. By exploring alternative ways to address the limitations of conventional observational and experimental methods, the authors of this book have used expert knowledge to complement poor data or replace missing empirical data, to cope with complexity that confounded the design and conduct of empirical studies, and to solve problems that required the coupling of knowledge generation with management or conservation decision making. The innovative and diverse array of methods illustrated in this book transcend our work in landscape ecology, providing tools and promoting insights that will be relevant within many other subdisciplines of applied ecology (Kuhnert et al. 2010; Orsi et al. 2011). Furthermore, the perspectives and breadth of studies these authors have presented support the growing consensus that the application of expert knowledge is no longer the domain of a few maverick ecologists working at the margins of methodological inquiry. The number of expert-based studies has more than doubled in the past 10 years, and expert knowledge is serving as a credible foundation for many of the most pressing and complex debates in applied ecology (e.g., O'Neill et al. 2008).

For many, their first foray into the collection and application of expert knowledge is a response to the challenge of having little or no empirical data to guide management and conservation decisions (e.g., Drew and Collazo, Chap. 5; Doyon et al., Chap. 10; Keane and Reeves, Chap. 11). When investigating a new research area or developing a decision-support tool, we direct our initial efforts towards identifying the relevant body of theory and collecting or incorporating empirical data. Often, however, we encounter data gaps that would limit the precision, accuracy, or applicability of the study or product, and we struggle with funding or time constraints that prevent the collection of empirical data to support our efforts. While wrestling with such challenges, we recognize that the professional experience and knowledge of our colleagues could potentially address many of the gaps in the theoretical or empirical knowledge. It is at these crossroads that we make the decision to either formally incorporate expert knowledge in our efforts or to initiate an empirical research program. If we decide on the former approach, a departure from the methods of spatial data collection and analyses that are familiar to landscape ecologists, the research becomes a study focused on the human subject – an area in which most scientists lack experience, and which lies outside our comfort zone. The case studies and discussions in this book will better prepare landscape ecologists to consider whether and how to incorporate expert knowledge in our research, and will better equip us to practice rigorous methods when eliciting this knowledge.

By presenting a diversity of projects, both theoretical and practical, this book offers insights that will allow ecologists to anticipate the potential applicability, advantages, and pitfalls of expert knowledge. All of the lead authors are landscape ecologists who have found themselves dependent on expert knowledge to supplement, complement, or even replace empirical data (Table 14.1). They have shared their experiences, both successes and failures, wrestling with how to elicit expert knowledge in a manner that meets scientific standards of transparency and repeatability. In this chapter, we will synthesize some of the common themes that have emerged from their experiences and highlight opportunities for further research and development (Table 14.2).

Table 14.1 Summary of the objectives and methods for eliciting expert knowledge

	Chapter and lead author									
	3-Low Choy	4, 9-Drescher	5-Drew	6-Moody	7-McNay	8-Johnson	10-Doyon	11-Keane	12-Williams	13-Kappel
Study context and justification										
Study location and topic	Australia; habitat	Canada; succession	USA; habitat	USA; habitat	Canada; habitat	Canada; habitat	Canada; succession	USA; fire	Australia; planning	USA; vulnerability
Expert knowledge supplemented (SP), complemented (CM), or provided the sole source (SS) of information	SS, CM	CM	SS	SS	CM	CM	SS	SS	SS, CM	SS
Pre-elicitation										
Number of experts invited/participated	11/9	18/13	4/4	278/53	n.a.	10/10; 10/6	10/7	~120/55	6	199/58
Experts = sages (SG), practitioners (PR), scientists (SC), or stakeholders (ST)	SG, PR, SC	SG, PR	PR	SG, PR, SC	SG, PR, SC, ST	SG, PR, SC	SG, PR	SG, PR, SC, ST	SG, PR, SC, ST	SG, PR, SC
Elicitation piloted using test subjects or reviewed by an elicitation specialist?	Y	Y	N	N	N	N	N	Y	Y	Y
Training provided to standardize terminology and mitigate biases?[a]	Y	Y	Y	Y	Y	Y	Y	N	Y	Y

(continued)

Table 14.1 (continued)

	Chapter and lead author									
	3-Low Choy	4, 9-Drescher	5-Drew	6-Moody	7-McNay	8-Johnson	10-Doyon	11-Keane	12-Williams	13-Kappel
Elicitation										
Method of elicitation	Interview; software & GIS	Focus group/ interview; software	Discussion/ map-based interview	Focus group	Focus group	Questionnaire; analytical hierarchy procedure	Questionnaire	Workshop	Focus group	Questionnaire
Knowledge elicited individually (I), in groups (G), or both (B)	I	B	I	G	G	I	B	B	G	I
Knowledge elicited in person (P) or remotely (R), or both (B)	P	P	P	P	P	R	P	B	P	R
Expert metadata collected?	Y	Y	Y	N	N	Y	Y	Y	Y	Y
Was consensus an objective?	N	Y (Chap. 4) and N (Chap. 9)	Y	Y	N	N	N	Y	Y	N
Elicited qualitative (QL), quantitative (QN), or both (B) forms of information[b]	B	QN	B	QL	B	QN	QN	QL	B	QN

Post-elicitation										
Was expert metadata used to weight or interpret the results?	Y	Y	Y	N	N	Y	Y	N	N	Y
Was knowledge uncertainty represented in the final output?	Y	Y	Y	Y	N	Y	Y	N	Y	Y
Validation of the expert knowledge using independent data (I), expert review (E), or no validation (N)	I, E	I	I, E	N	I	I	I	I, E	E	E

[a]Training consisted of formal instruction of the experts before the elicitation, including a detailed explanation of the project's objectives, discussion of key definitions and terminology, and explanation of the method of knowledge elicitation

[b]Qualitative information was considered to be a simple ranking of items or processes (using an ordinal scale), the use of non-numerical terms such as "better" or "worse," or the specification of key factors that might influence the process of interest; quantitative data allowed for the elicitation of magnitudes of differences among items or processes, including probabilities, the weighting of factors, or assignment of relative scores

Table 14.2 Summary of the chapter authors' experiences from case studies, and their suggestions for advancing the collection and application of expert knowledge

Chapter and lead author	Key lesson learned during the project	Element of the project in need of improvement	Recommendations and key areas for future research
3-Low-Choy	The elicitation method must be tailored to meet the needs of the experts	Assess the relationship between site-level knowledge of experts and the landscape scale of the supporting GIS data	Convince organizations to maintain and develop their corporate expertise strategically; clarify the relationship between site-level and landscape-scale knowledge
4, 9- Drescher	A single method of elicitation did not capture knowledge adequately due to differences among experts and institutional cultures	Incorporate uncertainty in expert knowledge directly into scenario simulations	Develop an efficient, effective, and explicit system to capture, evaluate, and use expertise and to quantify uncertainty within resource management organizations
5-Drew	Allocating adequate time to plan and practice elicitations before collecting data is critical to identify weaknesses in the elicitation design or expert pool	Necessary to develop ways to more effectively relate expert site-level knowledge to landscape-scale data to ensure the selection of appropriate land cover data and landscape metrics	Develop tools to identify an expert's domain of expertise, weight their knowledge accordingly, and scale this knowledge up to larger scales
6-Moody	Elicitation can serve as a rapid process for collecting qualitative data and measures of uncertainty from large groups of experts	Require more time to prepare for workshops with a particular focus on developing effective facilitation skills	Develop methods to integrate inter-expert uncertainty into spatial models and resulting products
7-McNay	Using a formal and transparent elicitation process enhanced the acceptability of the results to stakeholders	Measure and incorporate uncertainty in knowledge rather than developing only a consensus view	Develop measures to ensure that information truly represents expert knowledge and that the elicited knowledge is not biased by the elicitation process. Assess the degree of uncertainty by capturing the variation among experts

8-Johnson	Depending on the context, expert knowledge can be very predictive of ecological processes in other geographic areas	Both record uncertainty and find ways to incorporate that uncertainty into the modeling process	Develop methods for weighting knowledge based on the magnitude of the uncertainty
10-Doyon	Expert knowledge is valuable for directing future empirical studies	Field-based exercises should be used to help experts understand the context and scale of the questions used during elicitation	Develop metrics that quantify the level of agreement among experts
11-Keane	Expert knowledge can be biased and inconsistent	Better measures of the degree of bias in knowledge are required; ensure that elicited knowledge relates only to the individual's area of expertise	Develop methods that allow the identification of only expert knowledge, and that include measures of bias, uncertainty, and reliability
12-Williams	Bayesian models are a powerful tool for including regional heterogeneity and uncertainty, and for targeting future collection of expert and empirical data	More formal documentation of the conceptual model used by each expert as well as the variation in knowledge among experts is necessary	Refine the Bayesian model to allow incorporation of the feedback and uncertainty reported by an expert panel who will review the model's outputs. Develop techniques for characterizing the experts and documenting the elicitation process
13-Kappel	For complex systems and processes, data gaps should not halt planning initiatives; expert knowledge is very useful in such cases because it provides enough confidence to proceed	Greater incentives are needed to attract experts and maintain their participation; a standing scientific advisory process might improve participation and expert investment in the process	Find incentives to encourage participation by experts. Develop methods that will allow the characterization of uncertainty as well as the accuracy and reliability of experts

14.2 What We Learned

14.2.1 Broad Application and Acceptance of Expert Knowledge

Expert knowledge can no longer be considered a fringe or secondary information resource. Although there is a long track record for the application of expert knowledge in natural resource management, we are now observing a greater level of respect for this approach because of the increasing degree of rigor (Sutherland 2006). The broader acceptance and resulting scrutiny provided by the scientific community is as encouraging as the growth in application of expert knowledge. Elicitation and expert knowledge are now valid areas of investigation for researchers in the natural sciences. Recent studies, for example, have considered the existence of bias and uncertainty in knowledge (Johnson and Gillingham 2004; Czembor and Vesk 2009), the ability to generalize knowledge to different landscapes or time periods (Doswald et al. 2007; Murray et al. 2009), the merits and drawbacks of expert knowledge relative to empirical data (Johnson and Gillingham 2005; Pullinger and Johnson 2010), and effective practices for eliciting knowledge (Kuhnert et al. 2010). Also, such studies are being published in the most highly respected ecological journals (e.g., Low-Choy et al. 2009; Murray et al. 2009; Aspinall 2010). These are exciting and worthwhile investigations with broad application to pressing issues in landscape ecology, such as conserving threatened species and understanding the effects of climate change (O'Neill et al. 2008; Wilson et al. 2011).

The case studies presented in this book provide insights and guidance on the application of expert knowledge to specific policy and management challenges. Drawn from communities of landscape ecologists in Australia, Canada, and the United States, the authors have illustrated the application of expert knowledge across such diverse fields as wildlife management (Drew and Collazo, Chap. 5; McNay, Chap. 7; Johnson et al., Chap. 8), conservation biology (Moody and Grand, Chap. 6), risk and vulnerability assessment (Kappel et al., Chap. 13), land use planning (Williams et al., Chap. 12), forest landscape succession and modeling (Drescher and Perera, Chap. 9; Doyon et al., Chap. 10), and fire ecology (Keane and Reeves, Chap. 11). In addition, these chapters provide the reader with an overview of a wide range of elicitation methods (Table 14.1). Other authors focused less on the specific uses of expert knowledge, and instead report on the development of more effective methods to elicit and understand the uncertainty inherent in expert knowledge (Low-Choy et al., Chap. 3; Drescher et al., Chap. 4).

Chapter authors have used expert knowledge to address gaps in the available empirical data, characterize the full state of knowledge of a given system, and expedite the delivery of a decision-support tool or of management guidance where the collection of empirical data would be impractical (e.g., for future states or events, for very large and variable landscapes, when rapid decisions are necessary). For example, McNay (Chap. 7) used expert knowledge to parameterize a predictive model of seasonal habitat use by woodland caribou. Although he had access to a considerable amount of empirical data, the key drivers of seasonal distribution and future habitat were complex and interrelated in ways that were not fully understood. In this situation, expert knowledge appeared to be the best basis for forming hypotheses and for integrating and parameterizing the available knowledge to produce predictive models. In comparison, Drew and Collazo (Chap. 5) had no empirical data to describe the distribution of the King Rail. They relied exclusively on experts to develop a set of complex and interacting hypotheses to describe the habitat relationships of this bird species and to guide the collection of empirical data.

In several chapters, expert knowledge was used to prioritize conservation or land use objectives, particularly when managers believed that inaction while awaiting better empirical data was not an option. Moody and Grand (Chap. 6) worked with experts to identify focal bird species that would be representative of broader faunal associations and used these species to guide regional conservation efforts in rapidly changing landscapes. Kappel et al. (Chap. 13) worked with a large number of experts to identify marine ecosystems vulnerable to key drivers of change. However, some chapter authors, including (Keane and Reeves, Chap. 11), expressed concern over the exclusive reliance on expert knowledge as a substitute for empirical methods, especially when rigorous methods were not applied during the elicitation process.

14.2.2 Investigating Expert Knowledge and Developing Rigorous Methods

An overarching theme that characterized all of the chapters was the recognition that elicitation should promote and support transparent and repeatable methods and provide for an assessment of uncertainty and bias in results. Although experts have long contributed their knowledge to support modeling and planning projects, past applications to the natural sciences had limited utility or acceptance because of non-repeatable and poorly developed methods (Sutherland 2006). Too often, elicitation simply involved an open invitation to discuss a particular subject with little to no development of the approach, documentation of the participants or the elicitation process, or use of rigorous methods (Johnson and Gillingham 2004). Poor research design often results in knowledge that has little internal consistency or external validity, and this has harmed the credibility of experts as information resources and active participants in science-based decision making.

Working properly with experts is not necessarily a simple or inexpensive process from either a time or a financial perspective. Drescher et al. (Chap. 4) allocated 12 months to prepare for the elicitation and Kappel et al. (Chap. 13) invited 199 experts to participate in a survey of the effects of 58 human stressors on 15 marine ecosystems. Many activities occur during the design and implementation of expert-based studies. Common recommendations suggest allocating enough time to refine the research questions, identify and characterize the experts, draft the elicitation questions, test and revise (i.e., pilot) the elicitation process and materials before collecting data, develop a strategy to motivate and maintain participation through what is often a long and demanding process, and assess the uncertainty and perhaps even the validity of the elicited knowledge (Low-Choy et al. 2009; Knol et al. 2010). The increased emphasis on design and planning reflects a growing understanding and appreciation of expert bias, the chance of miscommunication, a variety of types of error, and participant burnout. By delving into the elicitation literature and collaborating with colleagues in the social sciences, landscape ecologists are discovering that many of these potential problems can be anticipated and mitigated through proper study design.

Careful attention to detail is required to develop effective and acceptable approaches for elicitation and for reporting the results. As is the case with the collection of empirical data, researchers and practitioners must develop methods that meet a high standard of scientific rigor. Chapter authors demonstrated a range of techniques that can be used to elicit and formally document expert knowledge. Some used computer-based methods to record the expert's knowledge (Low-Choy et al., Chap. 3; Drescher et al., Chap. 4), whereas others used more generic survey tools that included questionnaires (Doyon et al., Chap. 10; Moody and Grand, Chap. 6) or focus groups (McNay, Chap. 7; Table 1). Low-Choy et al. (Chap. 3) described innovative elicitation software that allowed the experts to relate their knowledge to a specific landscape and continually evaluate the consistency and logical validity of their responses. This work, in particular, highlighted the recent methodological advances that have increased the rigor of eliciting expert knowledge.

The literature provides some guidance on best practices, potential biases, and the general steps used for elicitation (Kadane and Wolfson 1998; Low-Choy et al. 2009; Knol et al. 2010; Kuhnert et al. 2010). Although these past works are a useful starting point, the studies in this book highlight the apparent need for an elicitation process and a method of analysis that meets the specific objectives of a project. Not by design, but by chance, we find that a diversity of methods were adopted in the studies described in this book. This variation is likely representative of the range of approaches available and the creativity being exercised by researchers and practitioners who are working with experts and applying their knowledge to answering difficult questions and solving difficult problems in landscape ecology. We believe strongly that the elicitation method should be crafted to meet study objectives; however, the existence of such a large number of approaches suggests the need for further research to improve our understanding of these methods and provide stricter guidance on the best elements to use in a given elicitation process.

Although the chapters in this book differ in the problem being studied, the geography of the study area, and the elicitation process (Table 14.1), there is nonetheless a set of consistent steps for developing a transparent and repeatable method for collecting and applying the expert knowledge. We present those steps in a generic framework (Fig. 14.1) that can serve as a starting point for inexperienced landscape ecologists who are interested in planning projects focused on expert knowledge or that include an element of expert knowledge. McBride and Burgman (Chap. 2) and the references therein expand on those steps with more detailed guidance.

Previous researchers have sometimes overlooked the need to clearly define the characteristics of an "expert" (e.g., Petit et al. 2003; Van der Lee et al. 2006), so

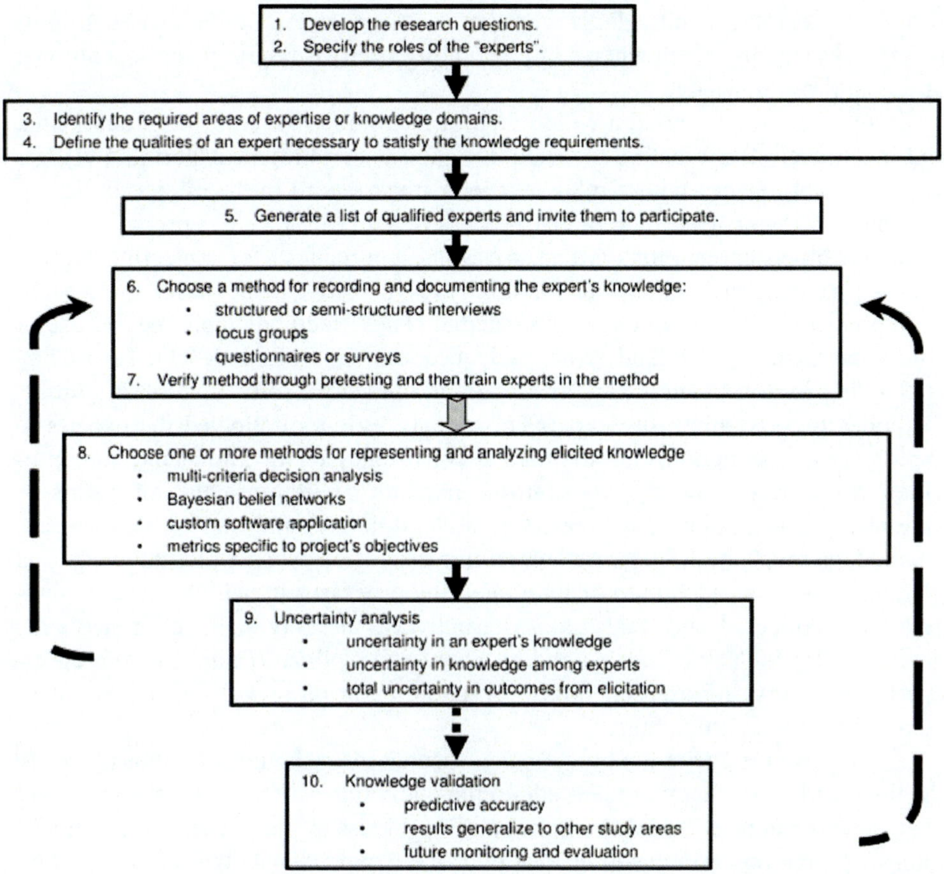

Fig. 14.1 A generic framework for study design and the elicitation of expert knowledge. The *grey arrow* represents linked processes for certain measurement techniques, the *dotted arrow* represents a process not appropriate for all study designs, and the *dashed arrows* represent feedback mechanisms that should be used in the presence of excessive uncertainty or weak validation

most chapter authors were careful to develop and document a clear definition of the "expert" and the domain expertise required to meet the project objectives. However, these definitions varied among studies; we found definitions based on the number of years of experience in a particular discipline or professional duty or study area, and definitions based on an index of expertise, such as the number of publications on a relevant subject. Other researchers have used even less direct measures of expertise, including membership in expert panels or committees (O'Neill et al. 2008).

Once defined, the experts must be sought out and invited to participate in the study. Authors in this book often used informal peer-nomination processes, such as recognition by colleagues or professional acquaintances. More formal approaches included chain referral ("snowball") sampling, in which the initial group of experts identified by the research team nominated additional participants (Chap. 8). One group of authors (Kappel et al., Chap. 13) used computer databases such as Google

Scholar to search for individuals who met their predefined definition. Such tools might be especially useful when a large pool of experts is required across a number of domains of knowledge.

Chapter authors reported a wide range of techniques for collecting, analyzing, and in some cases evaluating the reliability and uncertainty of expert knowledge. We found approaches with a relatively long track record in the elicitation literature, such as the analytical hierarchy process (Chap. 8), as well as project-specific computer-based applications (Chap. 4). Some approaches for collecting knowledge were more generic and were potentially less sensitive to the biases and sources of imprecision inherent to expert knowledge. These methods included the use of facilitated focus groups and structured questionnaires (Table 14.1). Low-Choy et al. (Chap. 3) discussed the application of an innovative software tool, *Elicitator*, for collecting and analyzing expert knowledge. This tool allowed the experts to explore their assumptions and the logical consistency of the knowledge they provided when describing the distribution and habitat requirements of plants or animals. The techniques for knowledge collection and analysis were sometimes coupled, as in Chap. 3, but were sometimes discrete. For example, the analytical hierarchy process and *Elicitator* integrated the processes by which expert knowledge was collected and analyzed. Alternatively, the Bayesian belief networks developed by McNay (Chap. 7) and by Drew and Collazo (Chap. 5) were developed using very different methods for eliciting prior probabilities from their respective expert participants.

Throughout the elicitation process, the research team should continuously verify the logic and consistency of the method, the elicitation scores, and the preliminary results. This can be accomplished through in-progress questionnaires or diagnostic tools that elicit process-related feedback from the experts and other project participants (e.g., research assistants, facilitators). The final step in the elicitation process is an assessment of the validity and uncertainty of the elicited expert knowledge. Uncertainty has a number of specific dimensions, as discussed by McBride and Burgman (Chap. 2), but generally represents the degree of variation in the answers elicited from a pool of experts as well as the resulting range in predictions or guidance provided by expert-based models or decision-support tools. Although there are some useful applications of consensus-based approaches for elicitation and decision making in landscape ecology, these approaches do not identify inter-expert variance, and there is growing agreement that this uncertainty should be documented rather than suppressed (Aspinall 2010). Validation is a comparison of the expert's individual or aggregate responses to some measure of truth where a measure of predictive accuracy is warranted. Verification and uncertainty are elements inherent to all expert-based processes and indeed to all empirical studies; however, validation is not always required or feasible (Fig. 14.1).

A major advance highlighted in this book was the improved degree of effort to identify, quantify, and account for the uncertainty inherent to elicited knowledge (e.g., Chap. 6; Chap. 8; Chap. 9). Several chapter authors noted the important distinction between aleatory uncertainty (i.e., uncertainty inherent in the nature of the system being studied) and epistemic uncertainty (i.e., uncertainty inherent in expert knowledge of the system). Recognizing this difference and its significance allowed them to develop methods to reduce epistemic uncertainty, primarily by paying much closer attention to the unique experience and judgments of individual experts (Chap. 9). A number of case studies also attempted to provide some means to formally evaluate the reliability of expert input both during and after the elicitation process (Chap. 5; Chap. 9; Chap. 10). Providing timely feedback to experts can allow them to correct their own responses (Murray et al. 2009). Techniques and tools are also available to improve the internal consistency between an expert's knowledge of system components and their expectations of the overall system behavior (Chap. 3).

Empirical data, where available, were used to validate the accuracy of expert knowledge; McNay (Chap. 7), Johnson et al. (Chap. 8), and Drescher and Perera (Chap. 9) made such comparisons. These authors assumed that the empirical data were obtained for situations similar to those on which the experts based their knowledge and that they were also precise and unbiased. In other chapters, cross-validation with empirical data was either unnecessary or impossible. For example, Drew and Collazo (Chap. 5) used expert knowledge to generate hypotheses about bird distributions and to design a population monitoring strategy. The data collected through annual monitoring were subsequently used to update and refine the model rather than to validate the model. Williams et al. (Chap. 12) and Kappel et al., (Chap. 13) used experts to address questions that focused on integrated socio-economic and ecological relationships and that considered many criteria, some of which were qualitative. There is no set of empirical observations that can serve to assess such complex or future processes, but model plausibility and internal consistency can nonetheless be verified using independent reviewers and other expert groups. Also, monitoring and active adaptive management experiments can both provide validation for expert-based decisions or predictions, but only once those data are collected.

The general steps for study design (Fig. 14.1) provide a robust starting framework, but more importantly, suggest that the elicitation and use of expert knowledge requires the same level of forethought and methodological rigor as empirically based studies. Indeed, many of the chapter authors implicitly or explicitly advocate for the development of better practices that can be used when eliciting and applying expert knowledge. Johnson et al. (Chap. 8) make such a plea when they report the results of a study that failed to adhere to any of the steps described in Fig. 14.1. To support such an approach, Low-Choy et al. (Chap. 3) provide a method and tool whose structure explicitly supports the use of good practices and that guards against many of the biases encountered during elicitation.

Beyond developing a defensible process for elicitation and meeting the direct objectives for using expert knowledge, a number of chapters highlighted methods designed to collect and analyze metadata that described the expert participants (Chap. 5; Chap. 8; Chap. 13). These ancillary data about the experts facilitated subsequent assessment of the elicited knowledge. Capturing the professional identity of the individuals allows modelers to explore the range and variability of the group's collective experience (Doswald et al. 2007). The research team can use these metadata to explore the reasons for outlier opinions, propose alternative hypotheses based on different groupings of each expert's unique perspectives or domains of experience, and assess the representativeness of the collected knowledge relative to the application setting.

14.2.3 Used Wisely, Experts Offer Valuable Contributions

Expert knowledge has often been thought of as temporary or substitute data for situations or questions in which empirical data are lacking. Although this is a valid and important use of expert knowledge, there are some applications in which expert knowledge may be more useful than empirical data. For example, experts offer many advantages for the modeling of complex systems, hypothesis generation, and reaching consensus decisions for management and conservation actions (see Fig. 1.1 in Chap. 1). In particular, the use of experts from a range of domains can reveal key aspects of a situation that were not known to the researchers. Landscape ecology addresses questions that pertain to broad spatial and temporal domains with many interacting cross-scale processes and elements. Such complex relationships can be difficult to quantify and understand using empirical data collected using traditional experimental design (Hargrove and Pickering 1992). If elicited carefully, expert knowledge can offer a broader geographic and temporal perspective than the typical 1- to 2-year studies that form the backbone of most empirical ecological and environmental data. Furthermore, experts can debate issues of environmental variability and data representativeness and can formulate hypotheses that conform to their broader combined experience so as to direct future investigations. However, as several chapter authors reiterate, the choice of experts for enlightening any of these processes is critical and should not be left to chance or opportunity.

Although we have emphasized the advantages of eliciting and using expert knowledge for applications in landscape ecology, expert knowledge is not without error, bias, and inaccuracy. Furthermore, expert knowledge is not a solution for all problems or an answer to all questions when empirical data are lacking: if there are no experts, there is no expert knowledge. Such was the finding of Kappel et al. (Chap. 13), who had too few experts to document the vulnerability of some marine ecosystems. Drew and Collazo (Chap. 5) also highlighted instances where the limited number of experts drawn from a narrowly defined domain (federal wildlife refuges) was not always sufficiently informative of the landscapes or species to be modeled. Where *knowledge* is lacking, experts might begin to contribute their *opinions*. The difference between expert knowledge and expert opinion (see Chap. 1) is not always obvious; even the authors within this book intermixed these terms, and experts themselves are not always aware of the limits of their knowledge until they are asked to quantify their degree of certainty. When elicitation focuses on a participant's domain of expertise, such that they reference events or processes that have occurred within their direct personal experience, then knowledge can be documented. However, when participants must extrapolate beyond their domain of expertise, either in time, space, or subject, their knowledge (by definition) incorporates more characteristics of conjecture, hypothesis, and opinion. This is not to say that expert opinion is of no value. Where direct knowledge is lacking, experts may still provide an educated and useful opinion on a particular question. However, landscape ecologists should carefully distinguish between knowledge and opinion in their analyses. This distinction is especially important because uncertainty will likely be much higher when based on opinion rather than knowledge. As an example, opinion may be ineffective for parameterizing quantitative models in which precision and accuracy are important to direct conservation activities (Chap. 8), but may be useful for developing hypotheses about ecological relationships (Chap. 5) or in risk analyses for future or unobserved events (Chap. 13).

14.3 Our Recommendations for Landscape Ecologists

Methods to collect and apply expert knowledge to questions and problems in landscape ecology are rapidly evolving. The contributions in this book demonstrate that the elicitation and use of knowledge of ecological systems and processes has moved from an *ad hoc* practice to a formalized and rigorous set of defensible methods. As with any scientific endeavor, however, there is room for refinement, improvement, and innovation (Table 14.2). Furthermore, the authors represented here constitute only a small portion of the researchers who are directly applying expert knowledge within the field of landscape ecology. We strongly suspect that many landscape ecologists remain unaware of the importance of rigor in the design of an elicitation study and of the basic elements of the process that are identified in Fig. 14.1. Also, although expert knowledge continues to play an important and growing role in the application of landscape ecological principles, its value remains ambiguous and its use remains contentious among the broader community of ecologists. To improve this situation, we have provided some guidance for best practices and have suggested areas of further research that will be necessary to improve the science of elicitation and the practice of application when developing studies or projects premised on expert knowledge (Table 14.2).

14.3.1 Become Informed: Review the Literature Prior to Eliciting Knowledge

Several authors in this book identified gaps in their own training, which left them unprepared for the complexity of designing, facilitating, and interpreting results from expert elicitations. Most people trained in the natural or life sciences have very little exposure to the assumptions that underlie research on human subjects and in the methods used to collect and apply expert knowledge. There is a wealth of existing literature, however, that provides well-founded guidance on defensible best practices for eliciting and using expert knowledge. Some of this work has focused on ecological applications (e.g., Low-Choy et al. 2009; Kuhnert et al. 2010), but other types of practitioners and academic disciplines have a longer history of using expert knowledge well. Thus, we urge the uninformed reader to explore the literature on statistics, health sciences, business, policy sciences, and psychology (Kadane and Wolfson 1998; Aspinall 2010; Knol et al. 2010). Many of the authors in this book have benefited from working directly with colleagues in the social sciences who might not have fully grasped the subject of their studies, but who understood very well the general process for effective elicitation. Just as we might seek out help from a colleague with advanced training in statistics, we must be open to the opportunities that experts in elicitation can provide, even if these experts are found in fields of study with few links to ecology. This message was delivered by McBride and Burgman (Chap. 2) and others (Table 14.2), who argued that improving the application of expert knowledge within landscape ecology will require greater awareness of the tools that are available, as well as the skills to select and tailor these tools to meet the needs of a given project.

For those wishing to learn more, the chapters and citations in this book identify many useful resources. Though this book is not a how-to manual for eliciting expert knowledge, each chapter offers valuable recommendations for motivating expert participants, improving communication, minimizing bias (Chap. 3; Chap. 4), documenting uncertainty (Chap. 6; Chap. 8; Chap. 9), and evaluating the accuracy of expert knowledge (Chap. 7; Chap. 8; Chap. 9). We have the following recommendations to improve the level of awareness and the capacity for self-learning by ecologists who are interested in applying expert knowledge to questions and problems in landscape ecology:

- Publication of special issues in journals of applied ecology to highlight and promote the effective and proper use of expert knowledge.
- Formal recognition of points of contact for both practitioners who have used expert knowledge and persons with expertise in eliciting expert knowledge. This "community of practice" would serve as a forum for discussing and guiding methods and for mentoring ecologists who are interested in applying expert knowledge.
- Development of a textbook or best practices manual that focuses on the most current methods for effectively working with ecological experts and eliciting expert knowledge. This text would consider all elements of gathering and using expert knowledge and would focus on the methodological hurdles or problem areas most likely to confront applied ecologists (Fig. 14.1). We suspect, however, that such a discipline-focused text is premature. We recognize that over the past 10 years ecologists have made significant progress in appreciating the complexity of expert-based studies, and applying better methods of elicitation. Also, there is a substantial literature from other disciplines that can direct ecologists in

the development and proper application of methods, and indeed, in understanding the nature of expert knowledge (Cooke 1991; Meyer and Booker 1991; O'Hagan et al. 2006; Collins and Evans 2007). However, considering the unique challenges faced by landscape ecologists, principally the interacting effects of spatial and temporal scale as well as process heterogeneity, we recommend the further refinement and testing of new and innovative methods (e.g., Chap. 3) and additional case studies and applications. Such work would allow a better understanding of the sources of bias and uncertainty inherent to landscape ecology and provide for a stronger foundation for a discipline-specific text. The increasing rate of publications in this area suggests that the science of expert knowledge, as applied to landscape ecology, may mature to a sufficient level to support such a text over the next 3–5 years.

14.3.2 Expand the Available Toolsets to Support Rigorous Elicitation of Knowledge

Authors in this book have demonstrated considerable innovation in developing methods that are effective for eliciting expert knowledge. Some notable advances of particular relevance to applications in landscape ecology include the improved integration of statistical analysis and GIS data within the elicitation process (e.g., Chap. 3). Such spatially explicit approaches will be more intuitive to landscape ecologists, making the knowledge reporting less abstract (Chap. 6). Despite these advances, however, all authors in this book concur that refinement and development of elicitation methods is a key area in need of further research (Table 14.2).

We noted considerable variation in the number of experts employed across the studies in this book and in previous research (e.g., Seoane et al. 2005 = 1; Chap. 13 = 58), and with the exception of certain knowledge areas in which no experts were identified, there was little justification of sample size. The social science literature provides some guidance on the best number of experts for an elicitation, but this advice is based on observations of group dynamics and biases (Aspinall 2010). In ecology, it seems likely that heterogeneity in the environments or ecological processes for which experts are knowledgeable, as well as variation in knowledge among expert participants, will affect recommendations for the minimum number of participants needed to meet the requirements for statistical rigor (Chap. 5). Also, the number of experts involved in a project is likely to require trade-offs among the breadth of the knowledge domain, the availability or number of experts working in that domain, and the effort and expense necessary to identify and recruit experts and to elicit their knowledge. Regardless of the practical limitations that confound the issue of sample size, some guidance is needed on when the use of too few experts threatens the validity of a study's conclusions.

Authors in this book were nearly unanimous in reporting the need to accurately identify and characterize expertise as well as the need for methods that better record and incorporate uncertainty during the elicitation process (Table 14.2). Clearly, useful expert knowledge is premised on identifying the correct pool of experts and differences in suitability among experts within that pool, but this step is often overlooked or is based on *ad hoc* criteria. Furthermore, there is less guidance on how to identify good experts *a priori* relative to the assessment and weighting of expert knowledge that occurs during and after elicitation. More research on understanding the implications of incorrect parameterization of the expert definition and of involving too few experts is clearly warranted. We have the following recommendations for further study:

- Development and testing of elicitation tools that allow experts to document their knowledge using an easy to understand, transparent, and repeatable process. These tools should accommodate all contexts of expert knowledge (Table 14.1, Chap. 1), including problem synthesis, hypothesis building, and model parameterization. Such tools should have inherent mechanisms that guard against bias and inconsistent logic or that allow experts to test for and correct such problems (e.g., Chap. 3).
- Development of tools or elicitation strategies that better match the spatial and temporal experience and knowledge of an expert to the proposed questions. For many problems in landscape ecology, we are seeking knowledge that informs our understanding of large-scale processes across landscapes or regions, whereas the experts may be more familiar with patch-level phenomena. Currently, we have a limited understanding of the implications of such scale mismatches or how to scale-up expert knowledge.
- Studies to understand the implications of the definition of an expert, the linkages between this definition and the problem being studied, the number of experts involved in a study, and the uncertainty in expert knowledge. A broader definition of the expert might result in a larger sample of experts, but would then result in a greater breadth of uncertainty in the expert knowledge. The implications of such decisions for the efficiency and reliability of the elicited knowledge should be investigated.

14.3.3 Continue to Critically Evaluate and Test Expert Knowledge

Ideally, every expert-based project should incorporate a critical analysis of the elicitation methods, the information elicited, and the reliability of the resulting decision-support products. Expert knowledge can be cross-validated against other expert sources (Chap. 13) or empirical data (Chap. 7; Chap. 8). However, there is no formal guidance for practitioners or researchers as to when validation of expert knowledge is necessary and, when it is necessary, how best to proceed with the validation. We suspect that the methods and rules for validation will prove to be project-specific, but some perspective on the scope of the available methods is necessary to help researchers understand this problem and choose an appropriate solution. Although validation may not be necessary or even possible in all cases, evaluation and verification of the elicitation process should be a structured component of all projects that depend on expert knowledge. Further research and debate is required to define the necessity and expected outcomes of validation.

Understanding the variation in expert knowledge and the possible biases that underlie this variation appears to be an area of increasing interest in the ecological literature (Doswald et al. 2007; Hurley et al. 2009; Chap. 13). The collection and analysis of detailed quantitative information from individual experts represents a large advance over approaches that capture only aggregate components of knowledge (e.g., consensus results). There are likely strong links between the selection of experts and the resulting uncertainty in knowledge. Thus, by characterizing individual experts and maintaining the ability to distinguish their personal responses within the pool of elicited information, we can better understand the sources of uncertainty (Chap. 8; Chap. 11). There remains, however, much room for innovation and further refinement of methods to evaluate the knowledge gathered through elicitation. Uncertainty is not an unknown concept to the practitioners and researchers who will apply expert knowledge. Indeed, uncertainty was well categorized (e.g. Chap. 2) or was at least recognized within many of the chapters. Despite the recognition of uncertainty as a pivotal concept in evaluating and applying expert knowledge, approaches for documenting uncertainty remain largely *ad hoc*. The authors represented in this book were nearly unanimous in reporting that the science of eliciting and using expert knowledge would be improved if the elicitation methods directly categorized, measured, and incorporated the uncertainty inherent in the knowledge (Table 14.2). Achieving this goal would require consistent and standardized measures of uncertainty in knowledge and a better understanding of the sources of this uncertainty, especially in the context of expert selection, and would require guidance on how best to manage and accommodate uncertainty. Specific recommendations include:

- Establish guidelines to characterize the reliability of expert knowledge. Recognizing that there are a broad range of applications and associated requirements for the precision and accuracy of knowledge as well as levels of involvement by experts, such guidelines would support the judicious application of expert knowledge to a given problem or question. An assessment system that positions the elicited information along a spectrum ranging from opinion to knowledge would be particularly valuable.
- Develop better methods and a consistent measurement scale for quantifying the degree of uncertainty in knowledge both among experts and within an individual's elicited responses. In theory, this measurement scale would partition uncertainty into the three main types: aleatory (due to the system's inherent complexity), epistemic (due to limitations of the expert's knowledge), and linguistic (due to the inherently subjective nature of the words an expert uses to describe their

knowledge). Such divisions would also improve the elicitation process. In addition, the quantification of uncertainty would allow researchers to weight the individual responses to account for their relevance in the context of a specific application or question.

Finally, through the diversity of the projects, the rigor of the different methods, and the insights of the authors, this book illustrates the exciting and valuable progress that is currently being made in the application of expert knowledge to answering the questions and solving the problems faced by landscape ecologists. Although there remains much room for innovation and improvement, the potential value of expert knowledge that is collected using a rigorous study design is high. This value is likely to be increasingly evident both in the short term, as a stop-gap measure when there is insufficient data and formal knowledge to support management decisions, and in the long term, as a way to complement and supplement empirical data and formal knowledge. This compilation of learning and experience suggests that there are few bounds to the effective and reliable application of expert knowledge. Where experts are available and a proper method is employed, neither the expert's discipline, the geography of the study area, nor the subject of study should prevent advancement of our understanding or the development of solutions to the complex problems faced by landscape ecologists. For these reasons, we are confident that applications of expert knowledge in landscape ecology will continue to expand and that the science of eliciting and using expert knowledge will continue to improve.

References

Aspinall W (2010) A route to more tractable expert advice. Nature 463:294–295

Collins H, Evans R (2007) Rethinking expertise. The University of Chicago Press, Chicago

Cooke RM (1991) Experts in uncertainty: Opinion and subjective probability in science. Oxford University Press, Oxford

Czembor CA, Vesk PA (2009) Incorporating between-expert uncertainty into state-and-transition simulation models for forest restoration. For Ecol Manage 259:165–175

Doswald N, Zimmerman F, Breitenmoser U (2007) Testing expert groups for a habitat suitability model for the lynx *Lynx lynx* in the Swiss Alps. Wildl Biol 13:430–446

Hargrove WW, Pickering J (1992) Pseudoreplication: a *sine qua non* for regional ecology. Landsc Ecol 6:251–258

Hurley MV, Rapaport EK, Johnson CJ (2009) Utility of expert-based knowledge for predicting wildlife–vehicle collisions. J Wildl Manage 73:278–286

Johnson CJ, Gillingham, MP (2004) Mapping uncertainty: sensitivity of wildlife habitat ratings to variation in expert opinion. J Appl Ecol 41:1032–1041

Johnson CJ, Gillingham MP (2005) An evaluation of mapped species distribution models used for conservation planning. Environ Conserv 32:1–12

Kadane JB, Wolfson LJ (1998) Experiences in elicitation. Statistician 47:3–19

Knol AB, Slottje P, van der Sluijs JP, Lebret E (2010) The use of expert elicitation in environmental health impact assessment: a seven step procedure. Environ Health 9:19

Kuhnert PM, Martin TG, Griffiths SP (2010) A guide to eliciting and using expert knowledge in Bayesian ecological models. Ecol Lett 13:900–914

Low-Choy S, O'Leary R, Mengersen K (2009) Elicitation by design in ecology: using expert opinion to inform priors for Bayesian statistical models. Ecology 90:265–277

Meyer MA, Booker JM (1991) Eliciting and analyzing expert judgment: A practical guide. Society for Industrial and Applied Mathematics and American Statistical Association, Philadelphia, ASA-SIAM Series on Statistics and Applied Probability 7

Murray JV, Goldizen AW, O'Leary RA, et al (2009) How useful is expert opinion for predicting the distribution of a species within and beyond the region of expertise? A case study using brush-tailed rock-wallabies *Petrogale penicillata*. J Appl Ecol 46:842–851

O'Hagan AO, Buck CE, Daneshkhah, et al (2006) Uncertain judgements: Eliciting experts' probabilities. John Wiley & Sons, New York

O'Neill SJ, Osborn TJ, Hulme M, et al (2008) Using expert knowledge to assess uncertainties in future polar bear populations under climate change. J Appl Ecol 45:1649–1659

Orsi F, Geneletti D, Newton AC (2011) Towards a common set of criteria and indicators to identify forest restoration priorities: An expert panel-based approach. Ecol Indic 11:337–347

Petit S, Chamberlain D, Haysom K, et al (2003) Knowledge-based models for predicting species occurrence in arable conditions. Ecography 26:626–640

Pullinger MG, Johnson CJ (2010) Maintaining or restoring connectivity of modified landscapes: evaluating the least-cost path model with multiple sources of ecological information. Landsc Ecol 25:1547–1560

Seoane J, Bustamante J, Diaz-Delgado R (2005) Effect of expert opinion on the predictive ability of environmental models of bird distribution. Conserv Biol 19:512–522

Sutherland WJ (2006) Predicting the ecological consequences of environmental change: a review of methods. J Appl Ecol 43:599–616

Van der Lee GEM, Van der Molen DT, Van den Boogaard HFP, Van der Klis H (2006) Uncertainty analysis of a spatial habitat suitability model and implications for ecological management of water bodies. Landsc Ecol 21:1019–1032

Wilson SK, Adjeroud M, Bellwood DR, et al (2011) Crucial knowledge gaps in current understanding of climate change impacts on coral reef fishes. J Exp Biol 213:894–900

Index

A.H. Perera et al. (eds.), *Expert Knowledge and Its Application in Landscape Ecology*,
DOI 10.1007/978-1-4614-1034-8, © Springer Science+Business Media, LLC 2012

T

Th

The